INSTITUT FRANÇAIS DU PÉTROLE

École nationale supérieure **du pétrole et des moteurs**

Centre exploration

Oil and Gas Exploration Techniques

SEISMIC SURVEYING AND WELL LOGGING

Sylvain BOYER
Log Analyst
Professor at École nationale
supérieure du pétrole et des moteurs

Jean-Luc MARI
Senior Research Geophysicist
Professor at École nationale
supérieure du pétrole et des moteurs

Translated from the French
by M.S.N. Carpenter
Reviewed by A. Rivollier

1997

ÉDITIONS TECHNIP 27 RUE GINOUX 75737 PARIS CEDEX 15

FROM THE SAME PUBLISHER

- Geophysics for Sedimentary Basins.
  G. HENRY
- Seismic Methods.
  M. LAVERGNE
- Seismic Well Surveying.
  J.L. MARI, F. COPPENS
- Full Waveform Acoustic Data Processing.
  J.L. MARI, F. COPPENS, PH. GAVIN, E. WICQUART
- Log Data Acquisition and Quality Control.
  PH. THEYS
- Wireline Logging Tool Catalog.
  Edited by M. VERDIER
- Acoustics of Porous Media.
  T. BOURBIÉ, O. COUSSY, B. ZINSZNER
- Formation Imaging by Acoustic Logging.
  Edited by J.L. MARI

IFP EXPLORATION AND PRODUCTION RESEARCH CONFERENCES

- Thermal Phenomena in Sedimentary Basins.
  Edited by B. DURAND
- Thermal Modeling in Sedimentary Basins.
  Edited by J. BURRUS
- Migration of Hydrocarbons in Sedimentary Basins.
  Edited by B. DOLIGEZ
- Petroleum and Tectonics in Mobile Belts.
  Edited by J. LETOUZEY
- The Potential of Deep Seismic Profiling for Hydrocarbon Exploration.
  Edited by B. PINET and Ch. BOIS
- Physical Chemistry of Colloids and Interfaces in Oil Production.
  Edited by H. TOULHOAT and J. LECOURTIER
- Subsurface Reservoir Characterization from Outcrop Observations.
  Edited by R. ESCHARD and B. DOLIGEZ

Translation of *Sismique et diagraphies*,
Sylvain Boyer, Jean-Luc Mari

Cover picture: Claude Laffont

ISBN 2-7108-0712-2
ISSN 1271-9048

OIL AND GAS EXPLORATION TECHNIQUES

Present-day exploration for oil and gas calls upon a wide variety of professional skills. Those require in-depth knowledge of the disciplines in Earth Sciences such as geology, geophysics and geochemistry, associated with engineering and data acquisition techniques such as well logging, seismic lines acquisition, treatment and interpretation, remote sensing, etc.

Powerful computers are used extensively in exploration, not only in the above mentioned techniques, but also for the management of huge quantities of collected data. They are also for modeling and simulation of natural phenomena such as the evolution of sedimentary basin architecture and fill, or petroleum formation through geological time.

The classical image of the petroleum explorer scouring a remote countryside with hammer in hand is therefore obsolete, even if field studies still play an important role in his training and sometimes in his professional activities.

The new know-how and techniques are also more and more frequently being used in what is called reservoir characterization which makes it possible to establish geological models to be used in the development and production of oil and gas fields.

For the success of the exploration assignements of oil companies, the staff depend on rapidly evolving technical know-how and are now obliged to constantly up-date their knowledge in the various exploration fields.

The aim of these volumes presented under the title of "Oil and Gas Exploration Techniques" is to provide a basis for such an updating. They have been prepared under the leadership of the *École Nationale Supérieure du Pétrole et des Moteurs (ENSPM)*, a world renowned educational institution in the field of petroleum, natural gas and internal combustion engines, and whose major characteristic is to be receptive to industrial needs and to incorporate and transmit the most recent information gathered, for professional development.

B. Durand
Head of the Center for Exploration
École Nationale Supérieure
du Pétrole et des Moteurs

ACKNOWLEDGEMENTS

We wish to express our sincere gratitude to the many persons who have reviewed this work and contributed numerous discerning comments.

In particular, we are indebted to the following:
Éric Blondin and François Verdier (Gaz de France),
Emmanuel Chevallier (Elf Aquitaine),
Françoise Coppens and Michel Lavergne (Institut Français du Pétrole),
Patrick Renoux (Compagnie Générale de Géophysique) and
Marie-Pierre Doizelet (Direction Départementale de l'Équipement des Vosges).

We also extend our warmest thanks to Institut Français du Pétrole, Éditions Technip and Mach 3 who ensured the practical side of producing this book and its attractive presentation.

FOREWORD

The present work is based on courses intended for students at the *École Nationale Supérieure du Pétrole et des Moteurs (ENSPM)*. Its aim is to show the respective contributions of downhole geophysical techniques (especially sonic logging) and borehole seismic surveying (such as Vertical Seismic Profiling) to the detailed matching of data provided by conventional seismics. This results in an improved understanding of reservoirs.

In the interface propagation mode, acoustic logs have a vertical resolution in the order of ten centimeters and a depth of investigation in the order of several centimeters. For refracted wave propagation, the vertical resolution varies from ten centimeters to about a meter, whereas values of the order of tens of meters are typically obtained in reflected mode. Sonic logging techniques can thus provide detailed information on the acoustic wave velocity and petrophysical properties of rocks as a function of depth. The use of reflected wave datasets yields a seismic image analogous to small-scale travel time sections which, under favourable conditions, enables the tracing of bed boundaries and the estimation of dips. A combination of density and sonic logging makes it possible to calculate the variation of acoustic impedance with depth; these data are then corrected, calibrated and matched before being converted to a time base.

The most common borehole seismic technique —vertical seismic profiling (VSP), with or without lateral offset of the source— has a vertical resolution of the order of one to ten meters, with the lateral investigation ranging from a few tens of meters to several hundred meters. It provides a time-based seismic image that is comparable to a trace or profile obtained from surface seismic reflection surveys. The VSP enables a fine adjustment of surface seismic data which can then be applied to greater depths. When constrained by well logging data (sonic and density), the VSP is able to provide an acoustic-impedance log for the well which may extend past the drilling depth.

The quality of the information derived from seismic surveys is highly dependent on the complexity of the geological media traversed by the well, the quality control of data acquisition, the efficiency of the algorithm used (as well as the computation assumptions), the resolution and the signal/noise ratio. In fact, the measurements acquired by well logging are indispensable both for the implementation of conventional seismic surveys and the processing of their results.

Thus, by taking well logging data into account, the many configurations used in field surveys (surface-to-surface, well-to-surface, well-to-well, 2-D and especially 3-D seismics) can yield a detailed structural image of the subsurface, especially the reservoir zones in a given field. Seismic surveys provide geologists and production engineers with detailed lithological information as well as various petrophysical parameters. To a certain extent, these survey techniques can lead to the location of hydrocarbon-bearing zones through the use of Direct Hydrocarbon Indicators.

FOREWORD

TABLE OF CONTENTS

FOREWORD 5

INTRODUCTION 11

Chapter 1. WIRELINE LOGGING TECHNIQUES

1.1 GENERAL REMARKS 15

1.1.1 Logging techniques and Logs presentation 15
1.1.1.1 Logging tools 16
1.1.1.2 Presentation of log data 16
1.1.1.3 The logging environment 20

1.1.2 Quality control and operating conditions 20
1.1.2.1 Raw log quality control 21
1.1.2.2 Checking of log data 22
1.1.2.3 Correction of measurements 22

1.1.3 Guidelines for the use of logs 22
1.1.3.1 Desired parameters 22
1.1.3.2 Environmental corrections 23
1.1.3.3 Interpretation of results 24

1.2 SONIC AND DENSITY LOGS 24

1.2.1 Acoustic well logging 24
1.2.1.1 Equipment and data acquisition 24
1.2.1.2 Quality control and corrections 41
1.2.1.3 Relations between acoustic wave propagation and media 44
1.2.1.4 Sonic log prediction 54

1.2.2 Density logs ($\gamma - \gamma$ logs) 55
1.2.2.1 Equipment and data acquisition 56
1.2.2.2 Quality control and data correction 57
1.2.2.3 Relations between measured density and the environment 59
1.2.2.4 Synthetic density log 60

1.3 INTERPRETATION AND APPLICATION OF WELL LOGGING RESULTS 60

1.3.1 "Quick Look" qualitative interpretation 61
1.3.1.1 The "Quick Look" method 61
1.3.1.2 Use of cross-plots 67

1.3.2 Quantitative interpretation 68

1.3.3 Other applications of wireline logs 68

1.3.3.1 Dip determination 68

1.3.3.2 Study of compaction 74

1.3.3.3 Study of fractured formations 75

1.3.3.4 Permeability 76

Chapter 2. SEISMIC WELL SURVEYING

2.1 DATA ACQUISITION 77

2.1.1 The well velocity survey technique 77

2.1.2 Vertical seismic profiles 80

2.2 PROCESSING OF WELL VELOCITY SURVEY AND VERTICAL SEISMIC PROFILING DATA 82

2.2.1 Processing of well velocity survey data 82

2.2.2 Data processing of vertical seismic profiling 89

2.2.3 Some applications of VSP 96

Chapter 3. SEISMIC AND WELL LOGGING SURVEY CALIBRATION

3.1 DISCREPANCIES BETWEEN SONIC LOG AND SEISMIC VELOCITIES 107

3.1.1 Preliminary remarks on geological media 107

3.1.2 Causes of discrepancies between seismic and well logging travel times 108

3.2 ADJUSTMENT OF SONIC INTEGRATED TRANSIT TIME USING SEISMIC TRAVEL TIME 109

3.2.1 Drift calculation between seismic time (TVC) and integrated sonic time (TTI) 110

3.2.2 Drift curve 111

3.2.3 Correction of sonic log to the check shot survey 112

3.2.4 Depth-to-time conversion of corrected sonic log data 114

Chapter 4. SYNTHETIC SEISMOGRAMS

4.1 CONVENTIONAL SYNTHETIC SEISMOGRAMS 121

4.2 VSP-TYPE SYNTHETIC RECORDS 123

4.2.1 Seismic traces and synthetic seismograms 126
4.2.2 Synthetic seismograms and attenuation effects 129
4.3 STRATIGRAPHIC DECONVOLUTION AND INVERSION PROCEDURES 130
4.3.1 Stratigraphic deconvolution 130
4.3.2 Inversion or stratigraphic deconvolution using constraints based on non-seismic data 134

Chapter 5. SEISMIC AMPLITUDE OBSERVATIONS

5.1 MAIN CAUSES OF AMPLITUDE ANOMALIES 140
5.2 AVO EFFECT-AMPLITUDE VARIATION AS A FUNCTION OF OFFSET 140
5.3 DIRECT HYDROCARBON INDICATORS (DHI'S) 144

Chapter 6. CONCLUSION AND CASE STUDIES

Determination of open fractures and assessment of the vertical resolution of surface seismic using sonic log data 149
Calibration of surface seismic data by means of well logs 153
Acoustic impedance in relation to porosity 157
High-resolution seismic well survey of a reservoir used for underground gas storage 159
Subwell prediction and detection of overpressure 165
Use of wireline logs in the interpretation of seismic lines across salt-bearing formations 171
Applications of VSP and dipmeter surveys in the development of a structural model 181

REFERENCES

REFERENCES 187

INTRODUCTION

The need to improve reservoir characterisation in order to optimise hydrocarbon extraction has led to the development of special techniques and increased collaboration of experts in many different fields, i.e.: geology, geophysics, logging and production engineering. The different types of information contributed by each of these approaches are complementary in nature, so it is essential to bring out links between the different disciplines.

The comparison between a seismic section (in two-way time) and an acoustic log (interval transit time versus depth: Δt) leads to questions about the relations between the two types of data and the possible combination of their corresponding datasets (Fig. 1).

The acoustic log provides an obvious link between geophysics, seismic and well logging data. Although covering different frequency bands (acoustic logs: in the order of 10 kHz; seismics: ranging from about 10 to 100 Hz), the two techniques are based on the same laws of wave propagation but with different methologies. Under a certain number of conditions, the seismic measurements collected at these different frequencies can be compared and used to improve knowledge of reservoir characteristics. Acoustic log has a very different vertical and lateral range of investigation compared with seismic surveys (surface or borehole).

It has a very good vertical resolution —of the order of tens of centimeters— but is limited to the

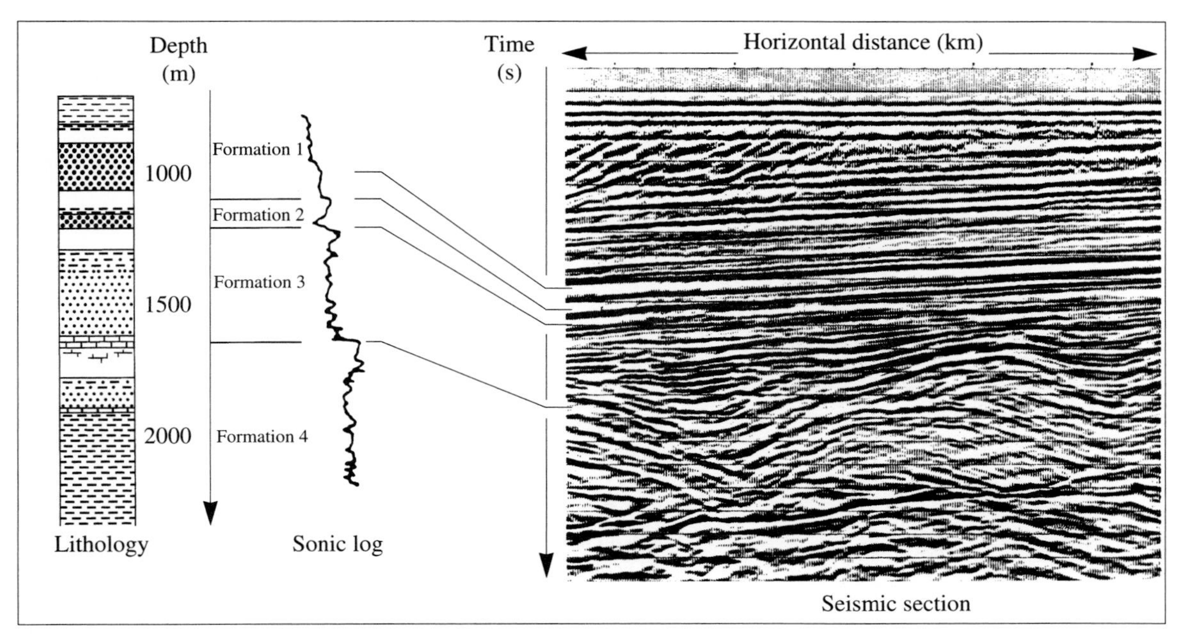

Fig. 1 Well and surface data matching.

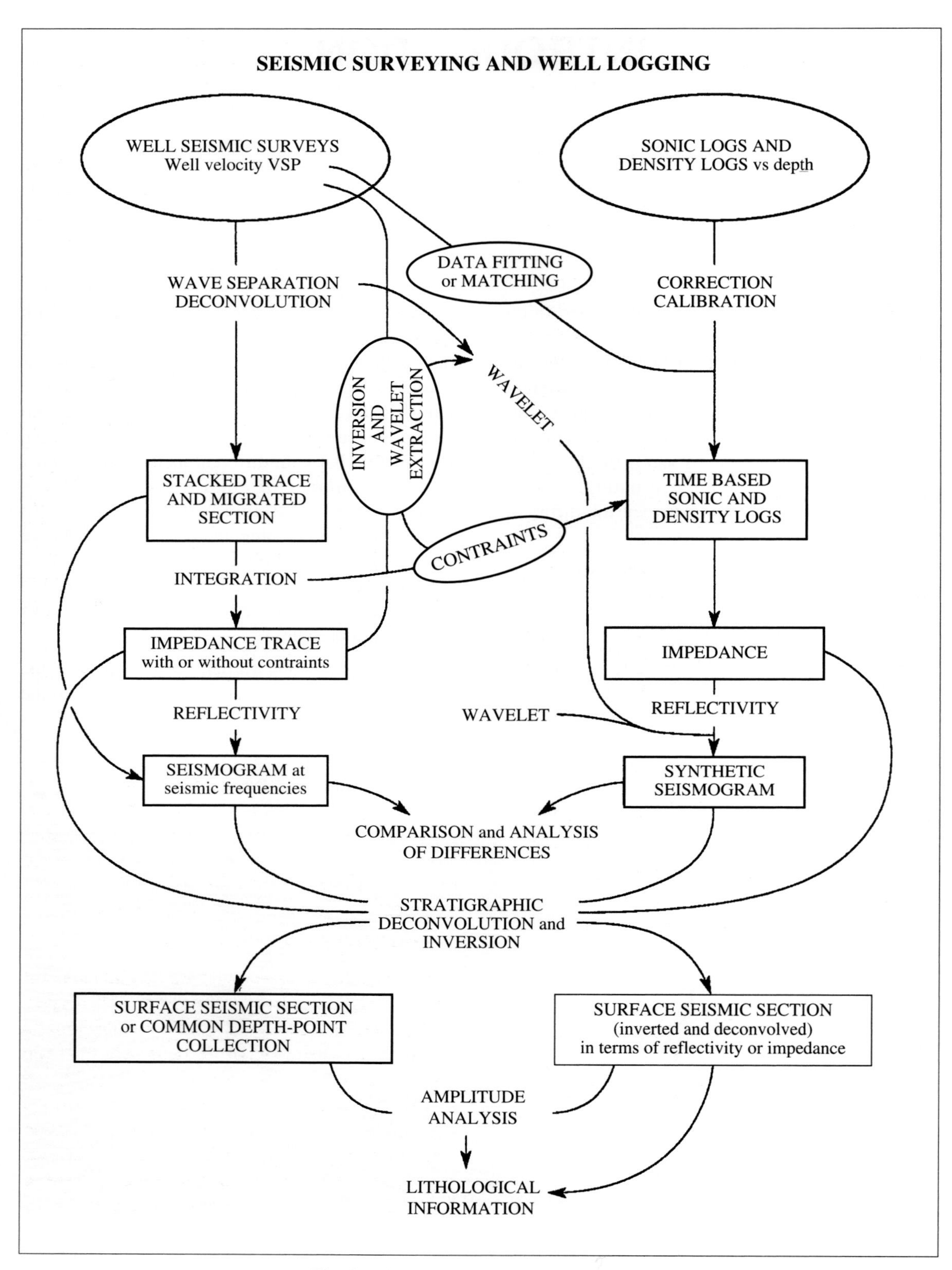

Fig. 2 Seismic surveying and well logging.

immediate vicinity of the borehole. By contrast, seismic profiling allows a more extensive spatial image, but the vertical resolution is no better than about ten meters.

Although the depth-to-time conversion of well log data is carried out using the acoustic velocities of formations obtained from acoustic logs (sonic logs), this method is insufficient to provide an effective comparison between seismic and logging survey datasets. There are discrepancies between the acoustic velocities derived from logging and seismic surveys, it is thus necessary to perform a sonic calibration for the depth-time conversion.

The sonic log calibration involves establishing a time-depth relation consistent with the seismic survey yielding the same vertical resolution provided by the sonic log. In other words, the sonic log measurements are recalculated to be compatible with variations in fluid and lithological composition, so the integrated travel time between two depth readings can be matched with the corresponding data from well velocity surveys.

A well velocity (or check shot) survey is carried out by measuring the travel times of head waves emitted from a surface shot by means of a geophone or an hydrophone placed at various depths in a well.

Check shot surveys are the predecessor of vertical seismic profiles. Vertical seismic profiles (VSP) may use more sophisticated tools to record the entire seismic wavetrain generated by surface source and transmitted through the earth filter downward. A VSP survey is usually recorded at a much higher density of depth points but may not cover the entire wellbore.

A high level of precision is required for the comparison of seismic and well logging datasets, particularly in the context of reservoir studies. Therefore, it is necessary to apply preliminary correction procedures before calibration of the sonic log to the check shot results or to well velocity survey.

The interval travel times obtained from the check shot survey and the sonic log are compared. The sonic log is then adjusted or calibrated according to well defined rules to match the check shot interval transit time.

Once the calibration has been carried out and a corrected time-depth relation established, it is possible to compare the well (logs) with surface seismic data. One technique employed for this purpose is the creation of a synthetic seismogram using density and acoustic velocity logs. Bulk density and acoustic velocity logs are used to create an acoustic impedance log. After depth-time conversion, the reflection coefficients derived from the acoustic impedance log are then convolved with an appropriate wavelet to produce the synthetic seismic section (often referred to as a seismogram).

Seismic data obtained from the vertical seismic profile (VSP) with or without source offset, are processed to provide seismograms at seismic frequencies that are directly comparable with synthetic sections and surface seismic sections. Even though these data have a poorer vertical resolution compared with well logging and a restricted frequency range, they can be used to adjust profiles obtained from seismic reflection surveys carried out at the surface. In addition, borehole seismic surveys can be used for defining appropriate operators for stratigraphic deconvolution and converting seismic sections to acoustic impedance sections or logs.

The combination of borehole seismic surveying with well logging techniques (see Fig. 2) enables a considerable improvement in the study of reservoirs. In some cases, the seismic and downhole logging datasets may appear inconsistent with each other. In the present study, an attempt is made to explain the various reasons for the observed discrepancies between these two approaches.

This text is divided into six chapters:

Chapter One is concerned with wireline logging. The principal features of the main well logging techniques are given in Section 1.1 (General Remarks), along with details about their application to the characterisation of geological formations. Particular emphasis is laid on the precautions to be employed in the use of well logs. The following Section (1.2) contains a detailed description of sonic and density logging tools as well as a discussion of their main fields of application. A brief summary of the interpretation of results is presented in Section 1.3, with special attention being nevertheless paid to the qualitative aspects, in order to bring out the importance of well logs in the main areas of interest to seismic surveying.

Chapter Two discusses conventional well velocity surveying and vertical seismic profiling (VSP). Both methods are presented with their associated data processing procedures, while some standard applications of VSP are also given.

Chapter Three deals with sonic calibration procedures leading to the definition of a Time-Depth relation ($T = F(Z)$), used to convert depth based log data to time based logs more easily comparable to seismic sections. Readers only requiring information about sonic log calibration methods may refer to this chapter. It is nevertheless recommended to read the preceding chapters as calibration requires a clear under-standing of the raw data quality control and editing procedures.

Methods for computing conventional seismograms (synthethic records) or VSP-type sections are presented in Chapter Four. The need for increasingly fine-scale data matching leads on to a discussion about stratigraphic deconvolution and inversion procedures — with or without the use of constraints — based on well logging and/or borehole seismic surveying measurements.

Chapter Five is concerned with the amplitude of seismic signals, outlining some of the causes of anomalies that arise during data matching. In addition, some observations are given on the significance of Direct Hydrocarbon Indicators (DHI's).

Finally, in Chapter Six, several case studies are given of the interpretations of seismic surveys and well logs.

Chapter **1**

WIRELINE LOGGING TECHNIQUES

1.1 GENERAL REMARKS

1.1.1 Logging techniques and Logs presentation

Well logs are obtained from tools introduced into a borehole on the end of a cable. These tools measure physical parameters such as current flow, naturel or induced radioactivity, wave propagation times and signal amplitude (acoustic or electromagnetic).

The petrophysical characteristics of the formations are inferred from calibrations carried out on the well logging tools.

Logging is generally implemented in open-hole sections, but is possible in cased holes when conditions are appropriate for the measurement of a particular parameter.

The measurements take account of a fairly small rock volume surrounding the tool: the vertical resolution (minimum measurable thickness) ranges from ten centimeters to one meter. The depth of investigation varies from a few centimeters to tens of centimeters; because of this fact, the environmental conditions in the well (e.g. presence of mud, mud cake, caved sections, fractures, micro fissures, etc.) can adversely influence the measurement.

Logging tools may or may not be positioned within the wellbore during the measurement. They may be centralised, eccentralised, eccentralised with a stand off, or in some cases the sensors are mounted on a pad type device to ensure borehole wall contact (see Fig. 1.1).

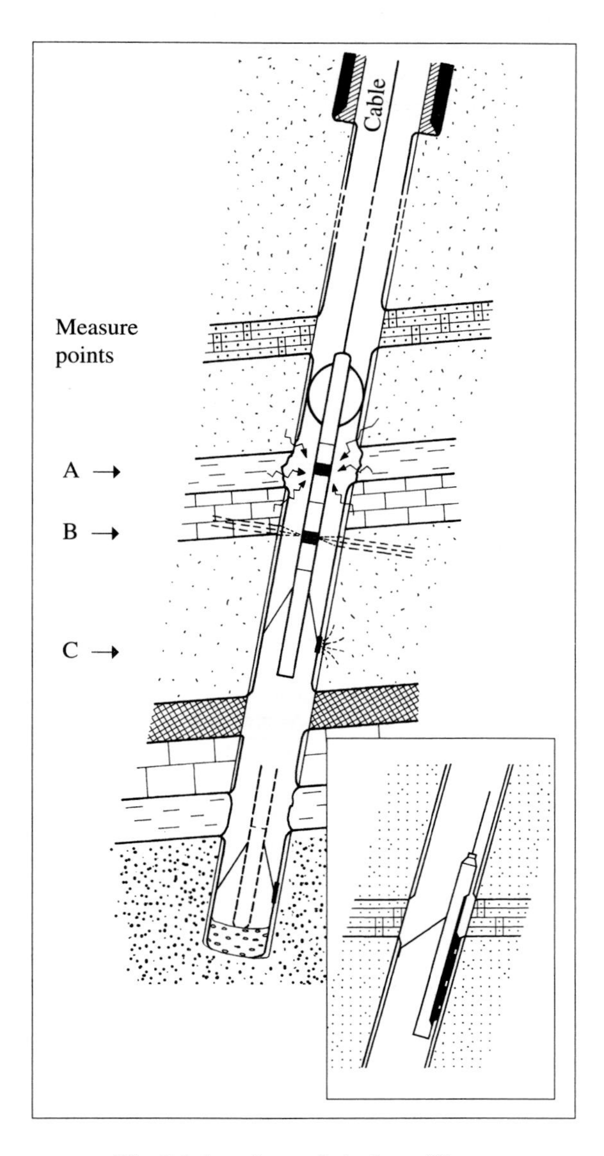

Fig. 1.1 Logging tools in the wellbore.

1.1.1.1 Logging tools

The principal families of well logging tools are:

(1) Caliper.

(2) Spontaneous potential (SP).

(3) Gamma Ray.

(4) Resistivity and conductivity tools (laterolog and induction logs), grouped together under the general term of resistivity logging:
- deep and shallow devices,
- micro-resistivity devices.

(5) Porosity logging tools, which are based on:
- induced radioactivity (neutron and density logs),
- acoustic wave propagation (sonic log),
- electromagnetic wave propagation (frequencies >1 GHz).

(6) Dipmeter.

1.1.1.2 Presentation of log data

The measured values are presented as curves plotted as a function of depth with respect to a datum. The sampling interval is generally 6 inches (15 cm), but 1.2 inches (3 cm) is sometimes used.

The zero depth datum can be set at the Rotary Table (RT), the Drill Floor (DF), the Kelly Bushing elevation (KB) or occasionally at Ground Level (GL). The zero datum is itself established with respect to a reference elevation (Datum Plane: DP), which generally corresponds to the Mean Sea Level (MSL).

Depth is recorded by measuring the cable being lowered into the well. The measurement accuracy after stretch and temperature corrections varies from 3 to 10 parts in 10 000. Measurements from combined tools are depth shifted by memorising the data of the uppermost sensors until the bottom sensor reaches the particular depth point.

TRADE NAMES OF MAIN LOGGING TOOLS

	HALLIBURTON		SCHLUMBERGER		ATLAS WIRELINE SERVICES	
Induction	Induction Electrolog Dual Induction Laterolog	IEL DIL	Induction Electrical Logging Ind. Spherically Focused Dual Induction Log Dual Induction SFL Array Induction Tool	IEL ISF DIL DIS AIT	Induction Electrolog Dual Ind. Focused	IEL DIFL
Resistivity	Dual Laterolog	DLL	Dual Laterolog	DLL	Dual Laterolog	DLL
Micro-resistivity	Microlaterolog Micro Spherically Focused	MLL MSF	Microlaterolog Micro Spherically Foc. Log Proximity log	MLL MSFL PL	Microlaterolog Proximity Log	MLL PL
Neutron	Sidewall Neutron Compensated Neutron Log	SNT CNS	Sidewall Neutron Compensated Neutron Log	SNP CNL	Sidewall Neutron Compensated Neutron Log	SWN CNS
Density	Compensated Density Log	CDL	Formation Density Litho Density Log	FDC LDL	Compensated Density Log Z.Densilog	CDL ZDL
Acoustic	Borehole Compensated sonic Long Spacing sonic	BCT BCT-EA	Sonic Log Sonic Long Spacing Sonic Digital Tool Dipole Sonic Imaging Tool	BHC SLS SDT DSI	Normal Space Acoustilog Long Spacing Acoustilog	AC ACL
Gamma ray	Gamma Ray Spectral Gamma Ray	UGR SGR	Gamma Ray Gamma Ray Spectroscopy	GR NGS	Gamma Ray Spectralog	GR SPL
Dipmeter	Four Electrode Dipmeter Omnigraphic Dip. Tool Six Arm Dipmeter	FED SED	High Resolution Dipmeter Tool Stratigraphic Dip. Tool Formation Micro Scanner	HDT SHDT FMS	Four-arm Dipmeter	DIP

- *The above list is not exhaustive and the abbreviations given may be subject to slight modifications.*
- *Other contractors offer comparable services under different trade names.*
- *Most of the commercial logging tools mentioned here can be used in combination with each other, under names and conditions given by the manufacturer (see catalogues for more details).*

Nevertheless, variations in the ascent rate of the tool, due to the condition of the borehole, may lead to incorrect depth of the tool and local deviations in the depth readings (i.e. "yo-yo" effects, sometimes attaining 1 m). The cable tension is recorded continuously, thus enabling the detection of such anomalies. In the case of separate logging runs, a poor setting of the zero depth datum may bring about a systematic and constant error in the readings.

Calibration marks on the margin of the log are used to check the cable speed (i.e., the theoretical ascent rate of the tool).

A typical log description is presented in the following paragraphs. The presentation of dipmeter data is treated in a different manner (see notes on use of dipmeter logging surveys).

A. Log header

The header includes data concerning the well itself (drilling and measured depths, depth reference used for datum, borehole diameter), the drilling conditions (mud, bit size, etc.) and information on the logging tools with their mode of operation (Figs. 1.2 and 1.4). In addition, some observations are made about the operation itself.

B. Overlap section (Fig. 1.2)

Logging is carried out with an overlap of about 30 m with the previous run in order to confirm the depth matching between the two runs and check the repeatability of the measurements. If the base of the previous run corresponds to a cased zone (as is usually the case), measurements that can be made through casing (radioactive or acoustic) will retain their general character but will be different due to casing and cement effects. Despite these discrepancies, the relative variations in the two logs are still sufficiently comparable to put the log "on depth". Checking of the depth values can also be performed with reference to the casing shoe, whose depth is measured on surface by different means.

HEADER

OVERLAP SECTION

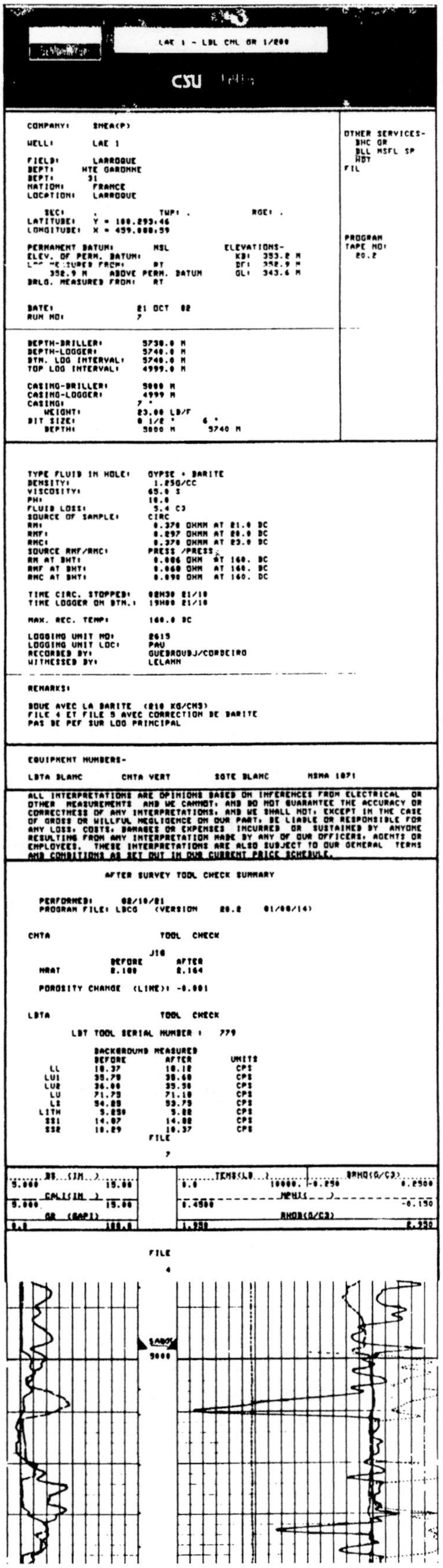

Fig. 1.2

Log presentation.

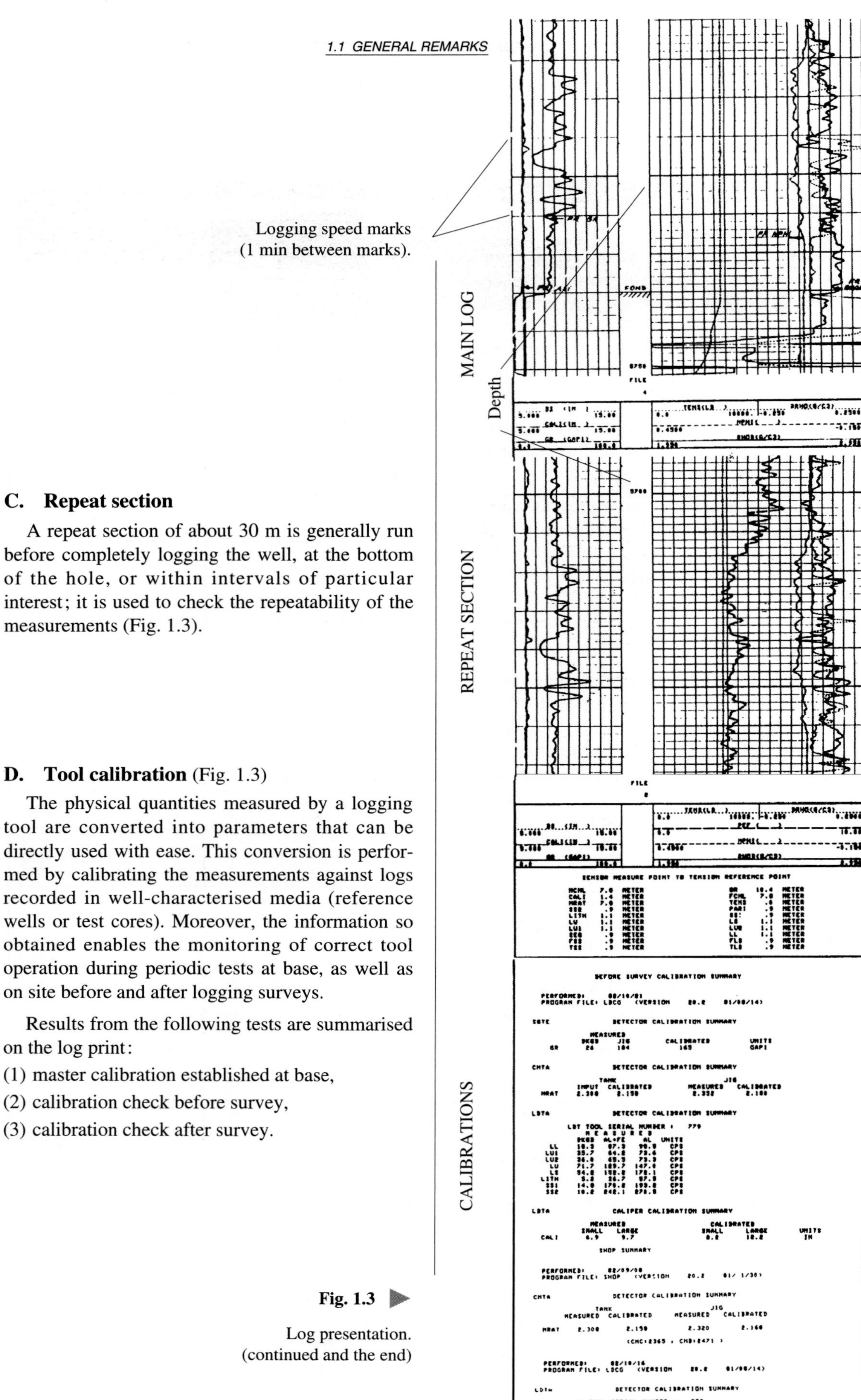

C. Repeat section

A repeat section of about 30 m is generally run before completely logging the well, at the bottom of the hole, or within intervals of particular interest; it is used to check the repeatability of the measurements (Fig. 1.3).

D. Tool calibration (Fig. 1.3)

The physical quantities measured by a logging tool are converted into parameters that can be directly used with ease. This conversion is performed by calibrating the measurements against logs recorded in well-characterised media (reference wells or test cores). Moreover, the information so obtained enables the monitoring of correct tool operation during periodic tests at base, as well as on site before and after logging surveys.

Results from the following tests are summarised on the log print:

(1) master calibration established at base,

(2) calibration check before survey,

(3) calibration check after survey.

Fig. 1.3

Log presentation.
(continued and the end)

Schlumberger

LAE 1 - LDL CNL GR 1/200

CSU Field Log

COMPANY: SNEA(P)
WELL: LAE 1
FIELD: LARROQUE
DEPT: HTE GARONNE
DEPT: 31
NATION: FRANCE
LOCATION: LARROQUE

SEC: . TWP: . RGE: .
LATITUDE: Y = 100.293,46
LONGITUDE: X = 459.088,59

OTHER SERVICES-
BHC GR
DLL MSFL SP
HDT
FIL

WELL LOCATION

PERMANENT DATUM: MSL
ELEV. OF PERM. DATUM:
LOG MEASURED FROM: RT
352.9 M ABOVE PERM. DATUM
DRLG. MEASURED FROM: RT

ELEVATIONS-
KB: 353.2 M
DF: 352.9 M
GL: 343.6 M

PROGRAM
TAPE NO:
20.2

DEPTH REFERENCES

DATE: 21 OCT 82
RUN NO: 7

DEPTH-DRILLER: 5738.0 M
DEPTH-LOGGER: 5740.0 M
BTM. LOG INTERVAL: 5740.0 M
TOP LOG INTERVAL: 4999.0 M

CASING-DRILLER: 5000 M
CASING-LOGGER: 4999 M
CASING: 7 "
WEIGHT: 23.00 LB/F
BIT SIZE: 8 1/2 " 6 "
DEPTH: 5000 M 5740 M

INFORMATION ABOUT STUDIED ZONE

TYPE FLUID IN HOLE: GYPSE + BARITE
DENSITY: 1.25G/CC
VISCOSITY: 65.0 S
PH: 10.0
FLUID LOSS: 5.4 C3
SOURCE OF SAMPLE: CIRC
RM: 0.370 OHMM AT 21.0 DC
RMF: 0.297 OHMM AT 20.0 DC
RMC: 0.370 OHMM AT 23.0 DC
SOURCE RMF/RMC: PRESS /PRESS
RM AT BHT: 0.086 OHM AT 160. DC
RMF AT BHT: 0.068 OHM AT 160. DC
RMC AT BHT: 0.090 OHM AT 160. DC

TIME CIRC. STOPPED: 02H30 21/10
TIME LOGGER ON BTM.: 19H00 21/10

MAX. REC. TEMP: 160.0 DC

LOGGING UNIT NO: 2615
LOGGING UNIT LOC: PAU
RECORDED BY:
WITNESSED BY:

MUD DATA

REMARKS:

BOUE AVEC LA BARITE (210 KG/CM3)
FILE 4 ET FILE 5 AVEC CORRECTION DE BARITE
PAS DE PEF SUR LOG PRINCIPAL

EQUIPMENT NUMBERS-

LDTA BLANC CNTA VERT SGTE BLANC NSMA 1871

ALL INTERPRETATIONS ARE OPINIONS BASED ON INFERENCES FROM ELECTRICAL OR OTHER MEASUREMENTS AND WE CANNOT, AND DO NOT GUARANTEE THE ACCURACY OR CORRECTNESS OF ANY INTERPRETATIONS, AND WE SHALL NOT, EXCEPT IN THE CASE OF GROSS OR WILLFUL NEGLIGENCE ON OUR PART, BE LIABLE OR RESPONSIBLE FOR ANY LOSS, COSTS, DAMAGES OR EXPENSES INCURRED OR SUSTAINED BY ANYONE RESULTING FROM ANY INTERPRETATION MADE BY ANY OF OUR OFFICERS, AGENTS OR EMPLOYEES. THESE INTERPRETATIONS ARE ALSO SUBJECT TO OUR GENERAL TERMS AND CONDITIONS AS SET OUT IN OUR CURRENT PRICE SCHEDULE.

Fig. 1.4

Log header.

1.1.1.3 The logging environment

A. Borehole conditions

The effects of caving or irregularities in well bore as well as the presence of fractures and micro fissures in the rock can all have some influence on the measurements obtained from different types of logging tools (i.e. according to depth of investigation, distance between source and receivers, presence of a sidewall pad, etc.).

B. Flushed and uninvaded zones (Fig. 1.5)

During drilling, the mud filtrate can penetrate the surrounding formations to distances of up to 1 m.

The rock immediately surrounding the borehole (a few decimeters) is flushed of its formation water and some of the hydrocarbons. This zone is defined as the "flushed zone". The "virgin zone" is that part of the rock completely unaffected by the invasion process.

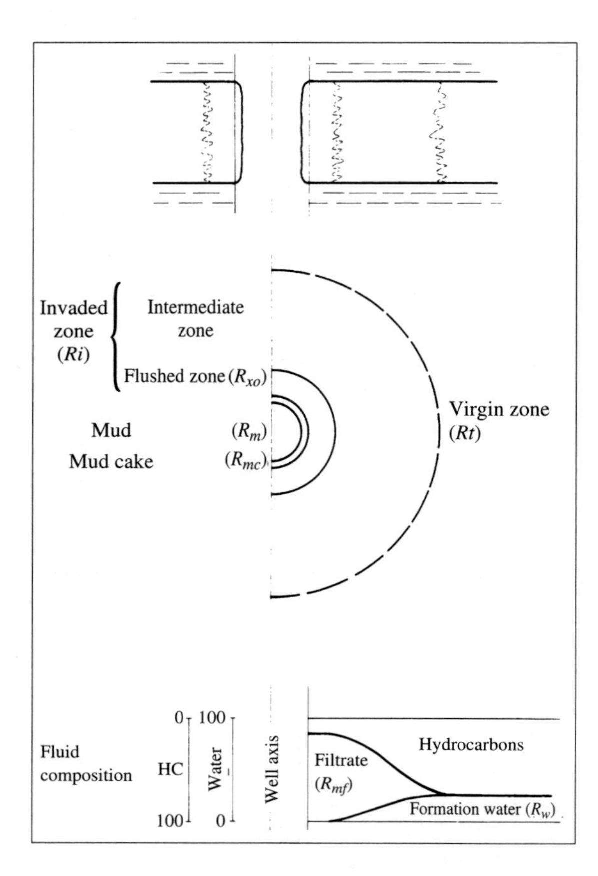

Fig. 1.5 Fluid invasion into porous and permeable formation containing hydrocarbons (R = resistivity).

The zone between the flushed and virgin zones is known as the "transition zone" where only some of the formation water and hydrocarbons are flushed by mud filtrate. The "invaded zone" is then defined as the flushed and transition zones combined.

Tools with a shallow depth of investigation, especially porosity instruments, read only within the flushed zone. Other types of tools provide logs that are influenced by physical characteristics in both the flushed and the virgin zones (e.g. resistivity logs).

Corrections and preliminary calculations are applied according to the type of log measurement in order to obtain parameters for the flushed/virgin zones.

1.1.2 Quality control and operating conditions

In the following paragraphs, a summary is given of the tests and corrections that are applied to raw data in order to obtain high-quality logs and minimise the possible errors arising from their use.

For a more detailed discussion of sonic and density tools, the reader is referred to the relevant sections of this chapter (1.2).

N.B.

- *Density logs are generally not recorded in cased wells.*
- *Acoustic logging tools can be run in cased wells. If the casing is perfectly cemented to the wall, it becomes transparent to acoustic logging. With poor cementation, however, casing waves appear in the recording. Acoustic Full Waveform recordings should be pre-processed (filtering of casing waves) before being used for characterisation of the formation. In cased wells, sonic logs may be used to assess the quality of the cementation (acoustic cement-bond logging).*

Identification of zones containing apparent anomalies

Whether or not the real cause of an anomaly is known (after simple inspection of the log or interpretation of the data) and irrespective of whether corrections can be carried out on the data, it is important to list the zones in the logs which contain anomalies. Among other applications, this list makes it possible to account for the deviations between sonic and seismic-survey travel-times. Such deviations can then be taken into account for improving the match between well logs and seismic surveys. Furthermore, the identification of anomalous zones enables the editing of sonic and density data with a view to computing synthetic seismograms.

1.1.2.1 Raw log quality control

A. Calibration procedures

It is indispensable to check whether the deviations between the values on the log calibration summary remain within certain tolerance limits (tolerance specifications are available from the logging survey contractor). Large deviations should lead to prudence in the use of the data and, if possible, a repeat of the log. With modern computerised systems, the calibration is performed by a calculator which stores the calibration parameters. In the case of a poorly calibrated log, a new calibration can sometimes be attempted by replaying the log with new parameters, thus avoiding a further downhole run.

B. Tool reliability check

The repeat section should yield values which are very close to the main log values. It should be noted that measurements from radioactivity logs are subject to statistical fluctuations.

C. Overlap between successive runs

It is very important to check the correct depth matching between two successive runs, and to check whether there are missing log data from one run to another.

D. Depth matching

It should be kept in mind that logging tools are run in a well on the end of a long cable subject to significant amounts of stretch due to tension and temperature effects. In normal and good borehole conditions, stretch is compensated for by the operator, however poor borehole conditions can easily cause depth errors that will need to be corrected for during any interpretation.

The depth matching between two successive logging runs is checked by means of overlap sections (see above).

The depth matching of two separately logs (e.g. sonic and density) should be checked in order to enable the subsequent calculation of a correct acoustic impedance log. A constant offset of whole sections of the log is readily detectable (differences of up to 1 m).

Recordings obtained in the same downhole run with combined tools are subject to the "yo-yo" effect (Fig. 1.6), leading to problems with depth matching that are more difficult to detect, unless cable tension is recorded. This type of effect can take place with an amplitude of about 1 m. Zones which exhibit strong contrasts in density and in acoustic propagation times have to be checked carefully before operating sonic calibration, as well as the tie between sonic log and seismic datasets.

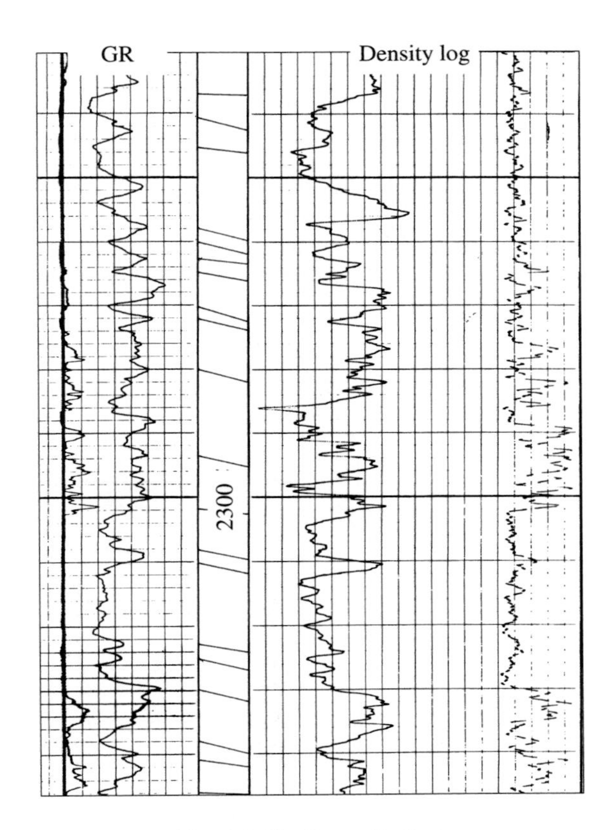

Fig. 1.6 The "yo-yo" effect.

E. Scale labelling

The log curve presentation scale should be verified as scales can be incorrectly labelled. Logs acquired and presented on computerised systems are generally immune to this type of problem.

1.1.2.2 Checking of log data

Errors arising from poor calibrations, incorrect scales or poor borehole conditions may lead to the use of bad data. It is thus important to cross check the data with methods for the interpretation of log data (e.g. cross-plots, etc.). Data should also be consistent with values in known and predictable marker beds.

The table below presents the values of interval transit-time (Δt) obtained by acoustic logging and the bulk densities (ρ_b) generally encountered in various materials:

	Δt (μs/ft)	ρ_b (g/cm^3)
Steel (well casing)	57	–
Anhydrite	50	2.98
Salt	67	2.05
Compact limestone	47	2.71
Compact sandstone	56	2.65
Compact dolomite	44	2.87

Environmental effects

Borehole conditions: the occurrence of caved sections, as detected by the caliper log (Fig. 1.7), can cast some doubt on the validity of the sonic transit-times (overestimated) or the densities (underestimated).

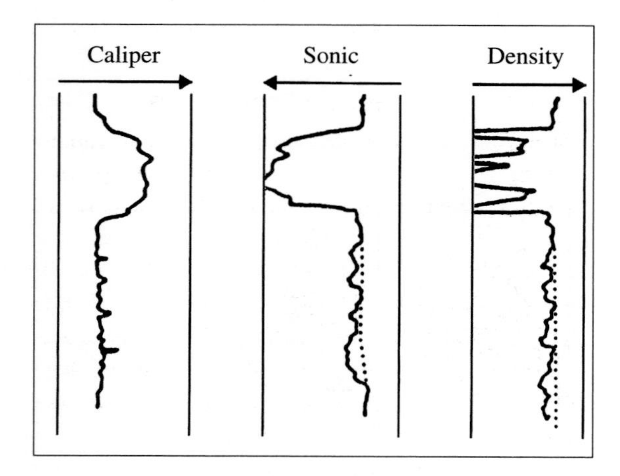

Fig. 1.7 Effect of wall caving on logs.

The presence of fractures or micro fissures in the borehole wall may also lead to an overevaluation of Δt and an underevaluation of ρ_b.

1.1.2.3 Correction of measurements

Measurements obtained from logging and seismic surveys do not consider exactly the same volume. In order to achieve a satisfactory comparison between these datasets, it may be necessary to apply corrections to the measurements.

A. Fluid invasion

Invasion corrections may be carried out during detailed quantitative study of the formations based on logging surveys and other types of well data (e.g. cuttings, tests, cores, etc.). These corrections are summarised for each of the logging tools in a later section of this chapter.

B. Lateral facies variations, anisotropy, dip effects

Sophisticated simulation techniques using seismic data as well as lithological and structural information (based on core and log analysis) are able to reduce the discrepancies observed between results obtained by borehole seismic surveying and logging. These costly and ponderous techniques are not in general use, being at least unnecessary for the simple matching of sonic against seismic data.

Some of the points mentioned here are further discussed in Chapter 4 (synthetic seismograms).

1.1.3 Guidelines for the use of logs

1.1.3.1 Desired parameters

A. Types of information sought

- Matrix density (ρ_{ma}), where the matrix corresponds to the solid portion of the formation without the shale fraction.
- Effective porosity (Φ), defined as the ratio between pore volume (exclusive of the shale fraction) and volume of the whole rock. This value includes all the free fluids.
- Water saturation (S_w) of the virgin zone, defined as the ratio between the volume occupied by free water and the volume corresponding to the

effective porosity. The difference with 100% is the hydrocarbon saturation (i.e. $S_{hc} = 1 - S_w$).

- Water saturation (S_{xo}) of the filtrate-flushed zone, which is the volume occupied by the filtrate water divided by the porosity of the formation. The difference with 100% is the residual hydrocarbon saturation (i.e. $S_{hr} = 1 - S_{xo}$).
- Shale content (V_{sh}), expressed as a percentage of the whole rock volume.
- Formation-water resistivity (R_w).
- Density of the hydrocarbon phase (ρ_{hc}).
- Rock mechanical parameters.

B. Data provided prior to logging

- Mud resistivity (R_m) and density (ρ_m).
- Mud filtrate resistivity (R_{mf}).
- Mud cake resistivity (R_{mc}).

C. Parameters obtained from measurement and calculation

- Natural gamma ray activity (GR).
- Static spontaneous potential (SSP).
- Resistivity of the uninvaded formation (R_t: approximated by deep reading resistivity values, e.g. ILd or LLd).
- Resistivity of the flushed zone (R_{xo}: approximated by shallow reading values, e.g. MLL ou MSFL).
- Bulk density of the formation (ρ_b).
- Photoelectric absorption factor (P_e).
- Neutron log porosity (Φ_N).
- Interval transit-time or acoustic propagation slowness (Δt).

The terms "neutron log", "density log" and "sonic log", are used to designate porosity type devices. "R_t" and "R_{xo}" correspond to resistivity logs best approximating the resistivities R_t of the virgin zone and R_{xo} of the flushed zone.

D. Basic equations

In a "clean" formation (i.e. without shale), the resistivity is related to the amount of fluid acting as an electrical conductor according to Archie's relation:

(1) in the uninvaded zone: $R_t = a/\Phi^m \times R_w/S_w{}^n$

(2) in the flushed zone: $R_{xo} = a/\Phi^m \times R_{mf}/S_{xo}{}^n$

(where in general, a = 1, m = 2 and n = 2)

Of the various equations that have been proposed for shaly formations, we consider here only the Poupon relation:

$$\frac{1}{\sqrt{R_t}} = \left(\frac{V_{sh}^{(1 - V_{sh}/2)}}{\sqrt{R_{sh}}} + \frac{\Phi^{m/2}}{\sqrt{aR_w}} \right) S_w^{n/2}$$

(for the flushed zone, it is simply necessary to substitute R_t, S_w and R_w with R_{xo}, S_{xo} and R_{mf}).

Knowing the effective porosity (Φ), the resistivities of the conducting fluids (R_w) for the formation water, R_{mf} for the filtrate, the resistivity of the shale (R_{sh}) and the shale content (V_{sh}) of the formation, the above equations make it possible to calculate S_w and S_{xo}.

With porosity logs, linear relations are conventionally assumed between the measured parameters and the percentage abundance of the different constituents (for more details, the reader should consult the section on sonic and density logging). The following are some examples of the relations used:

(1) density log:
$\rho_b = (1 - \Phi - V_{sh})\, \rho_{sh} + V_{sh}\, \rho_{sh} + \Phi\, \rho_f$

(2) neutron log:
$\Phi N = (1 - \Phi - V_{sh})\, \Phi N_{ma} + V_{sh}\, \Phi N_{sh} + \Phi\, \Phi N_f$

(3) sonic log (Wyllie):
$\Delta t = (1 - \Phi - V_{sh})\, \Delta t + V_{sh}\, \Delta t_{sh} + \Phi\, \Delta t_f$

where *ma* denotes the matrix (solid phase without shale), *sh* the shale present in the formation and *f* is the fluid included in the pore space volume. Note that the term "shales" includes the clay material, its associated porosity and bound water.

1.1.3.2 Environmental corrections

Logging does not provide a direct measurement of rock characteristics, and is influenced to a variable extent by the media which exist between the tool and the surrounding formations. Preliminary corrections for borehole effects (hole diameter, temperature, pressure, presence of mud and mud cake) and surrounding formations should be carried out before proceeding with quantitative interpretations. Equally, the consideration of such effects during the analysis of log data can contribute to an improvement of their quantitative interpretation.

1.1.3.3 Interpretation of results

Using high-quality measurements which have been corrected for environmental effects, it is possible —through the use of tool response curves and the various relations described above— to distinguish the different formations traversed by the well and also determine their mechanical and petrophysical properties. Simple on-site techniques (e.g. "Quick Look") and more sophisticated quantitative interpretation methods are used for the interpretation of the logs (see 1.3: Interpretation and Application of Well logs).

By means of simulations, it is possible to calculate the tool response under various conditions, i.e.:

(1) in a formation uninvaded by mud filtrate,

(2) in the same uninvaded formation, but with a different hydrocarbon composition,

(3) in a formation with slightly different characteristics (lateral facies variations, cementation, etc.).

It is even possible to calculate a synthetic log for a measurement that has not been recorded by well logging (e.g. pseudo-sonic and synthetic density logs).

1.2 SONIC AND DENSITY LOGS

1.2.1 Acoustic well logging

The use of acoustic well logging to determine the compressional wave velocity of a formation (Fig. 1.8) is a routine and relatively long-established practice (Summers and Broding, 1952; Vogel, 1952). More recently, however, the full waveform is recorded in order to determine the propagation velocities of the different types of waves and measure certain petrophysical properties with a view to obtaining lithological information (Arditty et al., 1984; Matthieu and Toksöz, 1984; Paillet and Turpening, 1984).

The different types of wave which make the wavetrain are as follows:

(1) **Body waves**:

- Compressional waves (also called *P* waves or longitudinal waves), where particle motion is co-linear with the direction of propagation.
- Shear waves (also called *S* waves or transverse waves), which have a vibration perpendicular to the direction of propagation.

Compressional waves travel faster than shear waves. The ratio of their velocities is a function of Poisson's ratio and is generally about 2 (found to be between 1.5 and 4). (*cf.* 2.2.3: Some Applications of VSP).

Shear waves will only be generated in a formation if they are faster than the compressional wave velocity in mud. Such formations are termed "fast" (compared to "slow" formations).

(2) **Interface waves**:

- Pseudo-Rayleigh waves are reflected conical dispersive waves (Biot, 1952). At low frequencies ($<$5 kHz), their phase and group velocities approach the *S* wave velocity of the formation, while at high frequencies ($>$25 kHz) their propagation velocity becomes asymptotic to the compressional wave velocity of the fluid. This type of wave is only encountered in fast formations.
- Stoneley waves are scattered along interfaces; in fast formations, they show group and phase velocities at high frequencies that increase asymptotically towards the propagation velocity in the fluid. In slow formations, these waves are more highly dispersed and are more sensitive to parameters linked to *S* wave propagation. At low frequencies, Stoneley waves are analogous to the tube waves observed in VSP surveys.
- Fluid waves are guided (or channel) waves, showing very little scattering, which are propagated through the fluid located between the tool and the borehole wall.

1.2.1.1 Equipment and data acquisition

The different features of acoustic logging tools are outlined below:

(1) By using tools with multiple transmitters and receivers it is possible to create stacks similar to surface seismic reflection survey stacks; common shot point, common reflection point and common receiver point. These data can be processed in a similar way to seismic surveys.

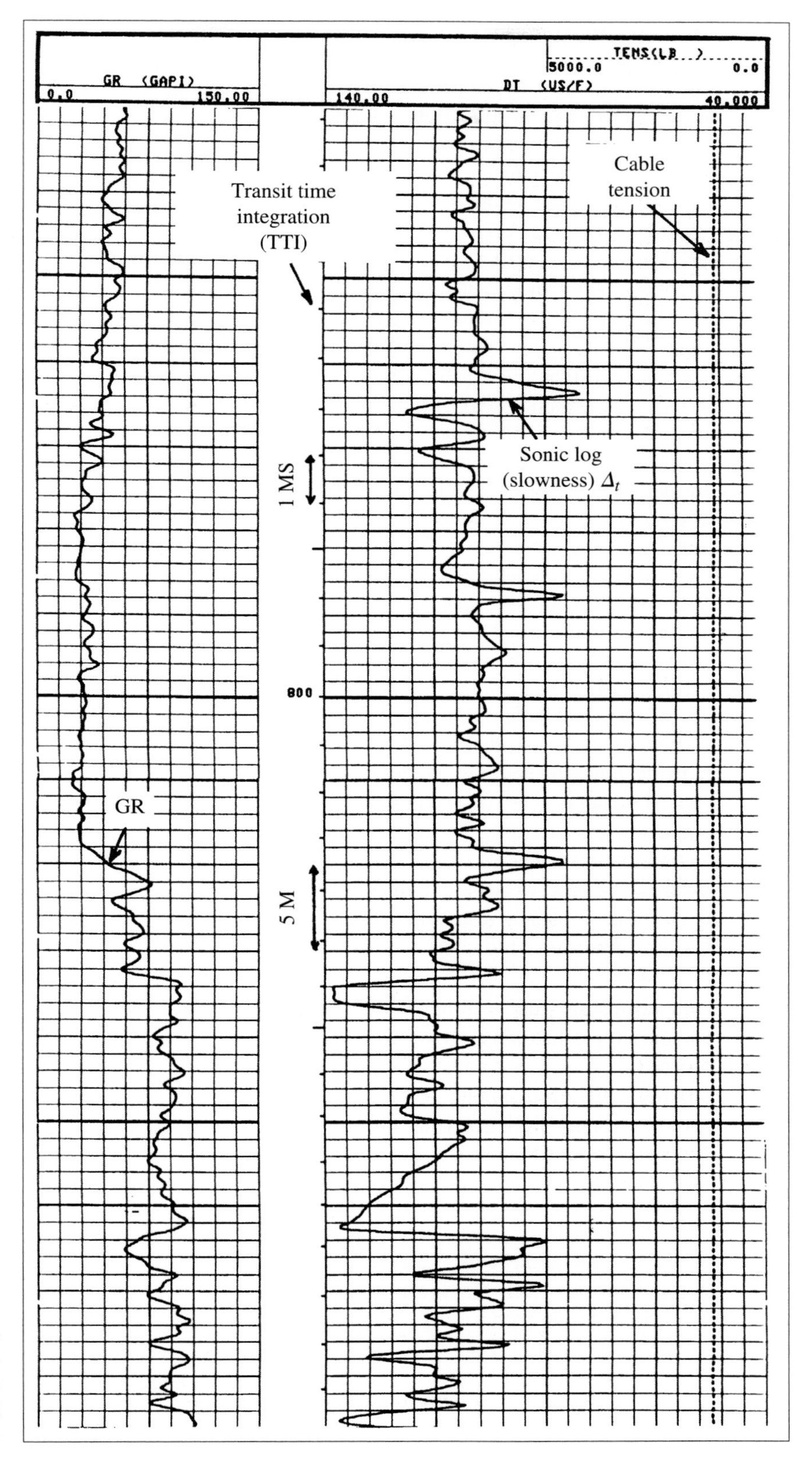

Fig. 1.8

Example of a sonic log, showing interval transit-times (Δt in µs/ft), combined with a GR log and cable tension curve.

(2) When tools are equipped with a large number of transmitters and receivers, data acquisition is performed sequentially by interrogating each transmitter-receiver pair in turn. With a limited number of transmitters and receivers, data acquisition can be carried out simultaneously from the same sonic pulse using several receivers.

(3) The transmitter-receiver "offset" and the inter receiver "spacing" determine respectively the depth of investigation and vertical resolution of the log. The vertical resolution is generally taken to be the receiver spacing (15-30 cm).

(4) The depth of investigation will be determined by optimising the path between transmitter and receiver in the mud column, invaded or altered zone and the virgin zone. Depending on the acoustic parameters of the media and of the tool geometric, the depth of investigation can then vary from 2 cm to 1 meter. Thus, in the usual case where the invaded zone is damaged and hence "slower", a short spacing tool will give a greater Δt than a long spacing tool which may read deeper into the formation.

(5) Sonic tools with long transmitter-receiver spacings enable a good time discrimination of the different arrivals, provided that the source is sufficiently powerful and the attenuation of the traversed media is not excessively high.

(6) The dominant frequency of the transmitter pulse (approximately 10 kHz) and the frequency bandwidth of the receiver (the ceramics used have a very wide frequency response: 100 Hz - 20 kHz) are important characteristics of the available tools.

(7) The time sampling step is generally a few microseconds for a listening period of a few milliseconds.

(8) Tools display different mechanical characteristics; some may be rigid (machined to avoid wave propagation via the tool body) while others are flexible.

A. Operations

For conventional operations and small diameter boreholes (<0.5 m) generally drilled in the oil industry, sonic logs are run with axially symmetric tools that are centred in liquid-filled wells (mud or water). The presence of gas bubbles in the mud usually leads to a mediocre quality of recording. In large-diameter boreholes, the tool is maintained in an off-centre position in order to avoid excessive dispersion of waves in the mud.

The logging speed is usually 10-15 meters per minute.

B. Calibration

Strictly speaking, the acoustic log does not require any calibration since the measurement of time is based on a quartz crystal with a precisely defined oscillation frequency, thus leading to almost no error in the calculated velocity. Several sequences of pulses are used in order to provide a measurement at 6 inch intervals. Even though the time measurement is relatively precise, the first arrival detection technique can lead to significant errors.

C. Types of tool

Monopole tools

The conventional sonic logging tool has an axial symmetry and is equipped with multidirectional receivers.

A compressional wave is generated in the fluid by the transmitter, thus giving rise to a compressionnal wave (P wave) and a shear wave (S wave) in the surrounding formation at the critical angles of refraction (see Fig. 1.9).

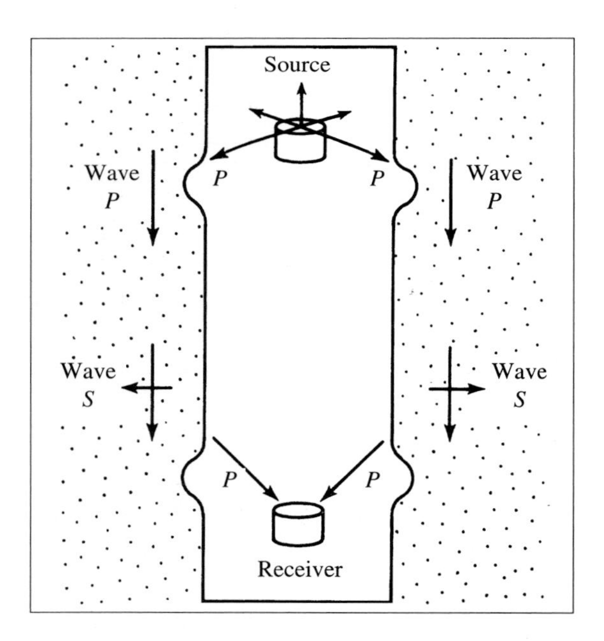

Fig. 1.9 Conventional acoustic logging: sketch showing generation of P and S waves. *(Williams et al., 1984)*

In a vertical well, this type of tool enables the recording of five modes of wave propagation (see following section, Presentation of acoustic data), viz.:

(1) refracted P waves,

(2) refracted S waves, only in fast formations,

(3) fluid waves,

(4) two types of dispersed tube waves, corresponding to the pseudo-Rayleigh and Stonely waves.

Estimation of S wave velocity using other types of waves

As mentioned above, shear waves can only be generated by refraction in seismically fast formations. The S wave velocity (V_S) of slow formations can be indirectly estimated in uncased wells using the Stoneley wave dispersion equation (Biot, 1952; Cheng et al., 1982). The main variables involved in this relation are borehole diameter, the density of wellbore fluids (ρ_f), the formation density of the formation (ρ_b), the Stoneley wave velocity (V_{St}) and the compressional wave velocity in the fluid (V_f). The dispersion equation for 0 frequency can be written as follows:

$$V_{St} = \frac{V_f}{\sqrt[2]{1 + \dfrac{\rho_f V_f^2}{\rho_b V_S^2}}}$$

At normal acoustic frequencies (3-20 kHz), it is necessary to independently measure six parameters in order to calculate the S wave velocity from Stoneley propagation modes. These parameters are:

(1) phase velocity of the Stoneley waves at a given frequency,

(2) fluid density,

(3) formation density,

(4) well diameter,

(5) compressional wave velocity in the fluid,

(6) compressional wave velocity in the formation.

Liu (1984) has proposed a method for the estimation of S wave velocities based on a comparison of these six measured parameters using theoretical charts deduced from synthetic acoustic logs. These logs are obtained from simulations using different values of shear modulus and different sets of frequencies, in addition to various densities, Poisson coefficients and tool/well diameters.

Tools with dipole-type source

Information on S wave propagation in both slow and fast formations may be obtained through the use of tools equipped with polarised transmitters and receivers. This type of tool generates P waves that are polarised at right angles to the axis of the well. The P waves create flexure modes at the borehole walls which give rise to pseudo shear waves travelling through the formation parallel to the axis of the well. Fig. 1.10 illustrates the principle of operation of this type of tool.

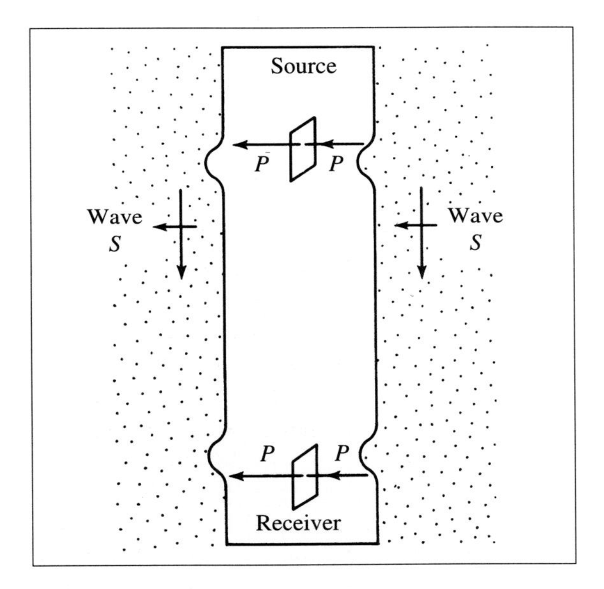

Fig. 1.10 Sketch showing asymmetric transmitter used for generation of S waves. *(After Williams et al., 1984)*

Mobil Oil has developed a dipole source tool known as SWAL (Shear Wave Acoustic Logging tool; Zemanek et al., 1984) where the transmitters and receivers are constructed using a technology based on the work of Sims (1968). In a SWAL tool, the transmitter and receivers are linked by a 7-conductor cable. The distance separating the transmitter and the nearest receiver may vary from 6 to 15 feet, while the distance between the two receivers is 3-5 feet. The transmitter frequencies are in the band 1-3 kHz.

Schlumberger proposes a tool of the same type known as DSIT (Dipole Sonic Imaging Tool) which has the unique feature of being able to operate in both monopole and dipole modes.

Figure 1.11 presents a comparison between logs obtained by LSAL (Long Spaced Acoustic Log - also a *Mobil* monopole tool) and SWAL tools in a seismically slow formation.

Different types of logging tools

The acoustic logging tools used by different contractors (Fig. 1.12) provide either real-time measurements of compressional wave velocity or are designed for full waveform data acquisition. The following examples are taken from the various available products:

(1) Monopole source or conventional sonic tools:

- the BHC sonic and LSS (Long Spacing Sonic) tools of *Schlumberger*, the Acoustilog of *Atlas Wireline Services* and the Full Wave Sonic Tool of *Halliburton Logging Services*,
- a sonde developed by the *Société d'Études de Mesures et de Maintenance (SEMM)*, which is a flexible tool equipped for simultaneous data acquisition on either the two nearest or the two farthest receivers,
- *Mobil*'s flexible LSAL tool (Long Spaced Acoustic Logging; *cf.* Williams et al., 1984), with transmitter-receiver and inter-receiver connections being made by cable,
- the flexible EVA tool, a product of *Elf-Aquitaine*,
- the Array Sonic (SDT-A/C), first proposed by *Schlumberger* in 1984 (Morris et al., 1984).

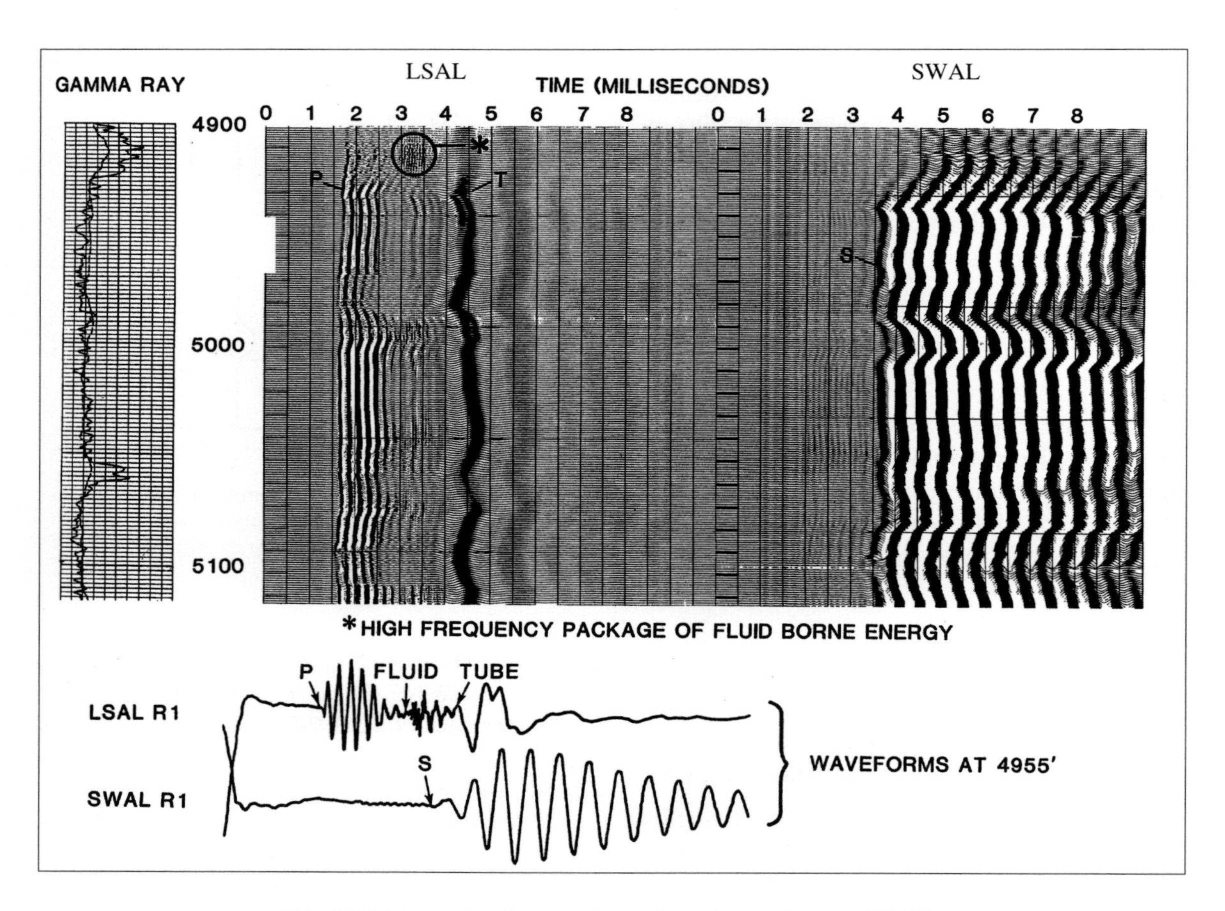

Fig. 1.11 Comparison between Long Spaced Acoustic Log (LSAL) and Shear Wave Acoustic Log (SWAL) recorded in unconsolidated Miocene formations. *(Williams et al., 1984)*

	SCHLUMBERGER			ATLAS	HALLIBURTON	ELF	SEMM	MOBIL	
	DSIT	SDT-C	DSWT	Acoustilog	FWS	EVA		LSAL	SWAL
		Rigid	Rigid		Rigid	Flexible	Flexible	Flexible (cable)	Flexible (cable)
Tool diameter	3" 5/8	3" 5/8	3" 5/8		3" 5/8	4"	48.3 mm		
Length	51'	37.9'			14.33 to 16.61 m	18 m	6.75 m		
Weight		350 kg			400 kg	300 kg	50 kg		
Pressure	20 000 PSI	20 000 PSI			20 000 PSI		300 bar	20 000 PSI	20 000PSI
Temperature	350°F	375°F			400°F		85°C	400°F	400°F
Recording speed	< 1800'/hr	1400-3000'/hr	900 or 1200'/hr	< 3600'/hr	30'/min (< 4000'/hr)		6-20 m/min		
Centring		centred			centred	"Moustache" centraliser	"Moustache" centraliser		
Data transmission	digital (12 bits)	digital	digital		digital	analogic	analogic		
Transmitters	2 dipoles 1 single-pole	2 magnetostrictive single-pole	2 single-pole	2 single-pole	1 single-pole piezoelectric device	4 single-pole magnetostrictive devices	1 single-pole magnetostrictive device	1 single-pole magnetostrictive device	dipole
Spacing between transmitters		2'	2'	2'		0.25 m		5'	3-5'
Frequency		5-18 kHz			13 kHz		17-22 kHz	15 kHz	1-3 kHz
Resonant frequency	1.5 kHz (single and dipole, LF) 10 kHz (single-pole, HF)								
Receivers	8 single/pole devices	2 + 8 piezoelectric ceramic devices	16 ceramic devices	12	4 piezoelectric devices	12 piezoelectric ceramic devices	4 piezoelectric devices	2 piezoelectric devices	
Spacing	6"	8 spaced at 6" 82 spaced at 2'	10 at 3" 5 at 6"	1/2'	1'	1 m	0.5 m (0.25 m)		
Band- pass	100 Hz - 5 kHz (single and dipole, LF) 8-30 kHz (single-pole, HF)	5-18 kHz			Flat from 07 to 30 kHz	1-25 kHz	1-30 kHz	30 Hz to 30 kHz	
Offset	9-12.5' (single-pole) 11-15' (dipole)	3-14.5' (8'-13,5')	5'-25'	up to 17.5'	10,11,12,13' From 10 to 20.5'	1-12.75 m	1 or 3 m	15-20'	6-5'
Vertical resolution	2' (→1/2")	2' (→1/2")		2'	2'	0.25 m	0.5 m (0.25 m)	5'	3-5'
Sampling step	512 samples/trace 10 µs (single-pole, HF) 40 µs (single and dipole, LF)	5 ms signal (to 10 ms) 5, 10 ou 20 µs		4 µs		2000 samples/trace from 5 µs	2000 samples/trace from 4 µs	2000 samples/trace from 5 µs	2000 samples/trace from 5 µs

1' (foot) = 0.3048 m   3"5/8 = 92 mm   LF = low frequency

1" (inch) = 0.0254 m   HF = high frequency

Fig. 1.12 Main characteristics of acoustic logging tools *(intended only as a rough guide).*

(2) *S* wave dipole emitter tools:

- *Mobil*'s SWAL tool.

(3) Mixed type tools (operating both in single-pole and dipole modes):

- *Schlumberger*'s DSIT,
- the MAC tool of *Atlas Wireline Services*.

Figure 1.13 illustrates the main features of some these tools (see manufacturers specifications for more detailed information).

D. Presentation of acoustic data

The standard way of presenting acoustic data is in the form of depth logs; the most common presentation is the Δt (interval transit time or slowness) versus depth for *P* waves, which can equally be used for *S* or Stoneley waves. In addtion, the integrated *P* wave transit time (also termed TTI: Transit Time Integrated; *cf.* Fig. 1.8) is also indicated on the log.

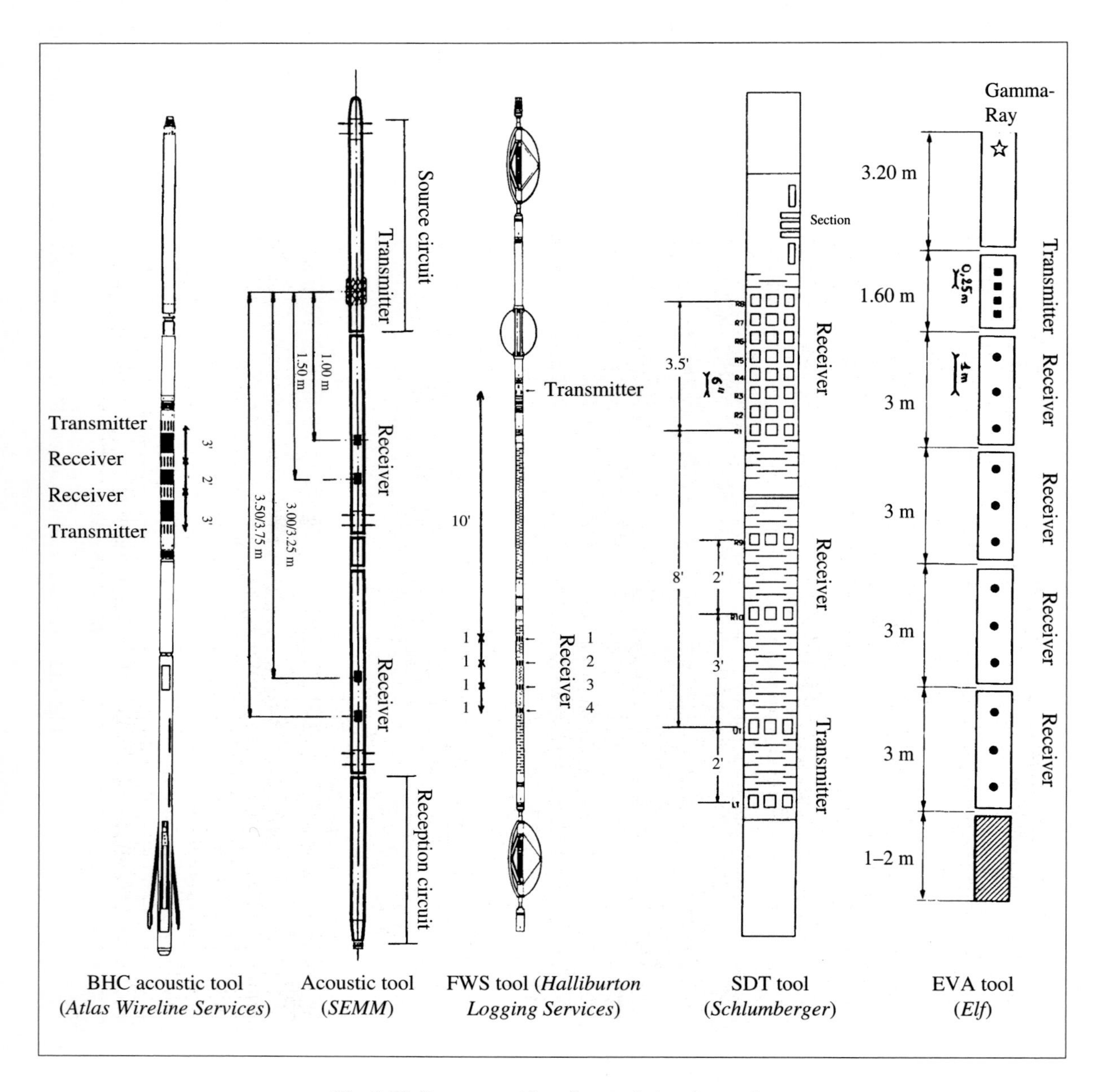

Fig. 1.13 Some examples of acoustic logging tools.

Full waveform data can also be plotted as common-offset sections, or alternatively as gathers of common source or receiver points, which are analogous to the sections utilised in seismic surveying. A common-offset section is made up of a set of sonic recordings plotted as a function of depth, the measurements being obtained with a constant transmitter-receiver distance (offset). Full waveform data may also be presented as a VDL (Variable Density Log) which is a continuous plot of the z-axis of the waveform with the amplitude variations being coded in grey scales.

The series of common-offset sections presented below serve to illustrate the different wave propagation modes.

With single-pole sources, the most commonly observed modes are due to refraction between the well fluid and the formation; in fast formations, compressional and shear waves are encountered as well as interface waves of the Stoneley or pseudo-Rayleigh type.

Figure 1.14 is an example of a common-offset section showing clearly distinct modes of propagation that was obtained with a 3 m transmitter and receiver distance.

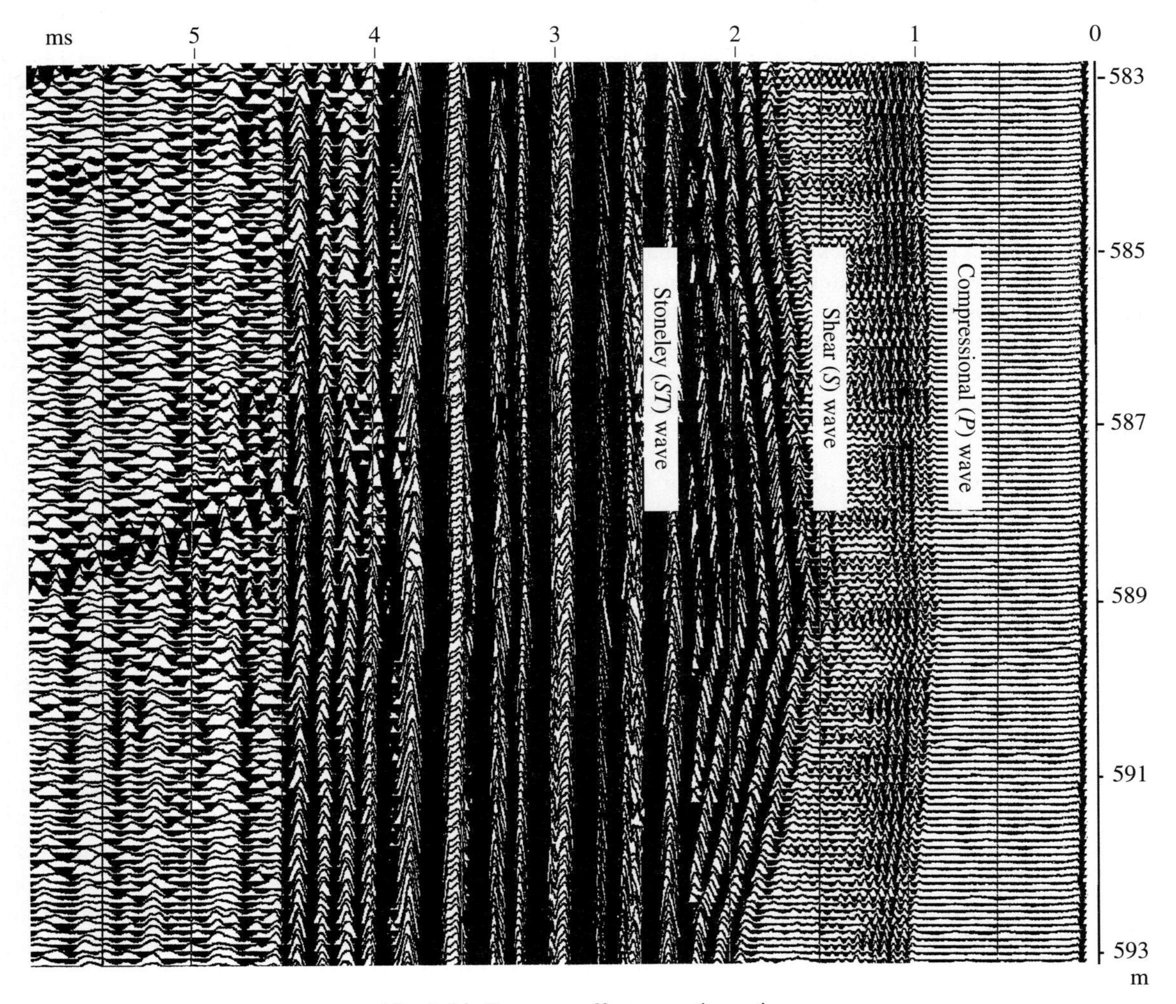

Fig. 1.14 Common-offset acoustic section.
(Courtesy of SEMM)

Some observations:

The acoustic logging tools integrate the geological information over a distance which is either the receiver spacing or the source-receiver offset (D).

In the case of acoustic data plotted as a common offset section (offset = D), the horizontal boundary between two semi-infinite formations appears as a ramp of length D. When transmitter T_x and receiver R_x are located in the same formation (1 or 2), the travel time obtained (T_1 or T_2) refers to the formation under consideration (1 or 2, as the case may be). However, if the tool is astride the formation boundary, the travel time T is intermediate between T_1 and T_2. The transition from T_1 to T_2 takes place over an interval of length D, while the start of the transition zone occurs at a distance of $D/2$ before the formation boundary.

Thus, the geological interface is situated at a distance which corresponds to half the offset away from the start of the sonic log transition zone, assuming the depth reference for the tool is taken at the mid-point between transmitter and receiver.

In the case of a layer of thickness H intercalated between two formations which are both thicker than D, the boundaries of the layer are always located at a distance of $D/2$ with respect to the extremities of the log transition zone (start or end, according to whether the tool is being pulled or lowered).

The travel time is not representative of the seismic velocity of the interlayer if its thickness is less thand D, as shown in Fig. 1.15.

Although the above remarks refer to travel times (integrated transit time over the source-receiver offset), they are equally applicable to the acoustic signal recorded by a receiver designed for full waveform response (Fig. 1.16). The same however applies to transit times with the difference that the receiver spacing becomes the limiting factor instead of transmitter-receiver offset.

Generally speaking, a common-offset section with spacing D has a vertical resolution that is less than a log obtained by comparing measurements from two receivers separated by distance E, the receiver spacing E (normally in the range 10-50 cm) is being generally less than D, which usually lies between about 1 and 10 m.

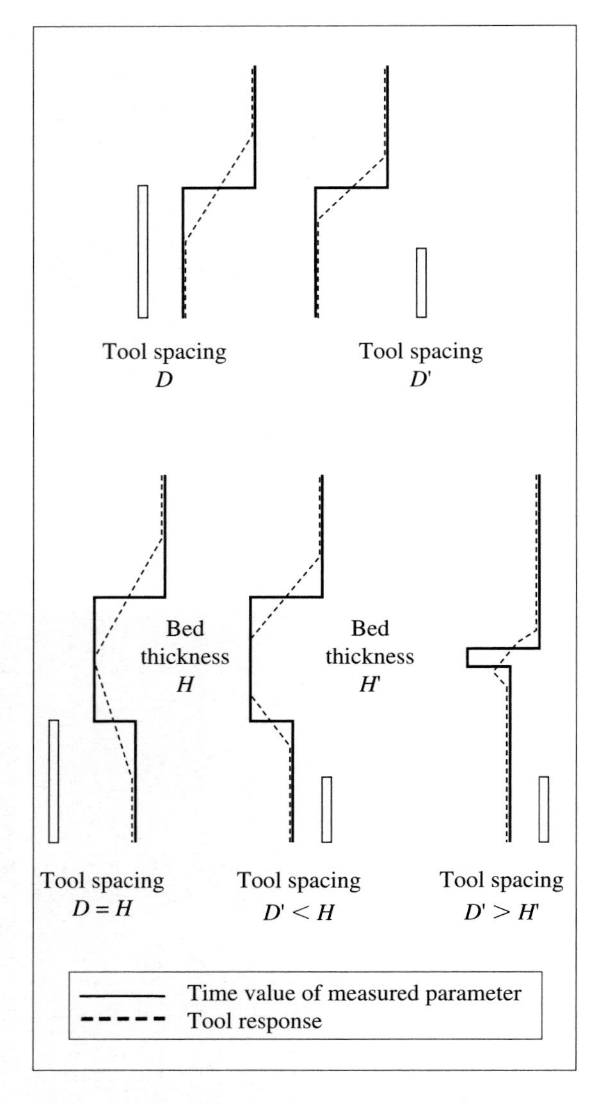

Fig. 1.15 Response of logging tools as a function of layer thickness.

When there is a contrast in acoustic impedance between two formations, the different types of wave (refracted body waves and interface waves) can be converted and reflected as shown in Fig. 1.17. This leads to the appearance on common-offset sections of V-shaped events (or criss-crosses) with low apparent velocity. When a refracted P wave is reflected off a formation boundary with a dip nearly perpendicular to the axis of the borehole, the apparent velocity of the reflected P wave is equal to the half-velocity of the incident P wave (Astbury and Worthington, 1986).

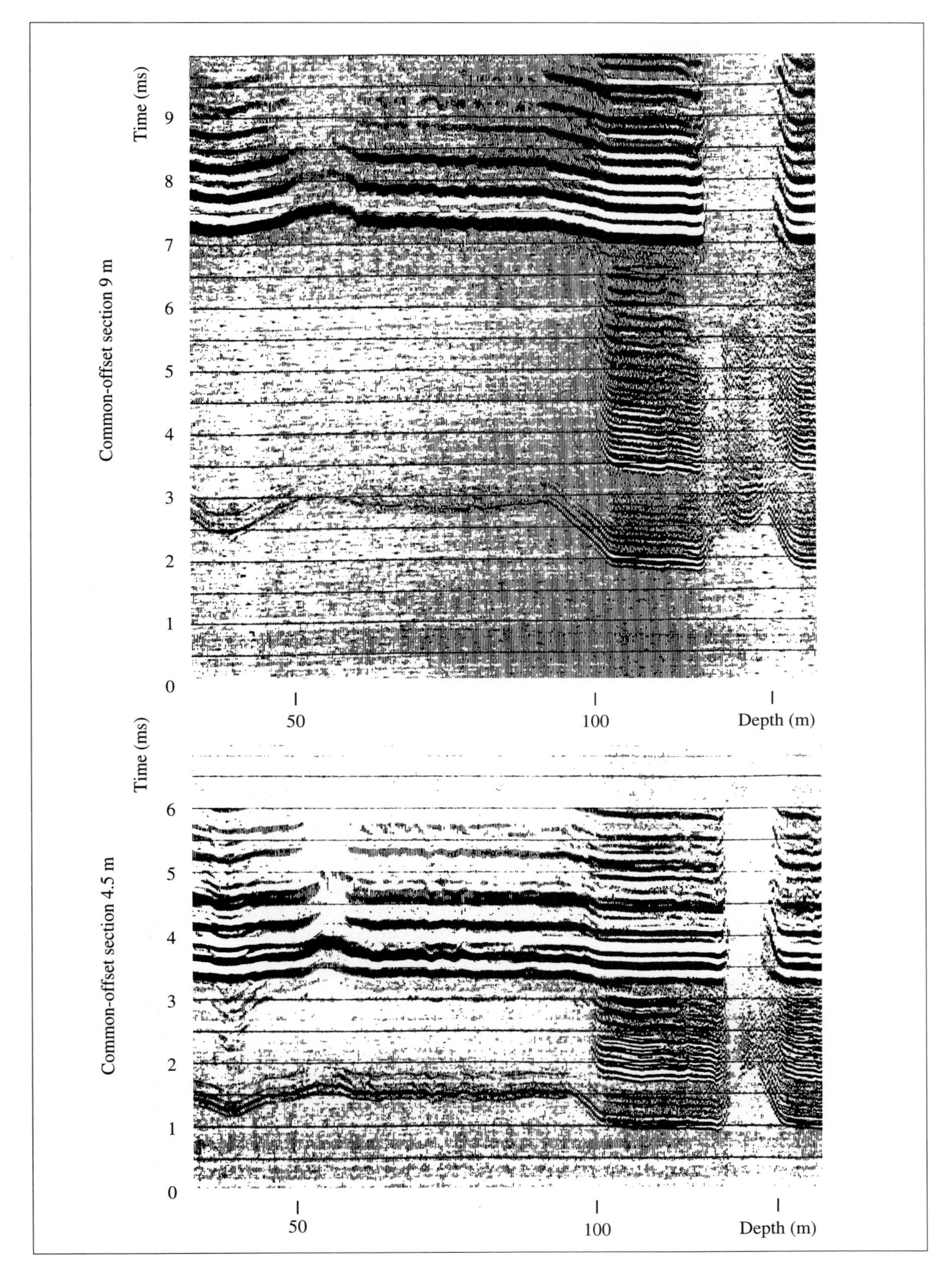

Fig. 1.16 Full waveform recordings (EVA tool). *(Courtesy of Elf)*

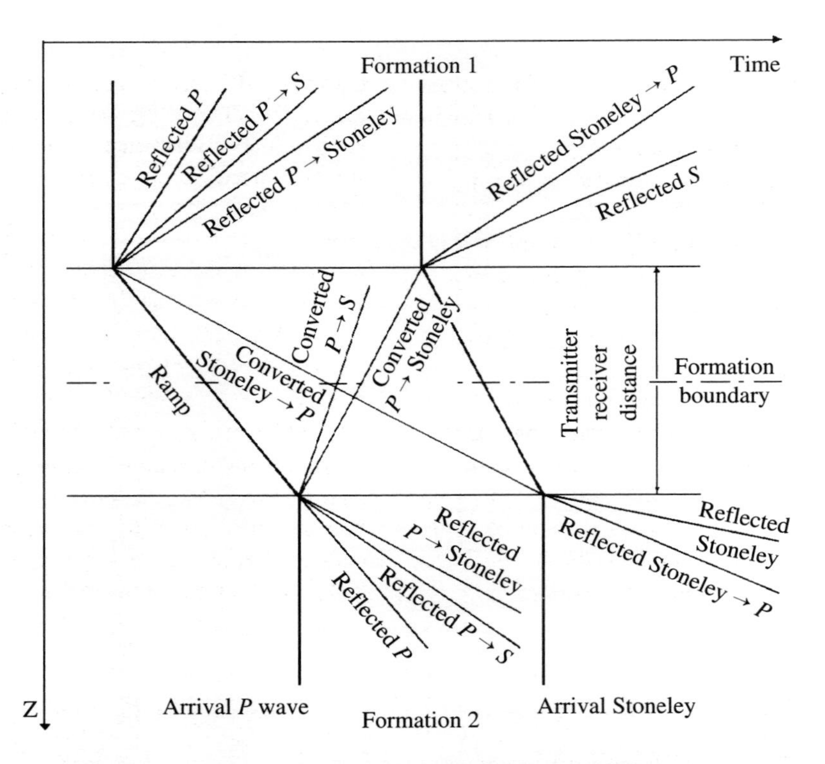

Fig. 1.17 Sketch section showing propagation modes generated by the passage of waves across a velocity discontinuity. *(Astbury and Worthington, 1986)*

The recording shown on Fig. 1.18 is a clear illustration of this type of propagation. In this section, it is possible to see the reflections of refracted *P* waves as well as various interface modes of the Stoneley type.

These reflections take place near the borehole wall and cannot under any circumstances provide information about deep heterogeneities in the vicinity of the well. Such reflections can also occur at casing joints, in the presence of caved sections and any other kind of borehole wall irregularity.

It is also possible to detect "leaky" compressional modes that are trapped in the well. These guided waves are created when the compressionnal body wave has an angle of incidence greater than the critical angle. At such angles, the incident energy is completely reflected in towards the well, thus giving rise to a channelled mode that is dissipated by conversion to shear waves at the well wall.

In general, these channelled modes are very strongly attenuated and are practically impossible to detect. Their attenuation is a function of the

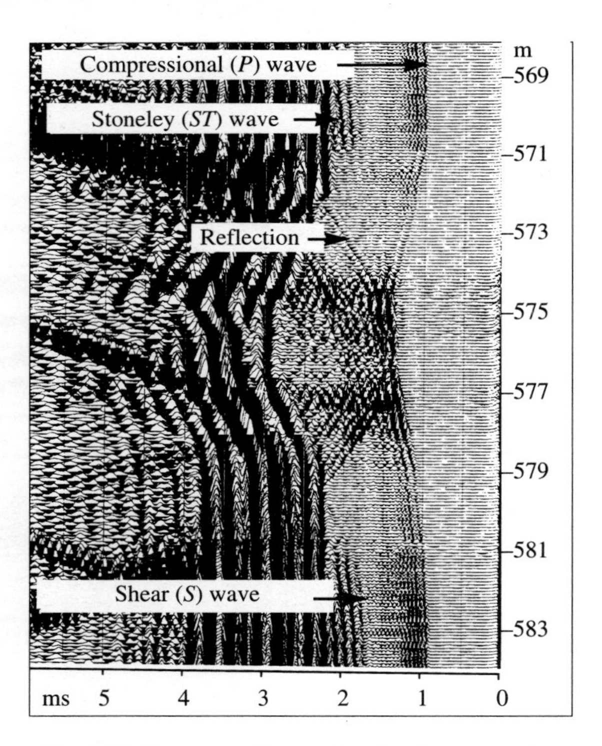

Fig. 1.18 Common-offset section showing conversion of propagation modes. *(Courtesy of SEMM)*

shear modulus of the formation, so, when the Poisson coefficient tends to 0.5, the formation appears as a fluid and such modes are no longer attenuated (Hornby and Chang, 1985).

This type of wave propagation is shown, for a depth interval of 993-1017 m on Fig. 1.19. Channelled waves are only slightly scattered, being propagated at an almost constant group velocity which is very close to that of the mud.

Using tools designed for full waveform acquisition, it is possible to obtain information on the characteristics of formations even when the well is cased. By the same token, sonic logs can be used to pick out poorly cemented zones that are characterised on VDL presentations by the presence of high energy arrivals (so-called tube waves) having infinite apparent velocity and long duration tending to mask the other waves.

A profile of casing wave energy versus depth is conventionally used as a cement-bond log. With good cement bonding of the casing, casing waves do not greatly hinder identification of the refracted body waves or interface waves. However, the Stoneley and pseudo Rayleigh waves loose their distinctive character and become no longer representative of the formation. In the case of formations with very high velocity ($\Delta t < 57$ µs/ft), the first arrivals from the surrounding formations can be confused with casing waves: leading to wrong interpretations (as an example poor cement bonding of the casing instead of fast formation and good cement bonding of the casing).

The acoustic section shown on Fig. 1.20 illustrates casing wave events in the depth interval 856-873 m, indicating the presence of a poorly cemented zone. Refracted *P* wave arrivals can be just picked out at 0.8 ms in this interval.

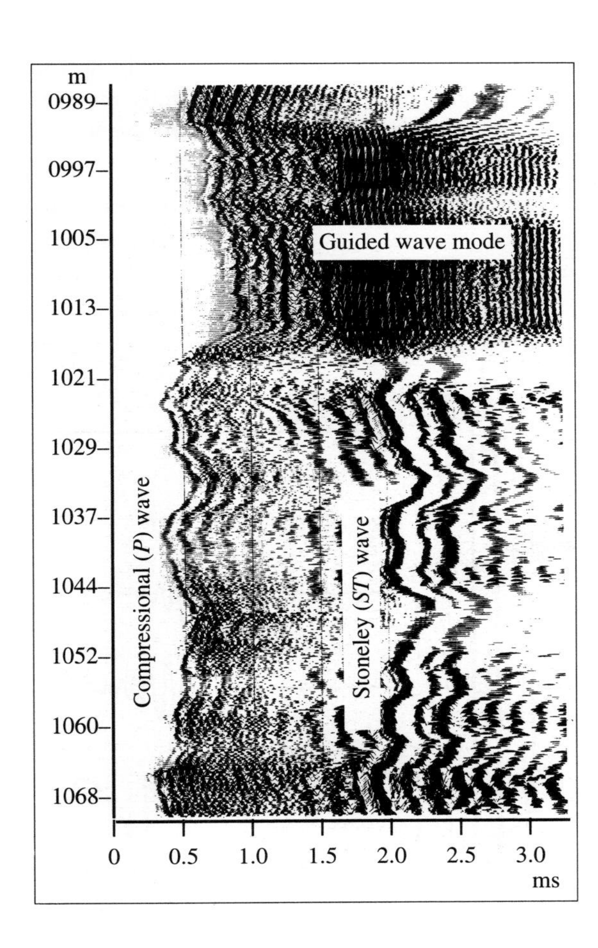

Fig. 1.19 Constant offset section showing channelled propagation modes. *(Courtesy of Gaz de France)*

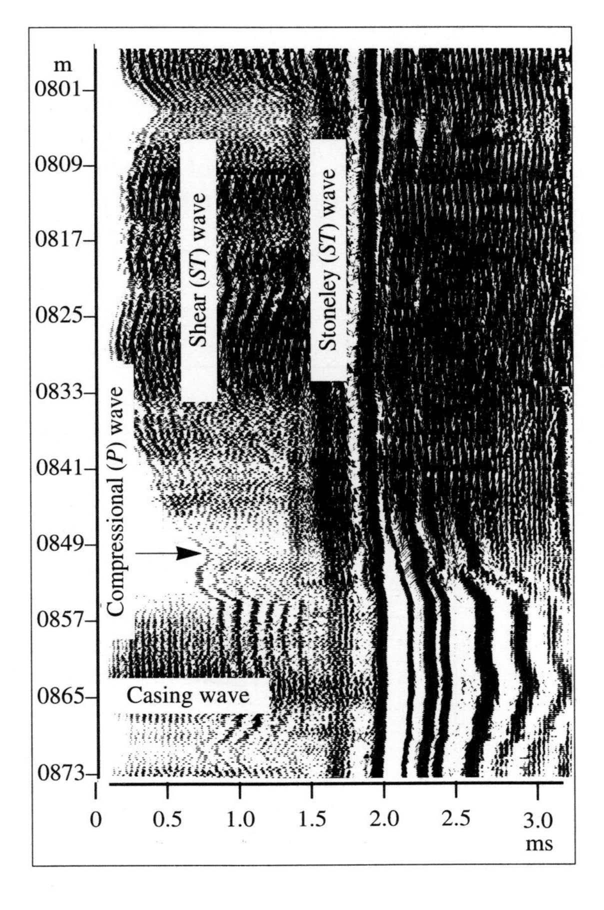

Fig. 1.20 Example of "casing" waves. *(Courtesy of Gaz de France)*

E. Obtaining acoustic parameters from log data

The following acoustic parameters are obtained from sonic logging surveys:

(1) propagation velocity,

(2) attenuation,

(3) frequency.

These physical quantities refer to the main wavesets observed, notably the refracted *P* and *S* modes (only in fast formations, recorded with single-pole axially symmetric tool) as well as Stoneley waves.

The measurements thus obtained from sonic logs can then be used to derive other parameters such as rock mechanical properties (e.g. Poisson coefficient, bulk and shear moduli, etc.) and porosity, as well as indications of permeability, degree of fracturing and lithology (see Section 1.3: Interpretation and application of well logging results).

The acoustic parameter conventionally acquired in real time is the *P* wave velocity, expressed in terms of transit time Δt (units of μs per foot), which is obtained by picking the refracted *P* wave arrivals.

a. Estimation of velocity and integrated travel time (TTI)

Measurement of the travel time between receivers enables a determination of the slowness Δt of a formation, the reciprocal of acoustic wave velocity.

The integration of Δt over depth yields the integrated travel time (also known as TTI: Transit Time Integrated), which corresponds to the time taken for a wave to travel over a certain distance (Fig. 1.21). It is generally represented by bars marking out 1 ms intervals of time, with 10 ms between longer bars.

α. Estimation of slowness (Δt or Dt) using standard tools

With conventional sonic tools, the transit time measurement of the *P* or compressional wave, Δt (μs/ft), is obtained by subtracting the travel times between a transmitter and 2 receivers usually spaced 2 feet apart. This technique theoretically removes the effects of the wellbore, provided the tool is centered in a perfectly cylindrical hole.

Detection of the first arrival (generally belonging to the *P* waveset) is performed using a minimum energy threshold (bias) (Fig. 1.22).

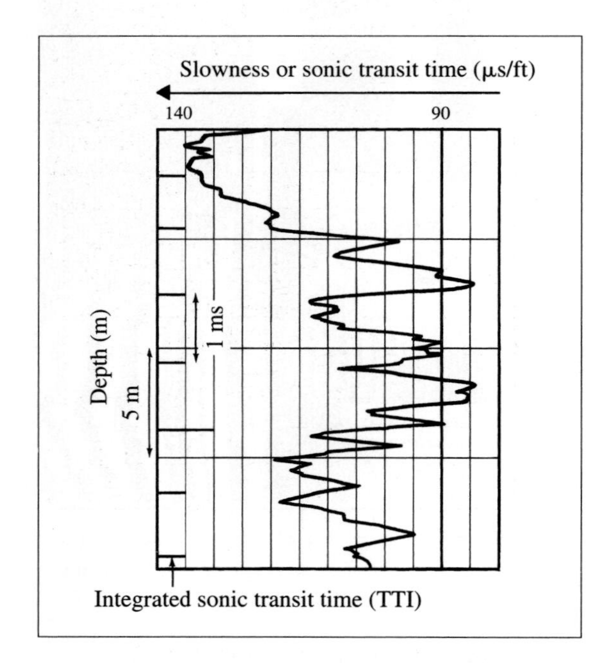

Fig. 1.21 Integrated travel time indicated on sonic log.

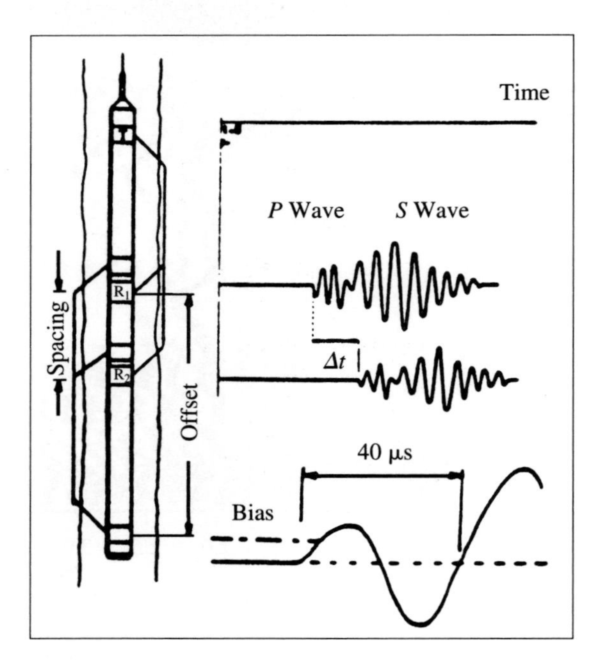

Fig. 1.22 Detection of arrival using threshold filter.

In practice, conventional systems are compensated for borehole effects and for the tool-hole configuration, so transmitter-receiver pairs (or a system incorporating automatic storage of transmitter-receiver transit times) are used to ensure a more reliable measurement of travel time. The compensation procedure involves averaging the transit times over the same depth interval from different receiver pairs.

Problems arising from the detection system

Any phenomenon involving noise or attenuation of the acoustic wave may cause errors in the measurement of interval transit time. In the case of tools using a bias system, errors in the estimation of transit time (Δt) are generally introduced by three factors: noise, cycle skipping and signal stretching. The correct choice of tool configurations and operating procedures will generally eliminate must of these problems, particularly with surface computer signal detection techniques. Zones with such problems as well as recording conditions must be identified in the remarks section of the log header.

Noise (see Fig. 1.23), with an amplitude higher than the bias level, occurring on a nearby (or distant) receiver will lead to an overestimation (or underestimation) of Δt. Generally speaking, the noise appears as isolated peaks and can be eliminated by editing.

Cycle skipping (see Fig. 1.24): signal attenuation may be linked to poor centring of the tool in the borehole, the presence of gas in the mud, etc., and can lead to the skipping of one or several cycles, thus lengthening the transmitter-receiver travel time (dT). As the frequency of the wavetrain is approximatively 25 kHz and thus corresponds to a period of 40 μs, this travel time dT is increased by a multiple of this signal period.

For a tool with two receivers and two sources (Fig. 1.22), the slowness Δt is given by the following relation:

$$\Delta t = \left[\frac{\frac{(T_2 - T_1)}{e} + \frac{(T'_1 - T'_2)}{e}}{2} \right]$$

where T_1 and T_2 are the direct travel times to receivers R_1 and R_2, T'_1 and T'_2 are the reciprocal travel times and e is the distance between the two receivers.

With a compensation system, assuming that cycle skipping occurs only on one source-receiver pair, the error in the determination of Δt is thus equivalent to $\Delta t/e$. For $e = 2$ ft and $dT = 40$ μs, the error on the raypath parameter Δt is 10 μs/ft.

Occasional cycle skipping gives rise to isolated peaks which have little influence on the calculation of integrated travel time. Nevertheless, if the cycle skipping persists through a sufficiently thick interval, entire sections of the log may be affected (Fig. 1.25), thus leading to an overestimation of Δt values as well as the integrated travel time.

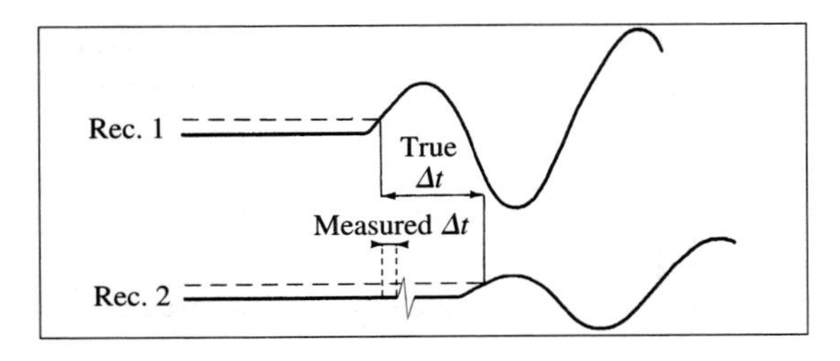

Fig. 1.23 Noise detection.

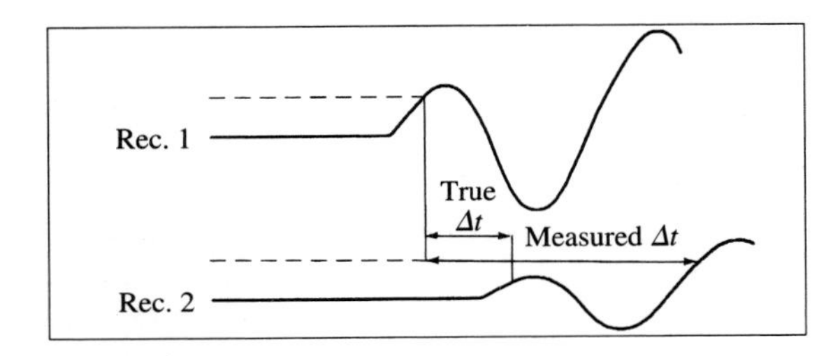

Fig. 1.24 Cycle skipping.

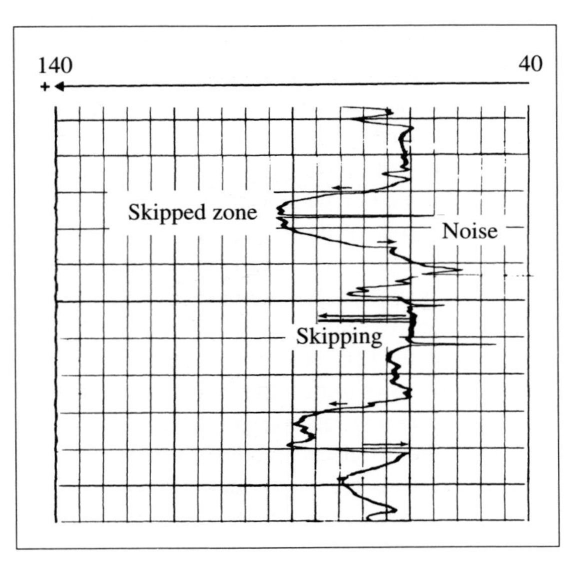

Fig. 1.25 Sonic log interval showing effect of cycle skipping.

Δt stretch (see Fig. 1.26): under the conditions described above, a rather weaker attenuation of the signal may lead to an increase in the measured time (up to a maximum of 40 μs divided by 4), thus associated with a maximum error on Δt of 2.5 μs/ft.

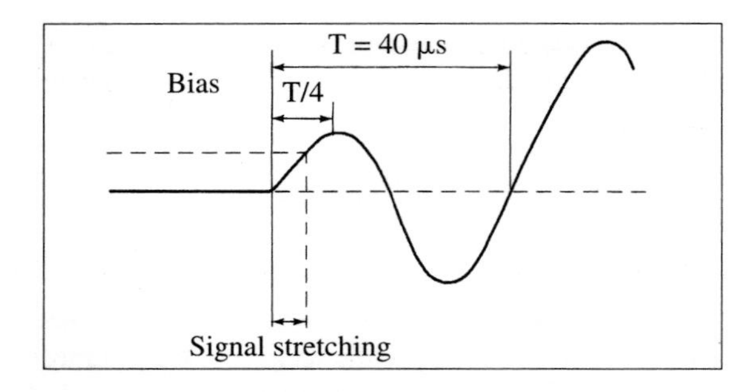

Fig. 1.26 Δt stretch effect.

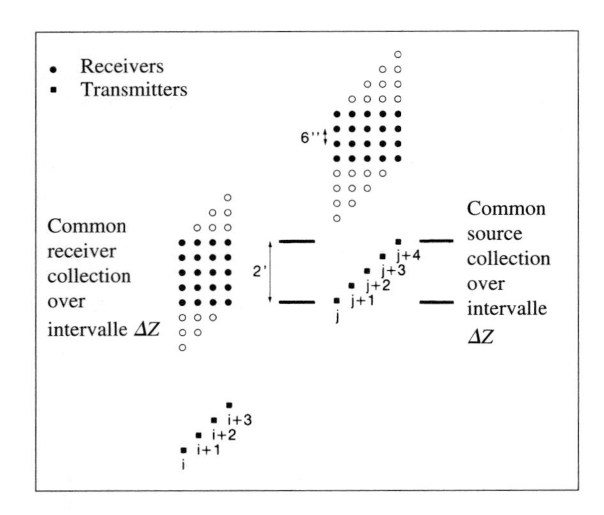

Fig. 1.27 Data gathering with common receiver depth and common transmitter depth. *(Courtesy of Schlumberger)*

β. Estimation of slowness (Δt or Dt) using transmitter/receiver array tools

For the more recent types of full wave recording tools (e.g. Array sonic, EVA, etc.), signal processing techniques are used which are similar to those employed in seismic data processing. These techniques make it possible to attribute an interval transit-time (or slowness) to each of the wave propagation modes, i.e. refracted P and S waves, and Stoneley-type interface waves.

The mean slowness of propagation of a wave across a given interval corresponds to the time delay acquired by the wave over this interval. The delay can be calculated by measuring the different arrival times at each receiver (or from different transmitter positions) located in the depth interval of interest for a common depth of transmitter (or receiver). As a consequence, the slowness of a particular formation may be estimated by measuring the delay in wave propagation by making use of sorted acoustic waveform data. This can be achieved by gathering data derived either from a common transmitter point or from a common receiver point. In this way, the average of the two delays provides a slowness value which is compensated for borehole effects.

Figure 1.27 illustrates the common receiver depth and common transmitter depth data gathering modes for a tool composed of one transmitter and eight receivers. In this particular case, the slowness of the formation is estimated over an interval equivalent to four times the spacing between two receivers (i.e. usually 2 ft for a spacing of 6" between two sources or receivers, as the case may be).

For a common transmitter (or receiver) depth gather composed of N recordings, where transmitter-receiver distances are increased regularly by the spacing between two consecutive receivers (or transmitter), the measurement of slowness for different waves can be carried out in various ways (Mari and Coppens, 1992):

- **Application of a threshold**

As in the case of conventional tools, the picking of arrivals is carried out by setting a detection threshold within a time window that is chosen for each recording. Slowness is the mean delay in wave propagation between two consecutive receivers (or transmitters), thus corresponding to the difference in arrival times recorded at each of the receivers. This method is very commonly employed to estimate the compressional wave velocity of formations.

- **Velocity analysis**

This method consists of establishing an energy map —or measure of multichannel coherence— as a function of arrival time recorded at the nearest receiver (with respect to a given transmitter) and measuring the delay between two consecutive receiver positions i and $i + 1$ (or two different transmitter positions, as the case may be). For a time t_0 and a given slowness Δt, each recording at position i

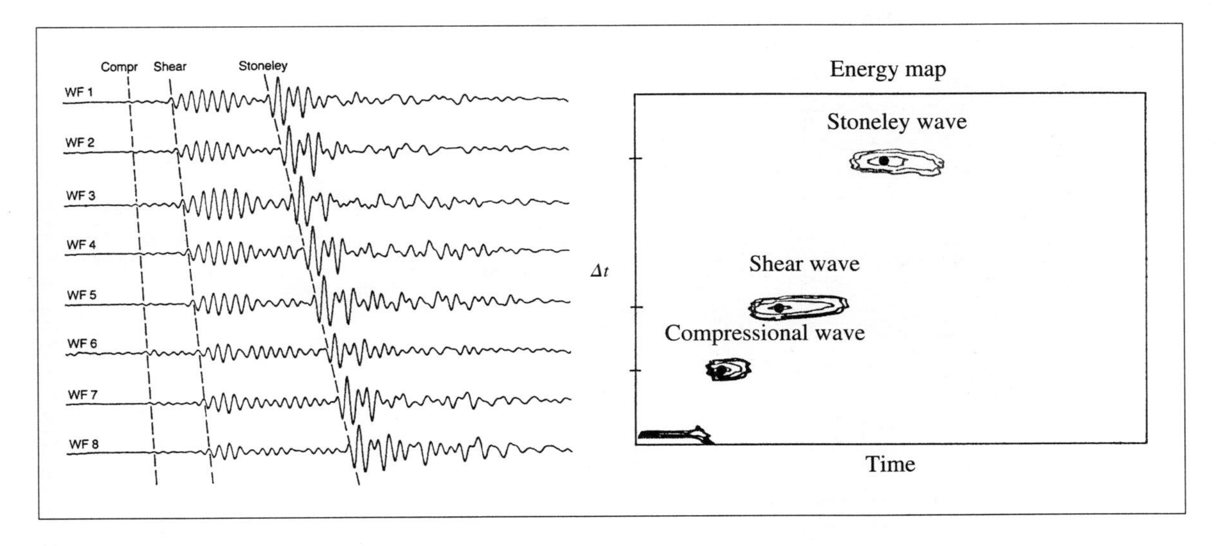

Fig. 1.28 Example of Δt determination using velocity analysis. *(Courtesy of Schlumberger)*

is shifted statically on the time axis by an amount $(i + 1)\ \Delta t$ in order to readjust all the recordings against t_0 before stacking. On the basis of the stacked trace, and using a short time window centred on t_0, it is thus possible to calculate an energy or coherence coefficient, more especially since the slowness value is chosen to ensure a correct line-up of arrivals observed from one recording to another.

Figure 1.28 illustrates a collection of traces composed of eight recordings, associated with a slowness vs. time plot (i.e. an energy or coherence map). The local maxima observed on a coherence map enable the identification of arrival times at the nearest receiver, thus providing the interval transit-time for each mode of propagation (compressional, shear and Stoneley wavesets can be clearly distinguished on such a diagram).

- **Cross-correlation method** (see Fig. 1.29)

In theory, this method is scarcely different from the velocity analysis approach. From each recording, a time interval is selected that is approximately centred on the wave under consideration. In this way, recordings of shorter duration are obtained which only contain a single type of wave propagation. By picking the maxima of intercorrelation functions taken consecutively in pairs (i with $i + 1$, and so on over the range 1 to $N - 1$), the mean delay between consecutive receivers may be directly obtained, and hence the interval transit-time or slowness.

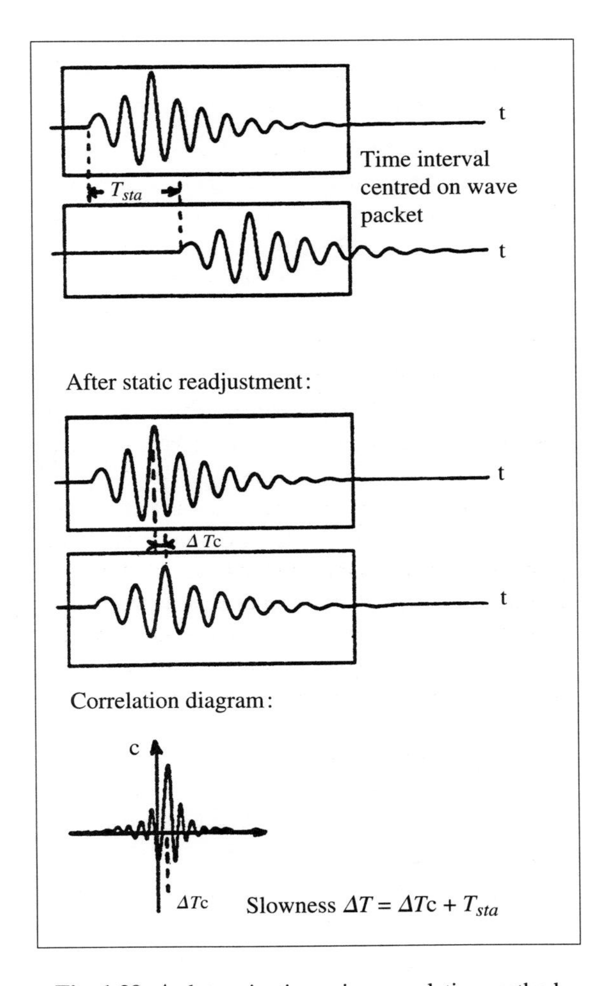

Fig. 1.29 Δt determination using correlation method.

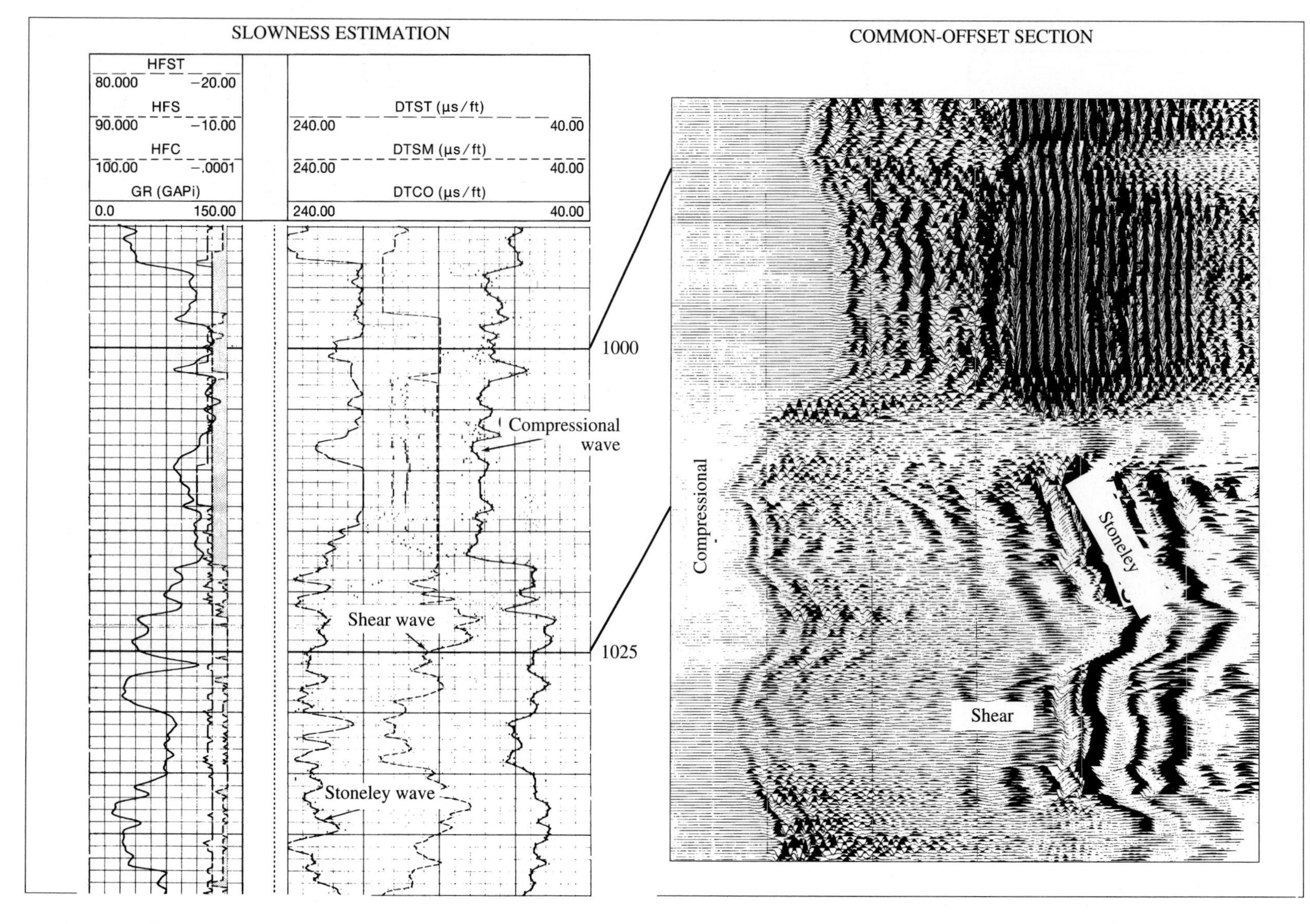

Fig. 1.30 Full waveform common-offset section compared with estimates of slowness for the same interval. *(Courtesy of Gaz de France)*

So as to avoid cycle skipping, the Δt value should be lower than the half-period of the signal pulse. In order to overcome this inconvenience, which may occur in very slow formations when the receiver or transmitter spacing is too large, the recordings are statically readjusted. The time readjustment T_{sta} is added to the slowness Δt obtained from the correlation procedure, thus yielding an estimate of Δt for the formation.

This method is applicable to all types of wave propagation. It is well suited for measuring the slowness of Stoneley interface waves, and is often used with a high-frequency cut-off filter (e.g. >5 kHz).

Figure 1.30 shows estimates of wave slowness for compressional (D_{tco}), shear (D_{tsm}) and Stoneley modes (D_{tst}) obtained with a Schlumberger Array Sonic (SDT) tool in the the depth interval 990-1045 m. These data are compared with a common-offset section recorded in the same formations, using a SEMM acoustic tool.

As regards the less conventional types of picking method, Mari and Coppens (1992) have proposed an approach based on artificial intelligence which incorporates rules for checking the coherence of selected arrivals from different receivers.

b. *Estimation of attenuation*

Several techniques are currently used:

(1) The decrease in amplitude can be calculated from equations of the type:

$$\frac{1}{d}\frac{A(R_i)}{A(R_j)} \text{ or } \frac{1}{d}\left(\frac{A(R_i)-A(R_j)}{A(R_i)}\right) \text{ or } \frac{20}{d}\log_{10}\frac{A(R_i)}{A(R_j)}$$

where $A(R_i)$ and $A(R_j)$ are the wave amplitudes recorded at receivers R_i and R_j and d is the spacing between R_i and R_j. d can be expressed as multiples of wavelength.

(2) The quality factor Q, a measure of the intrinsic attenuation of the medium, can be obtained using the following relations:

- the variation $\Delta\tau$ in rise time τ divided by the slowness Δt of the formation, i.e.:

$$Q = c\frac{\Delta t}{\Delta\tau}$$

where c is a constant (Fig. 1.31),

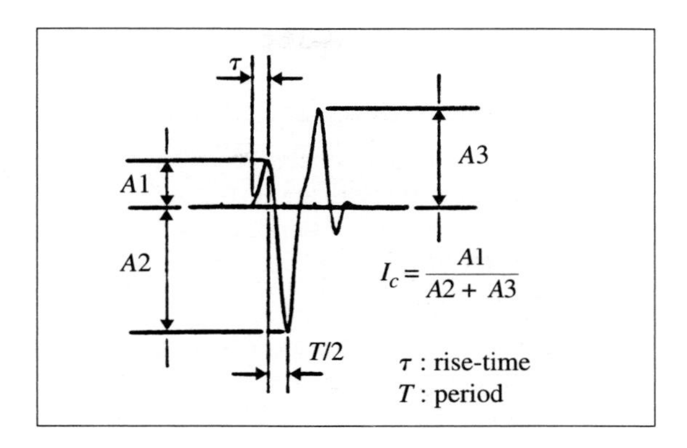

Fig. 1.31 Diagram showing rise time (τ), period (T) and shape factor (I_c) of a signal.

- measurement of the slope of the amplitude spectrum ratio, leading to an estimation of the attenuation coefficient α which is given by:

$$\alpha = \frac{\pi f}{QV}$$

where f is the frequency of the acoustic signal and V is the formation velocity,

- calculation of a shape factor (I_c; *cf.* Lebreton, 1978) based on the amplitudes of the first refracted P wave arrivals at a single receiver. Experimental trials have shown an almost linear relation between I_c and $1/Q$.

Attenuation is inversely proportional to the quality factor Q and can be taken as approximately equal to $27/Q$ (expressed in units of dB/λ).

c. *Estimation of the frequency*

This can be carried out in the frequency domain by spectral analysis, or in the time domain by measuring the apparent period of the signal.

1.2.1.2 Quality control and corrections

Anomalies in the interval transit-times (Δt) need to be detected and corrected, where this is possible, in order to prevent including them in the process of calibrating the sonic log to the well seismic data.

Some of these anomalies may be detected by simple examination of the sonic log, while others are revealed by cross-correlation of the different recordings. Finally, a certain number of anomalies can only be identified through a full interpretation of the log data from either a qualitative or a quantitative point of view.

Once the sonic log has been corrected for such anomalies, the interval transit-times can be integrated as a function of depth from the surface (zero depth datum) to yield the integrated travel time (TTI).

The transit time integration displayed ont the log should be verified in a zone of relatively constant Δt, such as casing.

On Fig. 1.32, it can be seen that seven intervals of 1 ms are indicated over a depth of 38 m. This calculates a slowness of 56 μs/ft, i.e.:

$$\Delta t = \frac{7 \times 1\,000}{38 \times 3.28} = 56\ \mu\text{s/ft}$$

The integrated travel time (TTI) is marked on the log by a small tick for each millisecond interval and a larger tick every ten milliseconds; it is recommended to check whether ten millisecond intervals really occur between each long bar, paying attention to the limits between runs, etc.

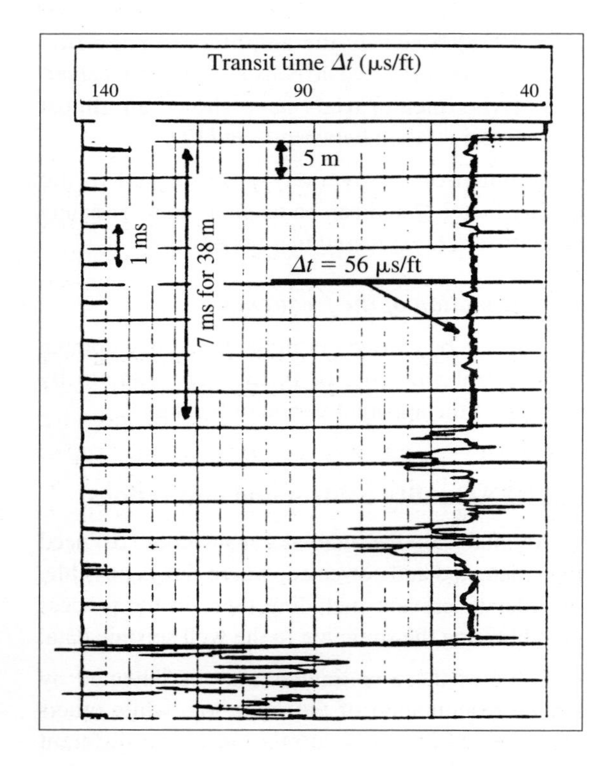

Fig. 1.32 Checking of integrated travel time.

A. Anomalies arising from detection system (noise and cycle skipping effects on sonic logs using bias system, see Fig. 1.33)

It is appropriate to take note of the remarks on the log header, which may identify problems during the logging run.

α. Noise

Noise is generally seen as isolated peaks of abnormal Δt, usually low; this effect can be easily removed. These peaks are very short (in depth) and not repeatable.

β. Cycle skipping

Cycle skipping causes isolated peaks with generally high Δt values that are increased by multiples of approximately 10 μs/ft; these peaks are readily eliminated under the conditions described previously (*cf.* Section: Problems arising from detection system; *cf.* Fig. 1.24).

Since thick zones may be affected, their proper identification depends on a thorough examination of the sonic log near the boundaries of these zones (i.e. looking for abrupt changes over distances less than the inter-receiver spacing) or a comparison with a full set of logging results (inconsistency of sonic log with interpretation based on other methods). In this way, the log measurements may be recalculated towards their correct values.

In the case of obvious cycle skipping, when there are numerous zones with significant effects,

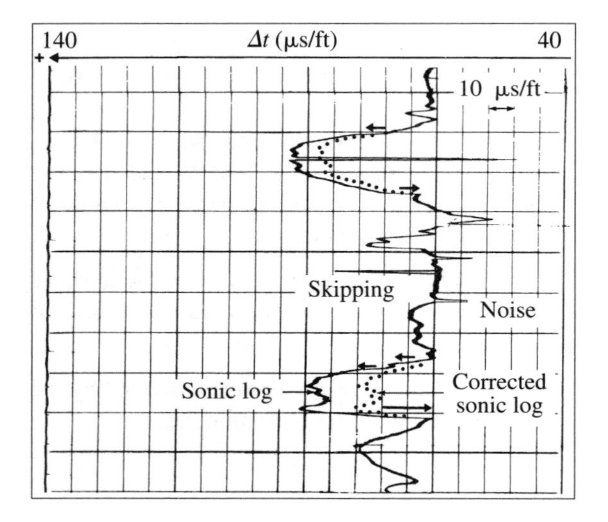

Fig. 1.33 Correction of cycle skipping effects.

an attempt at correction may be made by removing the observed shift so as to avoid major errors in the integrated travel time (TTI) (Fig. 1.33).

γ. Δt stretching

This effect is practically impossible to detect and cannot be corrected.

Noisy zones or zones with only localised cycle skipping are easy to detect. It is important to pick them out, even if their influence on integrated travel time TTI is small, so that they can be referred to if any anomalies remain after calibration (unseen cycle skipping zones).

B. Anomalies related to acoustic wave path

α. Caved sections (cf. Fig. 1.7)

The caliper log can detect caved zones, thus casting doubt on excessively high transit time readings Δt measured in certain sections.

β. Well wall deterioration (linked to drilling operations or fracturing; cf. Fig. 1.7)

These effects bring about Δt values that are higher than the real values of the undamaged formations.

γ. Raypath through mud (cf. Fig. 1.34)

With large diameter boreholes, slowness values close to 190 μs/ft may indicate a wave path through mud and not through the formation. A quantitative geological interpretation, or direct knowledge of the formations, is required to help in correcting the measured Δt value.

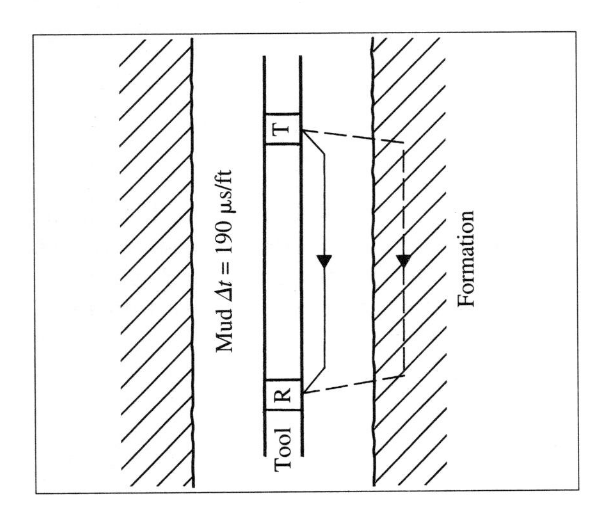

Fig. 1.34 Wave path through mud.

δ. Fluid invasion

The acoustic wave emitted by the transmitters generally propagated in the flushed zone of the formation. If it is necessary to know the travel-time of this wave in the uninvaded zone, a correction can be envisaged that makes use of the sonic tool response relations (see below for equations linking acoustic wave propagation to environmental parameters).

In the case of a water-invaded formation, a full logging survey interpretation can lead to a lithological characterisation as well as estimates of shale content, porosity and degree of hydrocarbon saturation, where these latter are not known. In fact, these data enable an approximate calculation of the Δt value of the uninvaded formation using, for example, the Wyllie time-average equation:

$$\Delta t = (1 - \Phi - V_{sh})\, \Delta t_{ma} + V_{sh}\, \Delta t_{sh} + \Phi\, \Delta t_{fl}$$

where *ma* denotes the matrix, *fl* the formation water, V_{sh} the percentage of shale present in the formation and Φ the effective porosity (in percent volume).

However, in formations where two fluids are present —particularly if gas is involved— it is far more problematic to establish the equations and corrections are more difficult to apply.

C. Anomalies related to geological formations

Some shale-rich or detrital formations with generally high apparent porosity show anomalously high Δt's that are incompatible with the known characteristics of the formation (as obtained from coring or other logs). Such anomalies may correspond to poorly compacted zones.

In carbonate-bearing or detrital formation with high porosity (especially if the formation pore-fluid has a high Δt, as in the case of gas), the occurrence of unusually low Δt values may appear to indicate the presence of vugs or large-diameter pores. However, it is generally impossible to perform a reliable correction, particularly because this phenomenon affects the transit times obtained both from surface seismic survey and from sonic log, but in different ways according to frequency, pore-size and nature of the formation fluids.

For more details on this phenomenon, the reader should refer to the section below (*cf.* Section: Influence of the raypath medium; see p. 51).

1.2.1.3 Relations between acoustic wave propagation and media

At the currently used acoustic log frequencies (10-40 kHz), and transmitter receiver spacings, it may be assumed the wave travels in a path very near the wellbore in the flushed zone. The fluids encountered are mud filtrate (saturation S_{xo}) and residual hydrocarbons (saturation $S_{hr} = 1 - S_{xo}$).

The following symbols are used to denote the different parameters involved (except where indicated otherwise):

Δt = interval transit-time of a wave (also designated as Dt or t)
V = seismic velocity of a wave
ρ_b = density of the formation
Φ = effective porosity of the formation, exclusive of the pore-space water associated with the shale fraction
Φ_t = total porosity, including pore-space water associated with the shale fraction
S = fluid saturation
S_{xo} = water saturation of the filtrate-flushed zone
S_w = water saturation of the uninvaded zone
μ = shear modulus (S wave propagation)
K = elastic modulus (body waves)
γ = P to S wave velocity ratio
E = Young's modulus
σ = Poisson's ratio (0.2-0.4)
β_B = Biot compressibility constant, i.e.: $1 - (C_{ma}/C_b)$
β_G = Gassman compressibility constant, i.e.: (C_{ma}/C_b)
C_b = bulk compressibility of the rock, i.e.: $1/K_b$
C_s = compressibility of rock-forming solid phases (matrix and dry clay fraction), i.e.: $1/K_s$
C_{ma} = matrix compressibility i.e.: $1/K_{ma}$ (for quartz $2.5 \cdot 10^{-13}$ Pa^{-1})
C_f = fluid compressibility
$C_f = (1 - S_w) C_{hc} + S_w \cdot C_w = 1/K_f$
C_{hc} oil = $1 \cdot 10^{-4}$ to $2 \cdot 10^{-11}$ Pa^{-1}
gas = 0.3 to $1.5 \cdot 10^{-9}$ Pa^{-1}
C_w water = $4.2 \cdot 10^{-12}$ Pa^{-1}.

The following subscripts are used as indices in the above terms:

P = compressional wave
S = shear wave
ma = matrix (solid phase exclusive of clay fraction)
sh = shale fraction
f = fluid
w = formation water
mf = mud filtrate
hc = hydrocarbons
hr = residual hydrocarbons
b = bulk rock (solid constituents and pore-fluids)
s = skeleton (solid phases and dry clay fraction).

A. Acoustic wave velocity and petrophysical properties

a. *Determination of rock mechanical parameters (cf. Fig. 1.36)*

The propagation velocities of P and S waves in elastic media are related to rock mechanical properties through the equations:

$$V_P^2 = \frac{K + 4/3\mu}{\rho_b} \quad \text{and} \quad V_S^2 = \frac{\mu}{\rho_b}$$

The ratio of P and S wave velocities is commonly written as:

$$\gamma = V_P / V_S$$

By combining the measurements of V_P, V_S and ρ_b it is possible to obtain the principal mechanical parameters K, μ and σ. Since it is independent of the density ρ_b, Poisson's ratio σ can be derived directly from γ:

$$\sigma = \frac{1}{2}\frac{\gamma^2 - 2}{\gamma^2 - 1}$$

In the case of shale free formations (i.e. a clean formation with $C_s = C_{ma}$, the Gassman's equation (1951) links the porosity to the mechanical properties and the compressional wave velocity of sedimentary rocks:

$$V_P^2 = \left[\frac{1}{\rho_b C_s}\left\{3\left(\frac{1-\sigma}{1+\sigma}\right)\beta_G + \frac{(1-\beta_G)^2}{(1-\beta_G) + \Phi\left(\frac{C_f}{C_s} - 1\right)}\right\}\right]$$

where $\beta_G = C_s/C_b$.

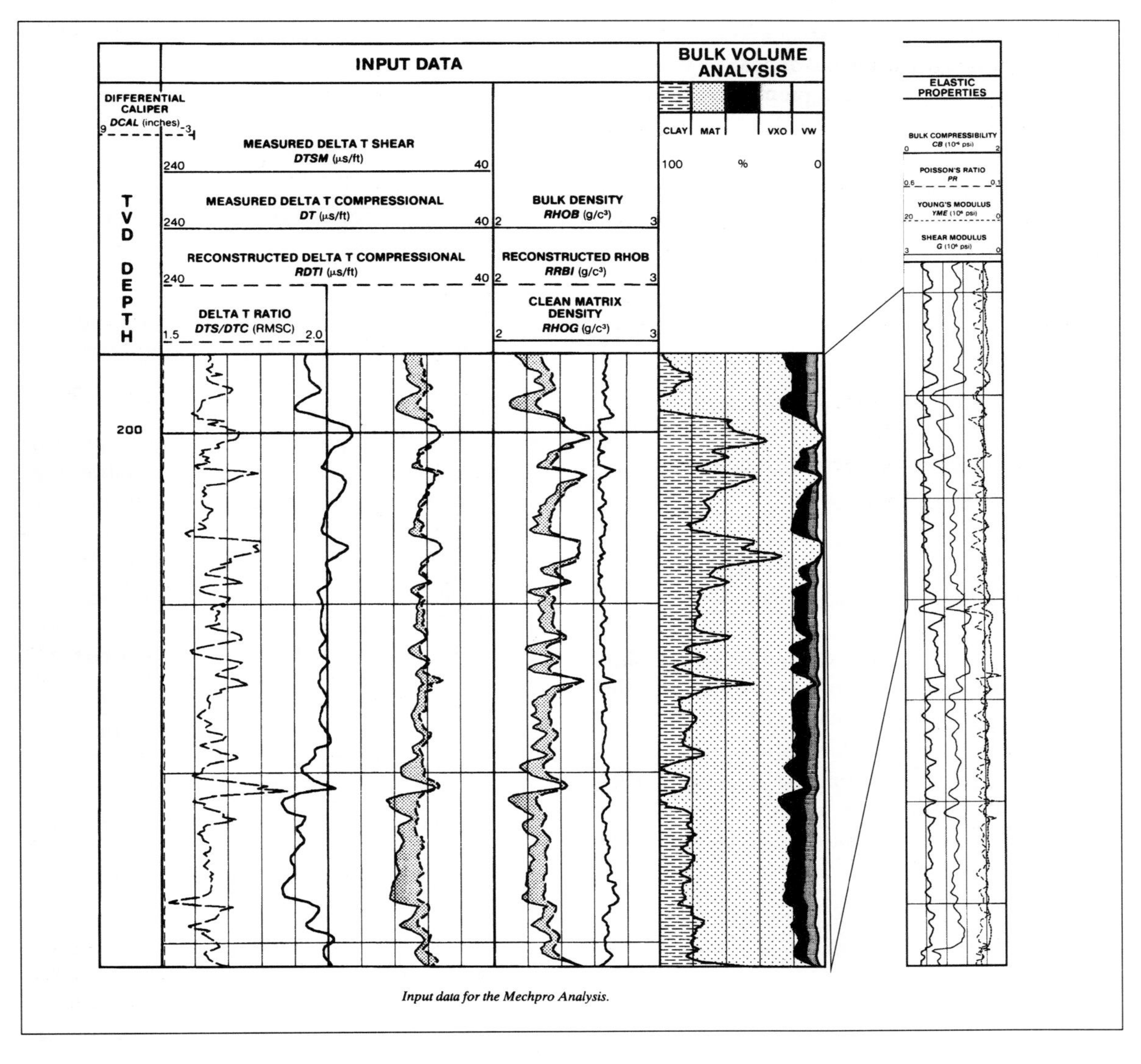

Fig. 1.36 Rock mechanical properties ("Mechpro" analysis by *Schlumberger*).

b. Determination of petrophysical properties

The quantitative interpretation of well logs currently combines acoustic measurements with Neutron and Density logs in order to determine the mechanical properties of formations, including the matrix characteristics, porosity and abundance of shale (see Section 1.3.1.2: Use of Cross-plots).

Full wave recording acoustic tools provide a large amount of information concerning the slowness (Δt) and amplitude (A) of P, S and Stoneley waves — where these occur— and make it possible to give a solution when other log measurements are missing.

α. Fluid saturated clean (shale free) formations

- Different formulae are used to describe the relations linking compressional (P) wave velocity with the characteristics of formations and their pore fluids. The best known of these relations is the Wyllie equation, which applies to formations saturated in a single fluid phase:

$$\Delta t_P = (1 - \Phi)\, \Delta t_{Pma} + \Phi\, \Delta t_{Pf}$$

For shear wave propagation, we may consider a time or velocity average equation to be limited to the solid phase constituents.

- Because the linear relationship between Δt and porosity does not always appear to be valid, other formulae have been proposed such as the set of equations illustrated in Fig. 1.37 (Raymer et al. 1980):
  - for the range of formation porosities usually encountered (i.e. <40%):

$$\frac{1}{\Delta t_P} = \frac{(1-\Phi)}{\Delta t_{Pma}} + \frac{\Phi}{\Delta t_{Pf}}$$

  or alternatively

$$V_P = \sqrt{\frac{\rho_{ma}}{\rho_b}}(1-\Phi)^{1.9} V_{Pma}$$

  - for formations with very high porosity (>50%):

$$\frac{1}{\rho V_P^2} = \frac{(1-\Phi)}{\rho_{ma} V_{ma}^2} + \frac{\Phi}{\rho_f V_{Pf}^2}$$

- Pickett (1963) demonstrated a relationship between P and S wave travel times and the matrix and porosity characteristics of rocks saturated with liquid (*cf.* Fig. 1.38).
- Theoretical and experimental studies (Morlier and Sarda, 1971) have identified the relation between acoustic attenuation and petrophysical parameters:

$$\delta = 1.2 \cdot 10^{-3} \frac{S}{\Phi}\left[\frac{2\pi k f \rho f}{\mu}\right]$$

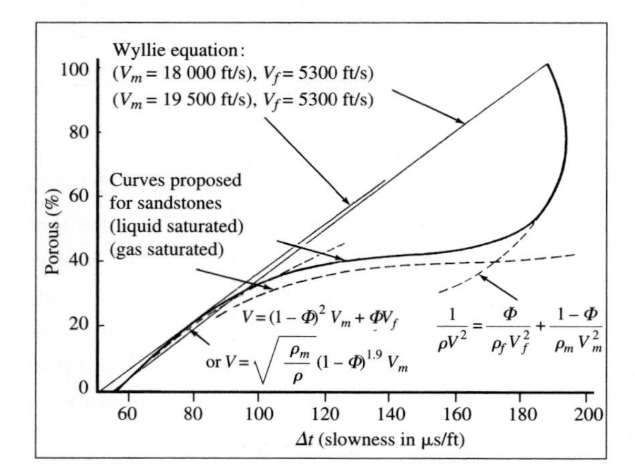

Fig. 1.37 Porosity vs. Δt curves for compressional waves.
(Raymer et al., 1980)

where: δ = is attenuation (dB/cm)
f = frequency (Hz)
ρf = fluid density
Φ = porosity (%)
k = permeability (mD)
S = specific surface area (cm^2 per cm^3 of rock)
μ = fluid viscosity (cP).

- More recent studies (*Elf*) also appear to indicate relations linking the amplitude and velocity of S waves with the formation characteristics (Fig. 1.39).
- Krief (1989), proposed a new type of chart (Fig. 1.40) which represents the relationships between body wave velocities and porosity /lithology of the formation. This set of curves was established using two types of equation:
  - in the following equations, the rock mechanical parameters μ and K are related to the acoustic velocities (V_P and V_S) as well as the density ρ_b:

$$V_P^2 = \frac{K + 4/3\,\mu}{\rho_b} \quad \text{and} \quad V_S^2 = \frac{\mu}{\rho_b}$$

$$K = \rho_b V_P^2 - \rho_b 4/3\, V_S^2$$

  - the Biot model equations are written as follows:

$$K = K_{ma}(1-\beta_B) + \beta_B^2 M_B$$

$$\mu = \mu_{ma}(1-\beta_B)$$

$$\frac{1}{M_B} = \frac{\beta_B - \Phi}{K_{ma}} + \frac{\Phi}{K_f}$$

  where K_{ma} and K_f are the elastic moduli for the matrix and the fluid, respectively and β_B and M_B are the Biot coefficients (Biot, 1952).

In the case of clean formations with water as the pore fluid, the Krief equations can be made consistent with the Raymer equations by introducing the following relation:

$$1 - \beta_B = (1-\Phi)^{m(\Phi)}$$

while substituting:

$$m(\Phi) = 3/(1-\Phi)$$

β. Clean formations (clay-free) containing gas (cf. Fig. 1.41)

It becomes more problematic to establish equations for formations containing two fluids, especially when there is a gas.

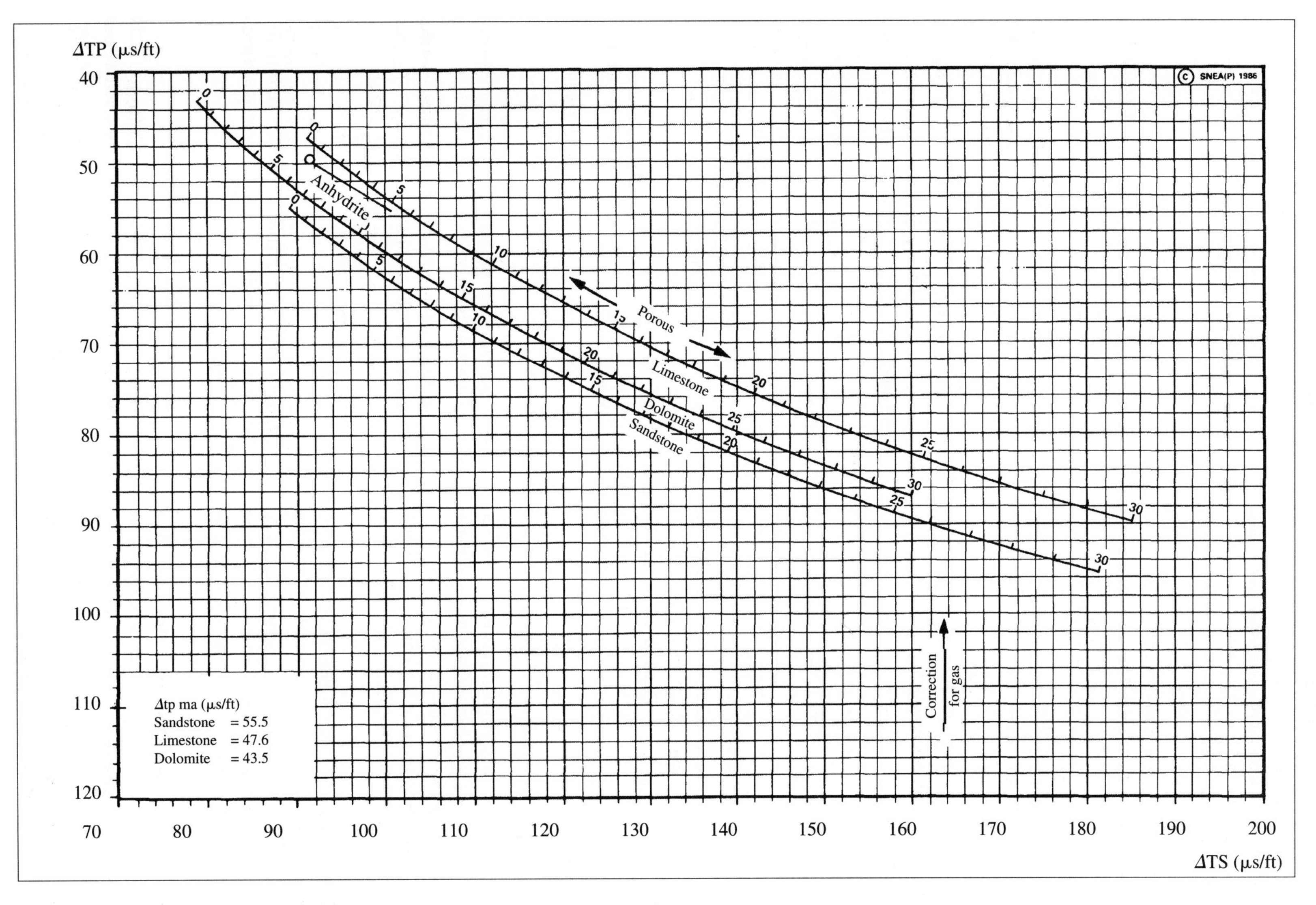

Fig. 1.38 Determination of porosity and lithology using EVA curves.
(Courtesy of Elf Aquitaine)

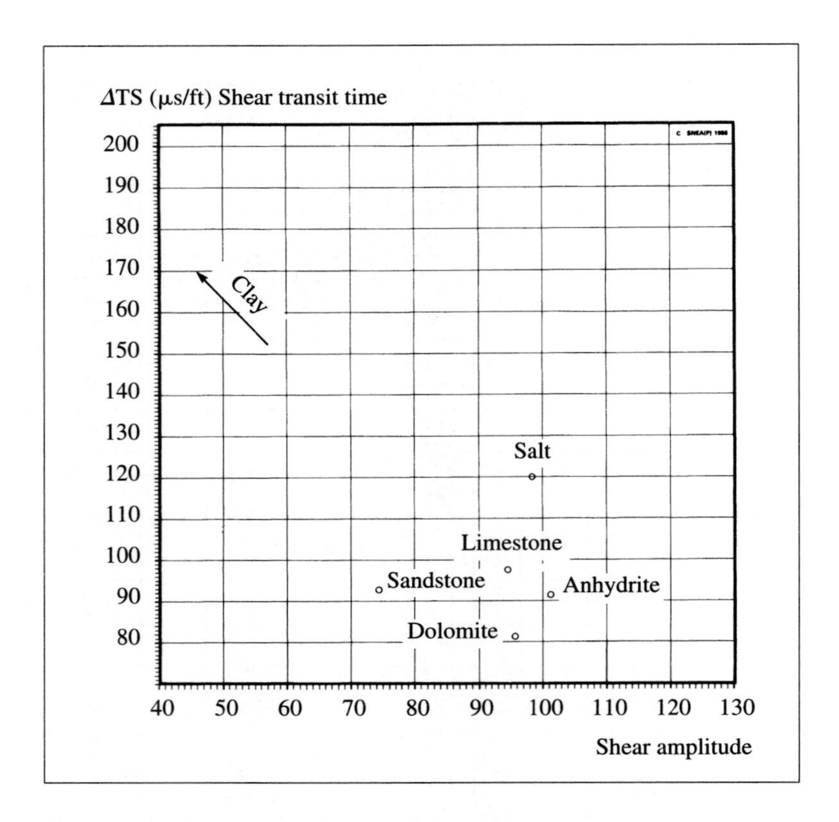

Fig. 1.39 Identification of matrix composition on the basis of EVA calibration curves. *(Courtesy of Elf Aquitaine)*

In fact, laboratory investigations show the strong influence of gas, even if present in minimal quantities, and the models currently used for determining petrophysical properties (Biot, Geerstma, etc.) cannot faithfully take account of this factor.

The presence of gas has a noteworthy effect on the velocity ratio (i.e. V_P/V_S). For low degrees of gas saturation (S_g in the range 0-20%, which corresponds to a water saturation S_w between 80 and 100%), this phenomenon mainly appears to have an effect on the compressional waves.

For very high degrees of gas saturation, even the shear waves would appear to be affected, as shown in Figs. 1.42 and 1.43.

Figure 1.43 shows the variation of P or S wave velocity as a function of water saturation in different formations, mainly of sandstone-type lithology. These diagrams also serve to compare the measured velocities with values predicted by the model.

The Krief equations can be rewritten to take account of the fluid constituents:

$$\frac{\Phi}{K_f} = \frac{\Phi S_{xo}}{K_{mf}} + \frac{\Phi(1 - S_{xo})}{K_{hc}}$$

Modified in this way, the equations reproduce fairly well the results of laboratory experiments (Fig. 1.42).

γ. Shaly formations

In shaly (or clay-bearing) formations, it is necessary to take account of the clay fraction and its particular characteristics. For example, the Wyllie equation may be written as follows:

$$\Delta t_P = (1 - \Phi - V_{sh})\, \Delta t_{Pma} + V_{sh}\, \Delta t_{Psh} + \Phi\, \Delta t_{Pf}$$

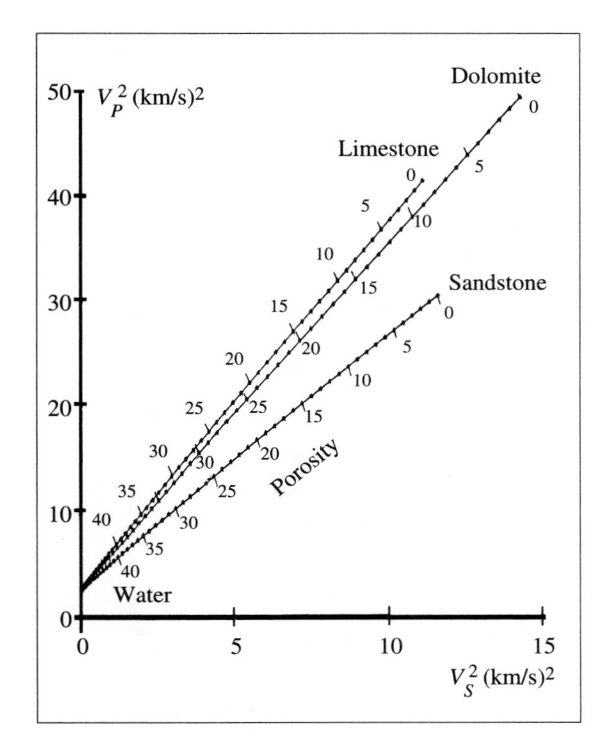

Fig. 1.40 Chart showing P and S wave square velocities as a function of lithology and water content. *(Krief, 1989)*

Fig. 1.41 Effect of the presence of gas at the top of the oil zone.

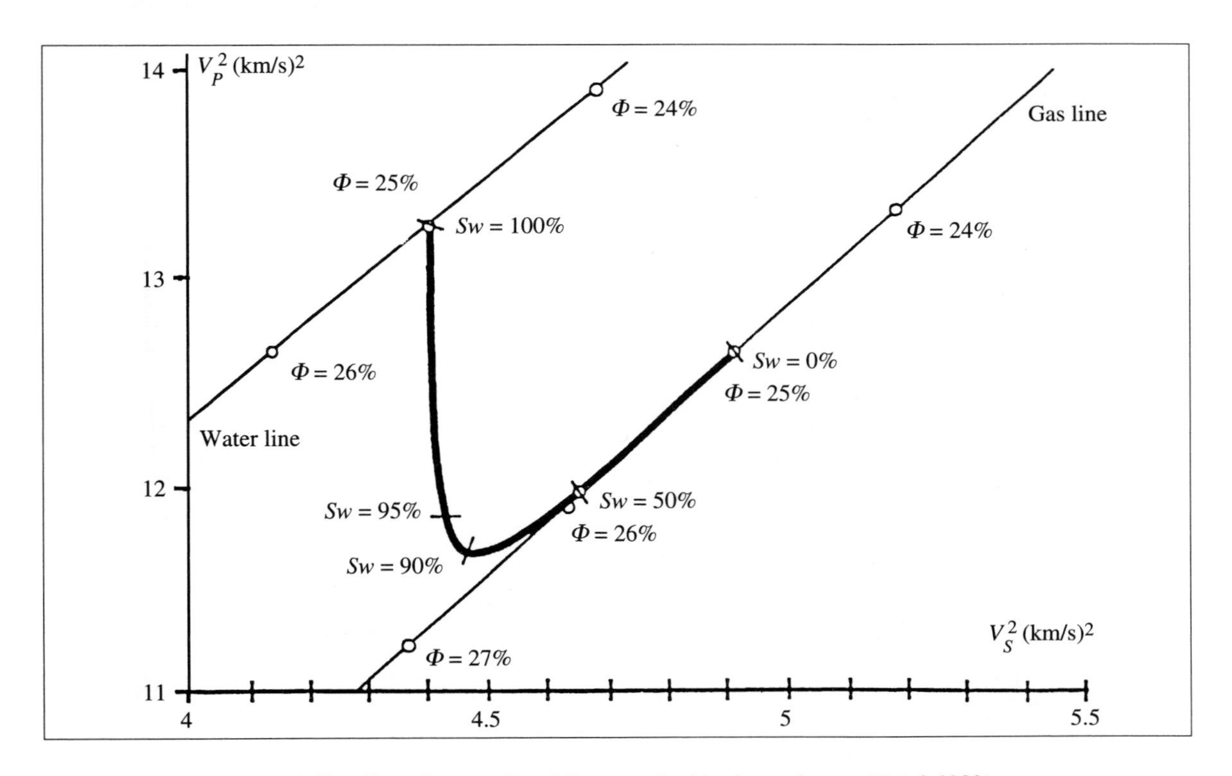

Fig. 1.42 Effect of gas on P and S wave velocities in sandstone. *(Krief, 1989)*

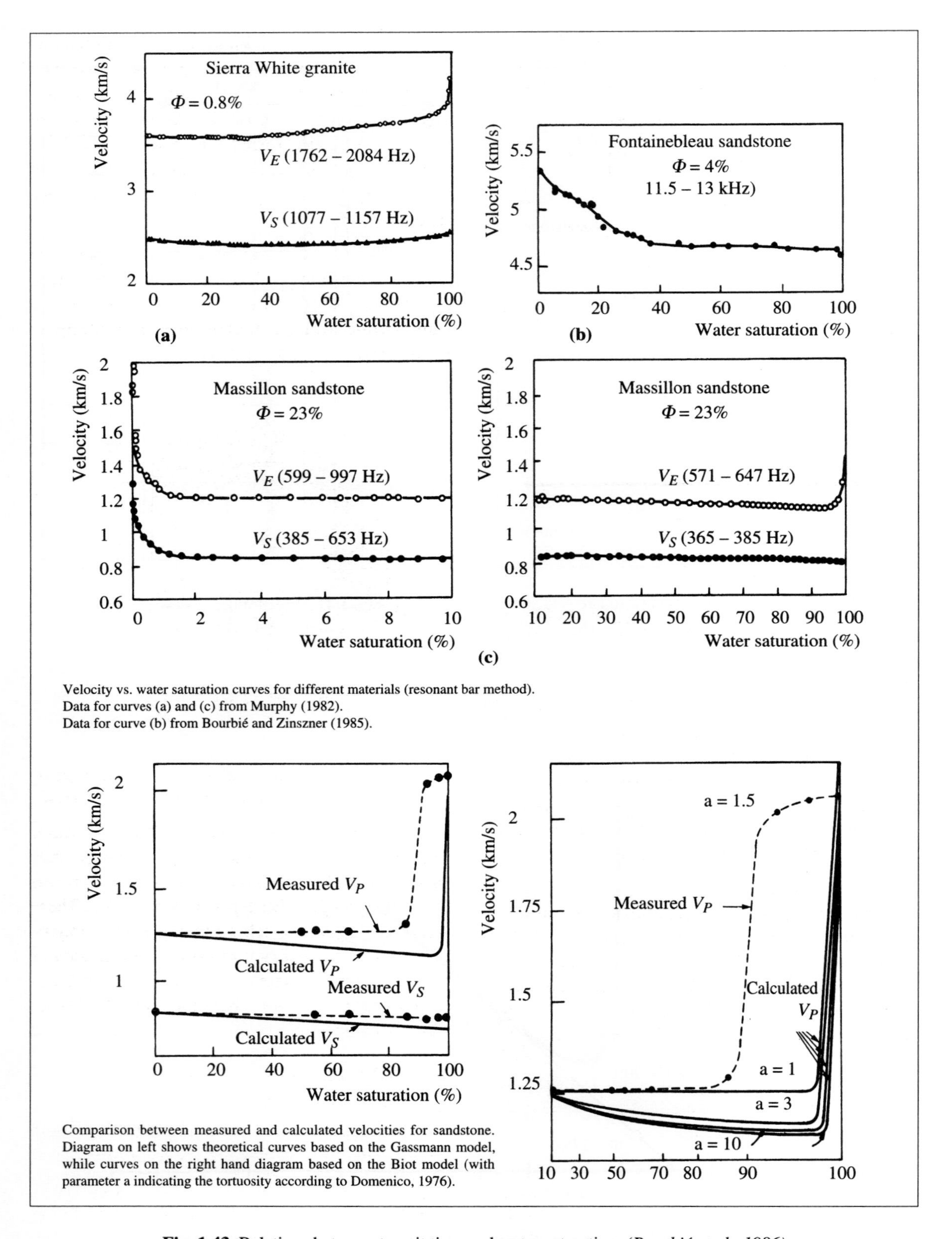

Fig. 1.43 Relations between transit time and water saturation. *(Bourbié et al., 1986)*

As regards the Krief equations (Krief et al., 1989), the relations are modified by assuming that the elastic moduli of the solids are harmonic mean of the respective constituents:

$$\frac{\Phi_t}{K_f} = \frac{\Phi S_{xo}}{K_{mf}} + \frac{\Phi(1 - S_{xo})}{K_{hc}} + \frac{V_{sh} - \Phi_{sh}}{K_{Wsh}}$$

$$\frac{1 - \Phi_t}{K_s} = \frac{1 - \Phi}{K_{ma}} + \frac{V_{sh}(1 - \Phi_{sh})}{K_{sh}}$$

$$\frac{1 - \Phi_t}{\mu_s} = \frac{1 - \Phi}{\mu_{ma}} + \frac{V_{sh}(1 - \Phi_{sh})}{\mu_{sh}}$$

where:

$$\Phi_t = \Phi + V_{sh}\,\Phi_{sh}$$

and making use of the following relations:

$$\frac{1}{M_B} = \frac{\beta_B - \Phi_t}{K_s} + \frac{\Phi_t}{K_f}$$

$$(1 - \beta_B) = \left(1 - \Phi_t\right)^{\frac{3}{(1 - \Phi_t)}}$$

δ. Influence of the formation

The propagation of acoustic waves depends on the arrangement of grains in the rock and their mineral composition, as well as the size and distribution of the pores, in other words, the elastic properties of the rock and its pore fluids. Frequency also has an effect on the mode of propagation, so that differences in travel time between waves measured by sonic logging (high-frequency) and seismic surveys (low-frequency) can be related to poor consolidation or the presence of vugs. Both of these phenomena are commonly encountered.

- **Poor consolidation**

In the case of poorly consolidated or slightly compacted formations (sands, marls, etc.) the measured interval transit-times Δt are much longer than the values expected from lithology, porosity or shale content. This is seen on both seismic surveys and sonic logs, but its effect varies as a function of acoustic wavelength. A compaction factor Bcp can be introduced to enable the calculation of porosity values that are consistent with known local conditions.

- **Vugs**

In the case of sonic logs, the ray paths appear to be deviated around the relatively large pores as seen in the case of certain carbonate formations (*cf.* Fig. 1.44), thus yielding low Δt's. Porosities calculated from volume-based measurements, such as Neutron or Density logs, yield values that are effectively close to those determined from core samples. On the other hand, the values derived from sonic logging seem to be far lower. This effect becomes even more important as the Δt of the pore-fluid increases.

From Fig. 1.44, it can be seen that raypath 1 is shorter than raypath 2, but with Δt in the fluid greater than in the matrix, it is possible to obtain a transit time along 2 that is less than along 1, i.e. $[(AB)\,\Delta t_{(fl+ma)}]_2 < [(AB)\,\Delta t_{(fl+ma)}]_1$.

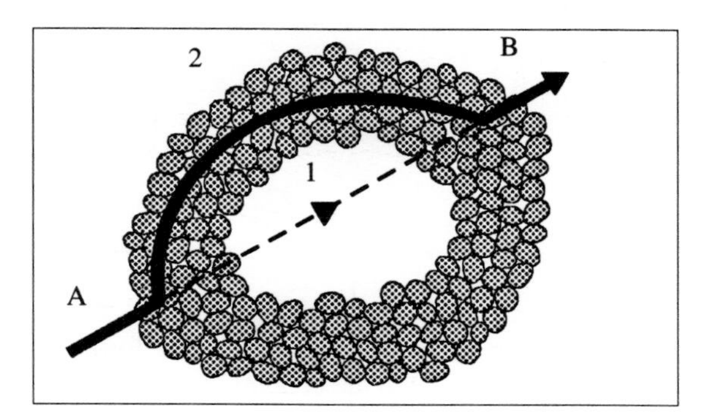

Fig. 1.44 Effect of vugs.

The critical pore size depends both on the type of fluid present and on the wavelength of the sonic pulse.

Studies which utilise all the available information (well logs, mud logs, cuttings, cores, geology, etc.) make it possible to identify inconsistencies between sonic logs and the other types of porosity log. These discrepancies may be related to vug effects (improperly termed secondary porosity) or cementation of the rock (study of abnormal trends on different types of porosity profile, e.g.: Neutron, Density, Sonic, Resistivity logs, etc.).

B. Bed boundaries and dips obtained from acoustic data

The use of sonic tools with multiple transmitters and/or receivers enables the recording of micro-seismic section in wells.

Figure 1.45 illustrates an example of a sonic log obtained in a deviated well. Under favourable cir-

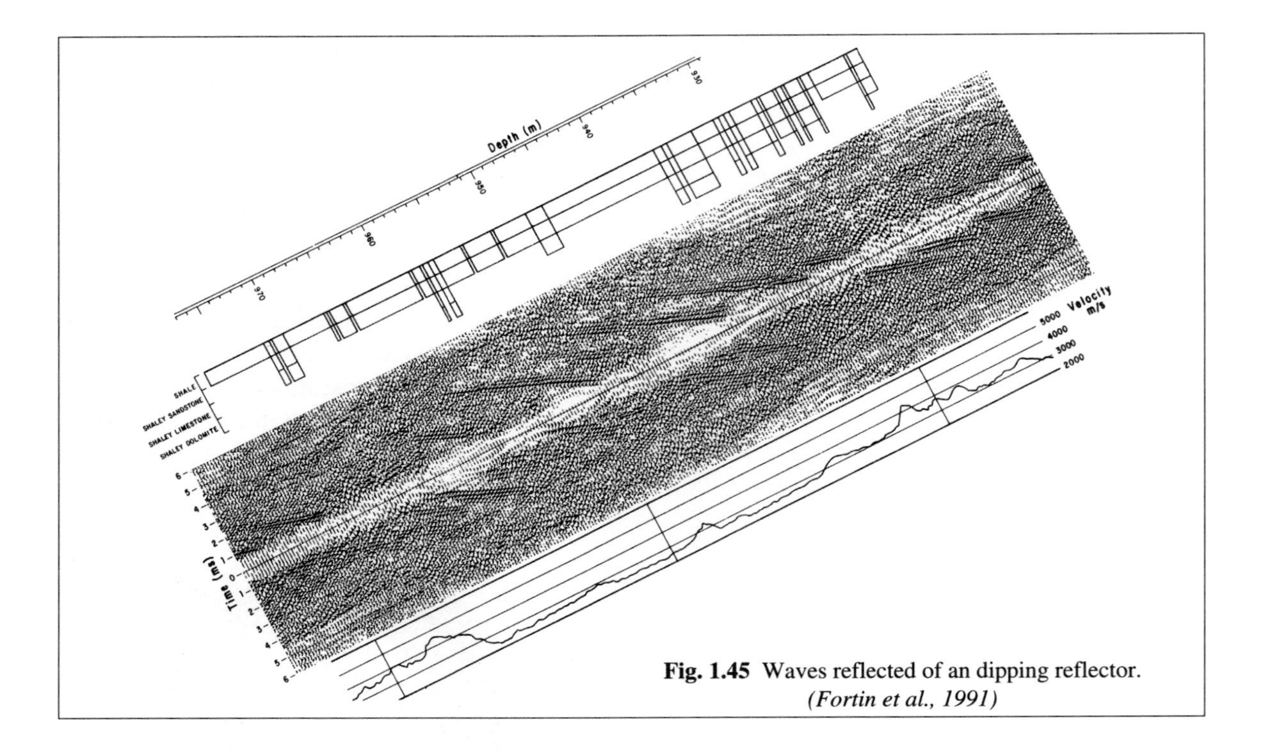

Fig. 1.45 Waves reflected of an dipping reflector.
(Fortin et al., 1991)

cumstances, i.e. when the axis of the hole forms an angle with the normal to the bedding plane (as in the case of strongly deviated wells or steeply dipping strata), it becomes possible to distinguish bedding boundaries and thus determine their dips. If the hole axis is kept close to the bedding plane (as in the case of controlled drilling), this may hopefully enable the tracing of such boundaries.

The lateral range of investigation away from the well is limited to about 20 m. In fact, investigation is a function of the attenuation of the medium, the characteristics of the transmitters and receivers, the tool geometry, and the apparent dip of the layers with respect to the tool axis (Fig. 1.46).

On sections obtained with data processing methods analogous to those employed in seismic surveying, after elimination of refracted and interface propagation modes, the breaks seen between the layers are related to the investigation limit and do not necessarily represent bedding discontinuities.

C. Permeability and Sonic logging

Many authors have sought to link permeability to the various measurements obtained from downhole tools, in particular the results of acoustic well

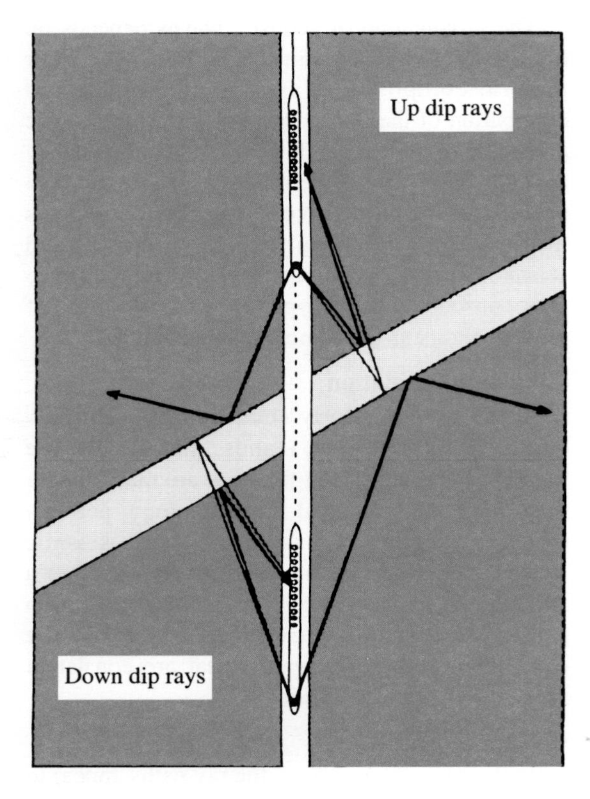

Fig. 1.46 Acoustic imaging.
(Courtesy of Schlumberger)

logging. Recent work is oriented towards the study of Stoneley and shear waves, in preference over compressional waves.

As regards compressional waves, Lebreton et al. (1978) have proposed use of a shape factor *Ic* (*cf.* Section 1.2.1.1. E. b.: Estimation of attenuation) to estimate permeability on the basis of an empirical relation of the type: $k = A \log Ic + B$, where A and B are constants determined according to the diameter of the well and the tool.

In accordance with the dispersion model proposed by Biot (1956) and Rosenbaum (1974), the variations of phase velocity and Stoneley wave attenuation are found to be a function of the petrophysical properties of the medium, notably the permeability.

In certain favourable cases, the inversion of phase velocity and attenuation carried out by means of the Biot-Rosenbaum model makes it possible to estimate permeability.

The figures presented below illustrate the application of this method to sandstone formations (ranging from very poorly to well consolidated) with permeabilities ranging from 10 to 3000 mD and having a porosity of the order of 30%.

The phase velocities (at 2 kHz) and attenuations (Stoneley wave amplitude ratios) are calculated using the Biot-Rosenbaum model and measurements of the petrophysical properties of the medium — particularly its permeability— that are derived from core analyses and log results (see Fig. 1.47). On these diagrams, the solid circles represent values calculated from numerical modelling.

Equally, the Biot-Rosenbaum model —when combined with logs of the Stoneley wave attenuation and phase velocity— may be used to obtain an estimate of permeability as shown on Figs. 1.48 and 1.49 (here, solid circles indicate the experimentally determined values).

The measurements of Stoneley wave attenuation (with a frequency centred on 2 kHz) are in good agreement with the dispersion model results, taking into account the intrinsic attenuation of the formation. Compared with attenuation, however, the phase velocity measurements are less sensitive to permeability of the medium.

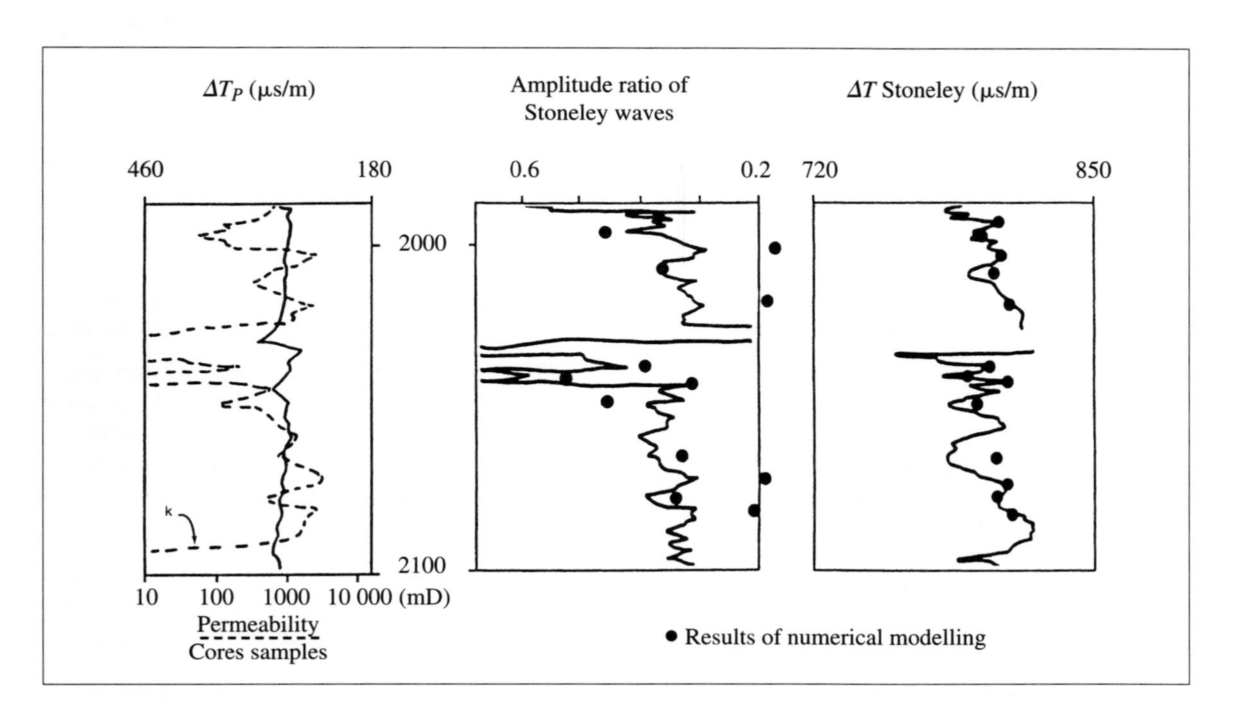

Fig. 1.47 Attenuation and slowness of Stoneley waves. Comparison between results and numerical simulations. *(Williams et al., 1984)*

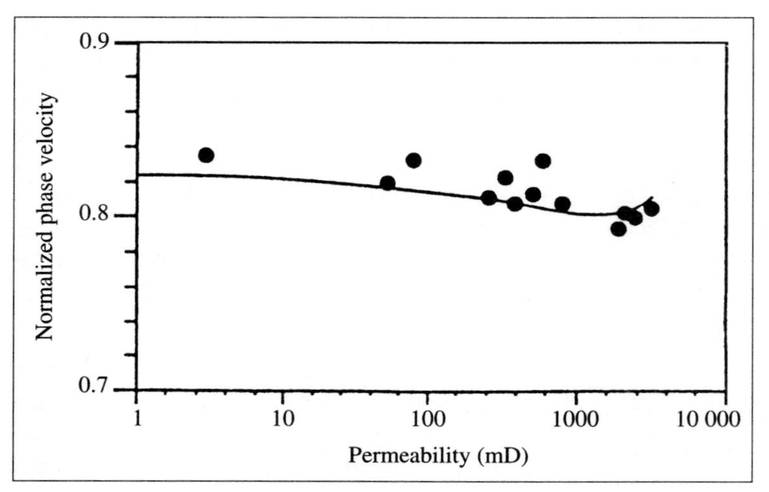

Fig. 1.48
Phase velocity of Stoneley waves plotted as a function of permeability, showing comparison between theoretical curve (Biot-Rosenbaum) and experimentally measured values. *(Williams et al., 1984)*

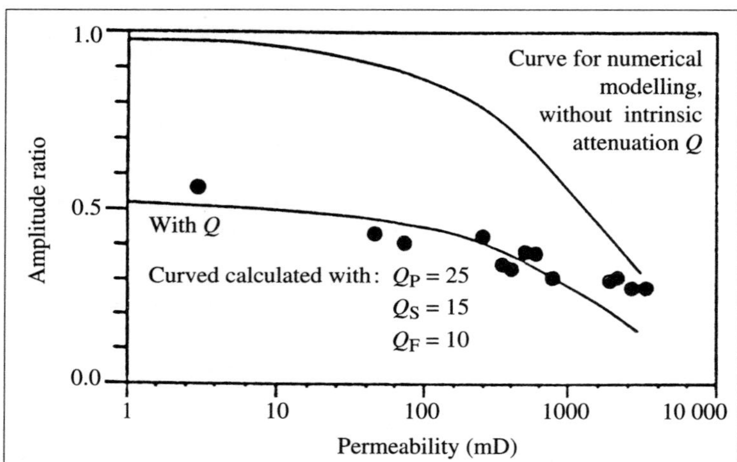

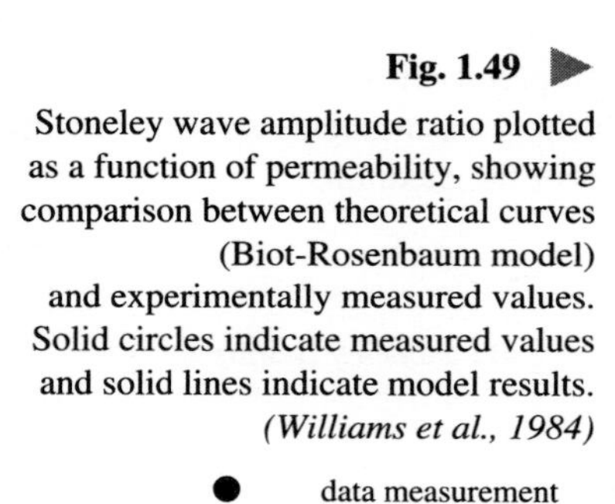

Fig. 1.49
Stoneley wave amplitude ratio plotted as a function of permeability, showing comparison between theoretical curves (Biot-Rosenbaum model) and experimentally measured values. Solid circles indicate measured values and solid lines indicate model results. *(Williams et al., 1984)*

At normal seismic frequencies (10-200 Hz), an analogous approach (Cheng et al., 1988; Mari, 1989) may be carried out using the dispersion model of White (1983) applied to tube waves.

1.2.1.4 Sonic log prediction

Sometimes a sonic log is not recorded. Since this information is essential in the depth calibration of seismic data, it is possible to reconstruct one on the basis of other logs.

When a relatively full set of logs is available for the estimation of porosity and shale content, or the determination of the nature of matrix and fluids, Δt values may be calculated from one of the equations mentioned in this text (e.g. Wyllie, Raymer-Hunt, Gardner, among others). It is quite clear that the choice of relation will affect the type of result obtained.

On some occasions, as in the case of overburden strata, only resistivity logs are run at depth along with a *GR* (Gamma Ray) and/or a *SP* (Spontaneous Potential) log. This means that a certain number of approximations have to be made:

(1) Although the V_{sh} value (shaliness) can be estimated from *GR* or *SP* logs, the result is less satisfactory if these logs are poorly representative of the true shaliness of the formation. This may arise, for example, in clastic formations containing heavy minerals, K-rich minerals, feldspars, micas, etc., or in the case of an SP response that is not linked exclusively to the shaliness abundance of shale.

(2) Appropriate slowness values can be assigned if cuttings are used to determine the nature of the matrix, i.e.:

sandstone $\Delta t_{ma} = 50$ µs/ft;
limestone $\Delta t_{ma} = 47$ µs/ft;
dolomite $\Delta t_{ma} = 40$ µs/ft.

(3) If the pore-fluid is water, then the slowness of the fluid (Δt_{fl}) can be fixed at between 189 and 210 µs/ft according to salinity. With hydrocarbon-bearing formations, this value is slightly lower than the true value.

(4) Porosity can be estimated from Neutron and Density logs by making assumptions about shaliness, matrix and fluids. Alternatively, it is even possible to evaluate porosity with the Archie equation using resistivity measurements:

$$Rt = FR_w / S_w^{\,n} \text{ where } F = a/\Phi^n.$$

Lastly, in the simplest cases, it is possible to apply equations which are adjusted to regional conditions and which incorporate only a limited amount of data. The two best known examples of such relations are given below:

- The Faust equation, based on resistivity measurements:

$$V = 1948\ (Z/\text{Res})^{1/6}$$

where V is the velocity in ft/s, Z is the depth in feet and Res is the formation resistivity in ohm-m.

- The Gardner equation, based on Density logs:

$$\Delta t = 10^6 / (\rho_b / 0.23)^4$$

where Δt is the slowness in µs/ft and ρ_b is the density in g/cm³.

1.2.2 Density logs ($\gamma - \gamma$ logs)

Assessment of the mass per unit volume or density of a formation (Fig. 1.50) is carried out through measurement of the electron density index ρ_{ei}, which is itself linked to the absorption of γ-rays emitted from a radioactive source placed in the tool. The index ρ_{ei} is related to the density ρ_b by the following equation:

$$\rho_{ei} = \rho_b\ (2\ Z/A)$$

where Z is the atomic number and A is the atomic weight.

Since the term 2 Z/A is approximately equal to unity, with few exceptions (in particular hydrogen,

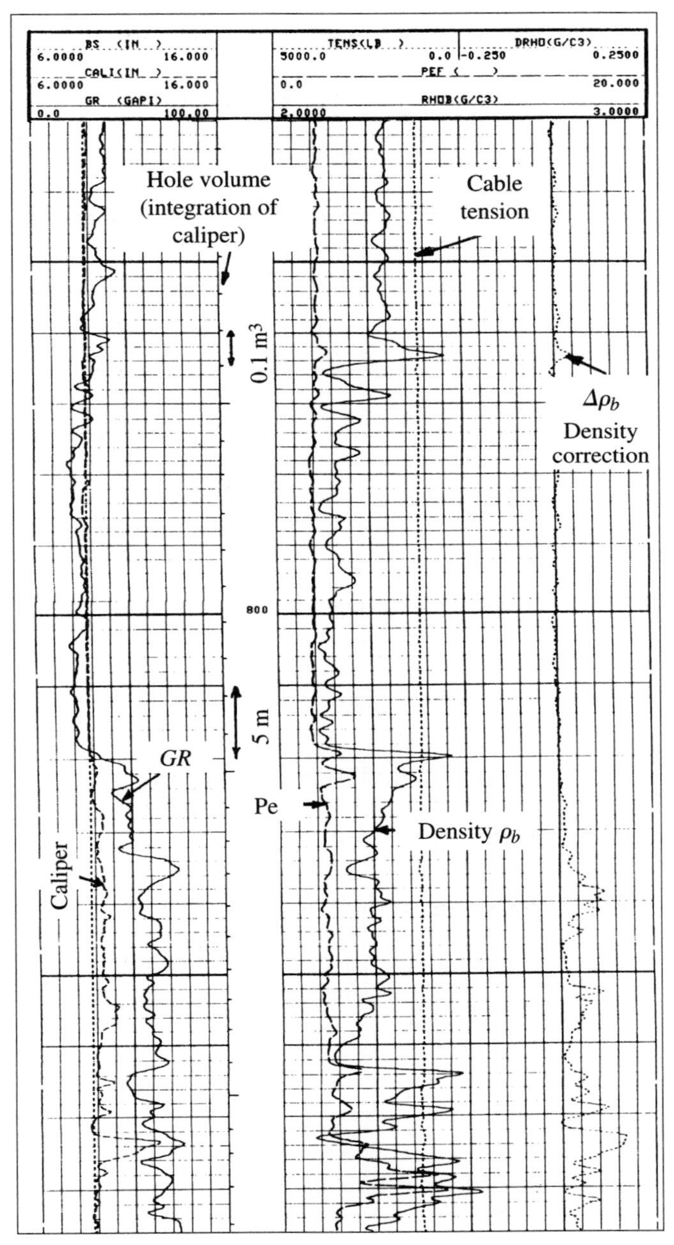

Fig. 1.50 ▶
Density log.

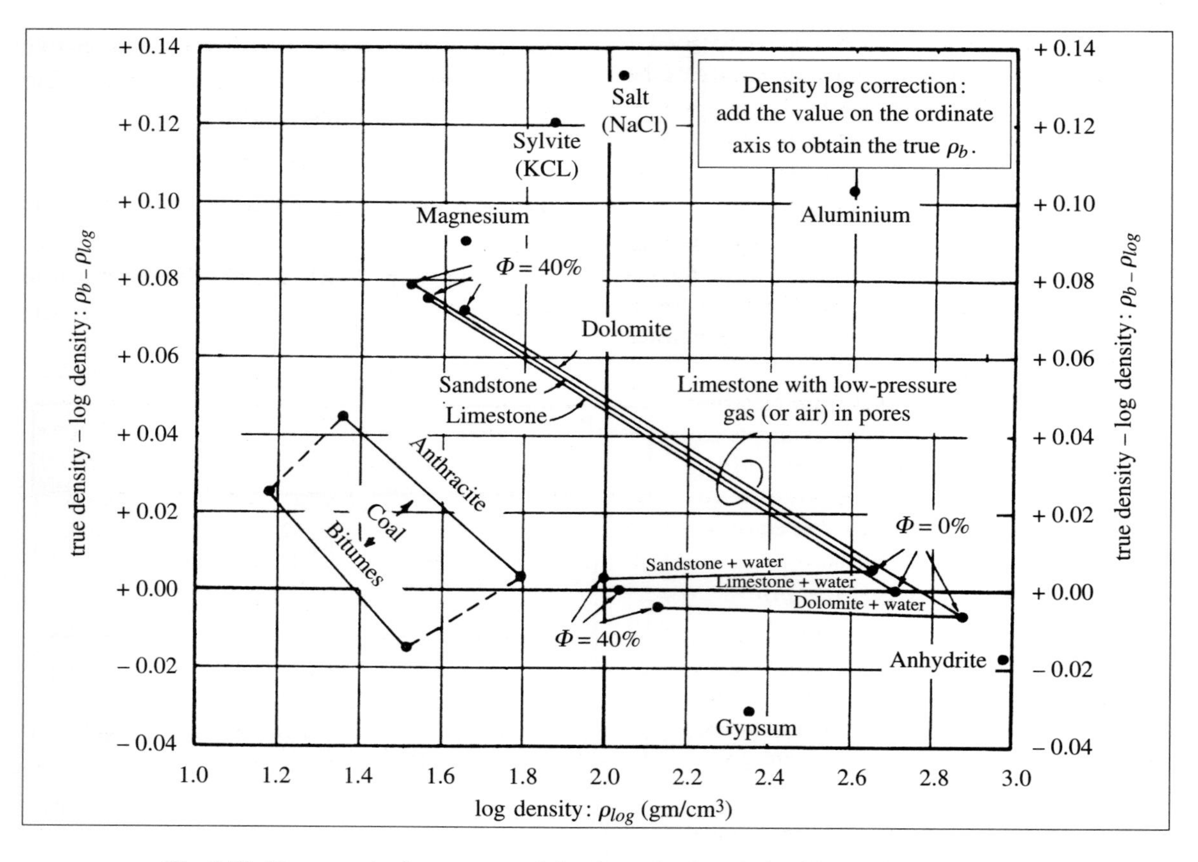

Fig. 1.51 Discrepancies between actual density and values derived from wireline logging. *(Courtesy of Schlumberger)*

where 2 Z/A = 2), the logging tool is calibrated to yield values close to the real densities of the most commonly encountered formations (e.g. limestone, sandstone and dolomite) containing water (Fig. 1.51).

1.2.2.1 Equipment and data acquisition

A. Operation

The conventional density tool makes use of a radioactive source (cesium-137) and two scintillation detectors. It is held against the well wall by a side-arm which also enables the measurement of borehole diameter (Fig. 1.52). Measurements are only rarely carried out in cased wells and cannot be used quantitatively.

B. Calibration

The density tool calibration is periodically checked at the operations base to make adjustments for individual source-detector characteristics

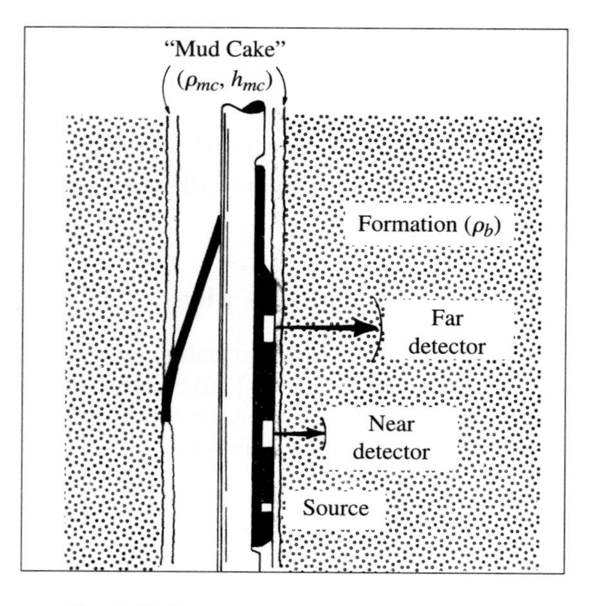

Fig. 1.52 Schematic drawing of a dual spacing Formation Density log.

enabling the computation from the count rate of 2 detectors to electron density. A wellsite calibration is performed to verify the integrity of the measurement system subsequent to transportation and once again after its descent into the well.

Any discrepancy between the value recorded by the Density tool and the actual density of the formation is due to the calibration ($\rho_e - \rho_b$) itself.

The table below reports the different density values for the main geological materials encountered in formations.

C. Recording

The γ radiation emitted by the source is measured at two different distances by means of the two detectors.

The measurement carried out on the far detector is corrected for the mud-cake effect using the response from the near detector. This corrected value is then converted into density and expressed in units of g/cm^3.

Thus, the correction so applied corresponds to the $\Delta\rho$ density correction curve on Fig. 1.50, which was used to provide a more accurate measurement of formation density (ρ_b).

Generally speaking, the extreme density corrections encountered in the presence of rough well walls or caved sections (>0.025 and <– 0.025 g/cm^3) can cast doubt on the values read off the density log.

N.B.

With the more modern tools, a supplementary measurement yields the photoelectric absorption factor (Pe, expressed in units of barns/electron). Since this quantity is practically independent of porosity and fluid, it is possible, under favourable conditions, to obtain a direct indication of the nature of the matrix.

1.2.2.2 Quality control and data correction

A. Calibration

It is always necessary to check the quality of calibrations. The calibration itself may be used to

COMPOSITION	FORMULA	TRUE DENSITY (g/cm^3)	2 Z/A	ELECTRON DENSITY	APPARENT DENSITY
Quartz	SiO_2	2.654	0.9985	2.650	2.648
Calcite	$CaCO_3$	2.710	0.9991	2.708	2.710
Dolomite	$CaMg(CO_3)_2$	2.870	0.9977	2.863	2.876
Anhydrite	$CaSO_4$	2.960	0.9990	2.957	2.977
Sylvite	KCI	1.984	0.9657	1.916	1.863
Halite	NaCI	2.165	0.9581	2.074	2.032
Gypsum	$CaSO_4\ 2H_2O$	2.320	1.0222	2.372	2.351
Anthracite		{ 1.400 1.800	1.030	{ 1.442 1.852	{ 1.355 1.796
Bituminous coal		{ 1.200 1.500	1.060	{ 1.272 1.590	{ 1.173 1.514
Fresh water	H_2O	1.000	1.1101	1.110	1.00
Salt water	200 000 ppm	1.146	1.0797	1.237	1.135
“Oil”	$n(CH_2)$	0.850	1.1407	0.970	0.850
Methane	CH_4	ρ_{meth}	1.247	1.247 ρ_{meth}	1.335 ρ_{meth} – 0.188
“Gas”	$C_{1.1}\ H_{4.2}$	ρ_g	1.238	1.238 ρ_g	1.325 ρ_g – 0.188

O. Serra, 1979.

establish corrections. In fact, in the case of certain formations such as halite or coal, corrections can be applied by reading off the true density on charts (Fig. 1.51).

In the case of gas-bearing formations, a correction must also be applied in order to create an acoustic impedance log, which makes use of true density values. However, such corrections are difficult to implement.

B. Anomalies arising from well conditions

It is essential to inspect the density values that are found in caved sections or in zones with strong density corrections ($\Delta\rho_b$). The occurence of extreme corrections (>0.025 and <– 0.025 g/cm^3) should lead to prudence in the interpretation of the corresponding densities read from the log (Fig. 1.53).

Another type of correction may be calculated on the basis of certain parameters and applied to the apparent density after subtraction of the $\Delta\rho_b$ value discussed above.

C. Corrections related to the zone of investigation

The density of a formation depends on porosity (containing water and hydrocarbons) and the properties of the pore-fluids (ρ_{mf} and ρ_{hc}). The ρ_b values indicated by the density tool refer to the flushed zone, which is invaded by the mud filtrate. Densities in the uninvaded zone (ρ_v) can be calculated by carrying out a full quantitative study of the formations based on logs and other well data (e.g. cuttings, downhole tests, core samples, etc.) (Fig. 1.54).

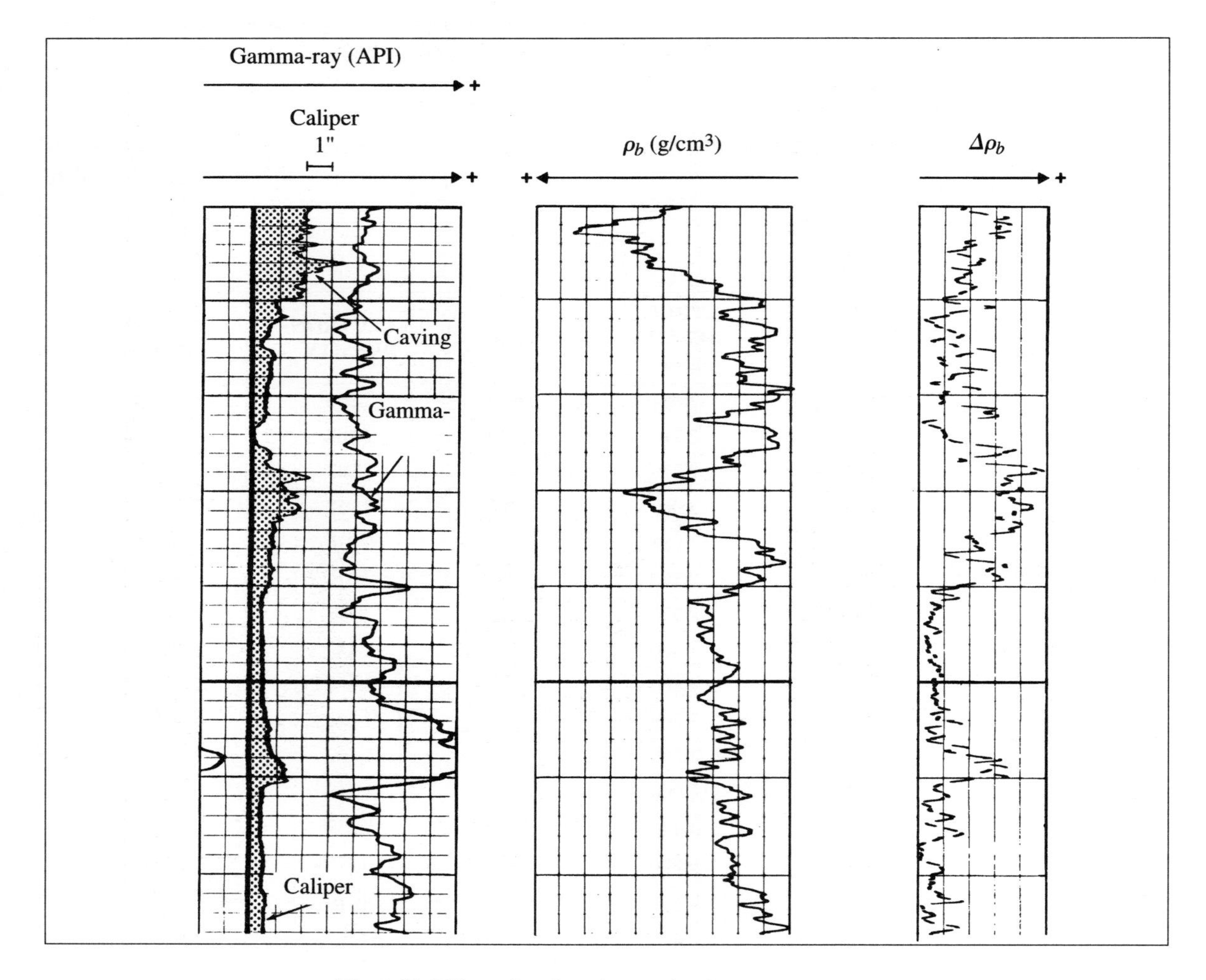

Fig. 1.53 Effect of well caving on density measurements.

As discussed below in Section 1.2.2.3, the correction for fluid invasion in the case of gas-bearing formations ($\rho_{hc} \approx 0.2$) with high porosity ($\Phi = 0.3$) and gas-rich fluids ($S_w \approx 0.2$ and $S_{xo} \approx 0.7$), can be written as follows:

$$\rho_b - \rho_v \approx \Phi (S_{xo} - S_w)(\rho_{mf} - \rho_{hc})$$

$$\rho_b - \rho_v \approx 0.1 \text{ g/cm}^3$$

1.2.2.3 Relations between measured density and the environment

The overall density of a formation depends on the matrix density (ρ_{ma}), the densities of the pore-fluids (ρ_{mf} and ρ_{hc}) and the porosity (Φ), as well as the water and hydrocarbon saturations (S_{xo} and S_{hr}, S_{hc}).

Taking the general case, for any given formation, we can write:

$$\rho_b = \sum_i \left(V_{ma_i} \, \rho_{ma_i} \right) + \Phi \rho_f$$

$$\rho_f = \sum_j S_{f_j} \, \rho_{f_j} \quad \text{and} \quad \sum_i V_{ma_i} + \Phi = 1$$

S_{f_j} is the fluid j saturation and V_{ma_i} is the percentage of matrix i, while ρ_{ma_i} and ρ_{f_j} are the density of matrix i and fluid j, respectively.

In most cases, the density measurements apply to the flushed zone, for which the following equations can be written:

(1) for clean formations:

$$\rho_b = (1 - \Phi)\, \rho_{ma} + \Phi (S_{xo}\, \rho_{mf} + S_{hr}\, \rho_{hc})$$

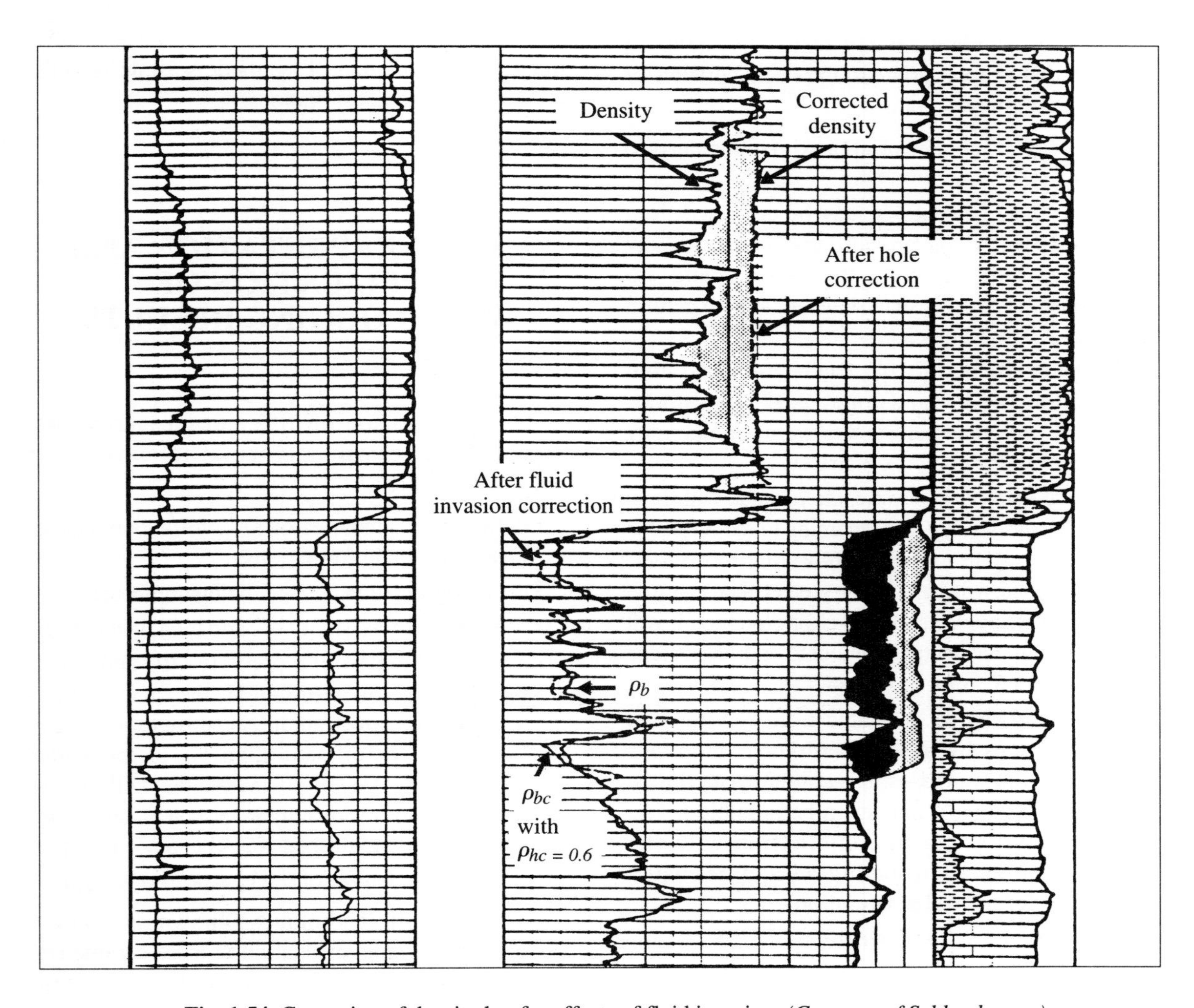

Fig. 1.54 Correction of density log for effects of fluid invasion. *(Courtesy of Schlumberger)*

(2) for shaly formations:

$$\rho_b = (1 - V_{sh} - \Phi)\,\rho_{ma} + V_{sh}\,\rho_{sh} + \Phi\,(S_{xo}\,\rho_{mf} + S_{hr}\,\rho_{hc})$$

In these formations, the density in the uninvaded zone may be obtained as follows:

$$\rho_v = (1 - \Phi)\,\rho_{ma} + \Phi\,(S_w\,\rho_w + S_{hc}\,\rho_{hc})$$

In order to convert the densities measured in the flushed zone to the values that they would assume in the uninvaded zone, the following correction must be applied to the log reading:

$$\rho_b - \rho_v = \Phi\,(S_{xo}\,(\rho_{mf} - \rho_{hc}) - S_w\,(\rho_w - \rho_{hc}))$$

$$\rho_b - \rho_v = \Phi\,(S_{xo} - S_w)\,(\rho_{mf} - \rho_{hc})$$

1.2.2.4 Synthetic density log

When a density log has not been recorded, it is possible to calculate density values by using an equation that is appropriate to the characteristics of the formation in question.

(1) In reality, a number of approximations need to be made:
- The shale content V_{sh} may be estimated from the *GR* or *SP* log, provided that no radioactivity is associated with the matrix and no constituents other than shale are able to perturb the *SP* measurements.
- The matrix density ρ_{ma} can be estimated from cuttings (e.g. sandstone: 2.65; limestone: 2.71; dolomite: 2.87). An average value may be taken for complex matrices.
- If the pore-fluid is water, the ρ_f value will lie in the range 1-1.1 g/cm^3 according to the estimated salinity. With hydrocarbon-bearing formations, this value will be slightly greater than the true fluid density. When gas is clearly present, it will occur in considerable amounts and an average value for S_w must be chosen.
- Porosity can be evaluated from Neutron and Sonic logs by making various assumptions about the matrix, shale content and fluids. Alternatively, it may even be possible to use resistivity data in combination with the Archie equation, taking an approximated value for the water saturation:

$$Rt = FR_w/S_w^2 \quad \text{where} \quad F = 1/\Phi^2$$

(2) Under the least favourable circumstances, it is possible to use empirical equations like those presented in the context of sonic log simulation (see Section 1.2.1.4). For example, the Gardner equation can be applied to sonic log data:

$\rho_b = 0.23\,[10^6/\Delta t]^{1/4}$, where Δt is in units µs/ft.

(3) Full waveform acoustic logging has the advantage of providing some information on the nature of the matrix (in typical reservoir rocks such as limestones or sandstones) as well as the pore-fluids (liquid or gas). This approach enables the calculation of porosity (see below Section 1.3), thus making it possible to solve the density equations and obtain values of ρ_b. Moreover, the combination of full wave sonic logging with *GR* provides an estimate of V_{sh} in the case of shale rich formations.

The uncertainties resulting from density log simulation —when performed using all the approximations mentioned above— do not generally hinder the calculation of synthetic seismic records used for depth time calibrations.

1.3 INTERPRETATION AND APPLICATION OF WELL LOGGING RESULTS

The combination of different measurements obtained from various types of logging tools makes it possible to:

(1) define the boundaries of reservoirs, compacted layers and shale formations,
(2) obtain information on the lithology of formations,
(3) determine the resistivity of formation water (R_w),
(4) estimate the effective porosity (Φ) and shale content (V_{sh}) of a formation, the quantities of hydrocarbon present in the uninvaded zone —through the oil saturation S_{hc} or the water saturation S_w ($S_{hc} = 1 - S_w$)— as well as the proportions of residual hydrocarbon S_{hr} and mud filtrate S_{xo} in the flushed zone (i.e. $S_{hr} = 1 - S_{xo}$),
(5) estimate the density of the hydrocarbons (ρ_{hc}).

Simple on-site techniques (known as "Quick Look") allow a rapid evaluation of these parameters by the straightforward comparison of logs, at least in the case of less complex formations (Fig. 1.58).

Quantitative interpretations based on the relations described previously — and also logging tool response curves — yield the percentage contents of the different constituents encountered in the formation (Φ, V_{sh}, S_w, S_{xo}, mineralogy and fluid composition). These data are presented as logs accompanied by a listing, thus enabling the rapid identification of interesting zones.

As pointed out in the previous sections on sonic and density logs, it is also possible to simulate a tool response curve by calculation in the following cases:

(1) in a formation that has not been invaded by filtrate,

(2) in the same uninvaded formation, but containing different hydrocarbons,

(3) in a slightly different formation due to lateral facies variations or changes in cementation.

It is even possible to calculate a synthetic log in cases where a particular measurement has not been run (e.g. pseudo-sonic and synthetic density logs).

1.3.1 "Quick Look" qualitative interpretation

1.3.1.1 The "Quick Look" method

All "quick look" methods are based on the following assumptions:

(1) constant water resistivity R_w,

(2) moderate, step profile invasion,

(3) constant lithology,

(4) clean formations (shale-free),

(5) a 100% wet zone exists,

(6) good hole conditions.

A. Identification of reservoir zones

This is carried out with the help of all the different types of wireline log data:

(1) Shale formations may be recognised on the logs by looking for the following characteristics, e.g.:
- high level of natural radioactivity,
- low resistivity,
- lack of fluid invasion,
- relative position of Neutron and Density curves,
- high values recorded by Sonic and Neutron logs.

(2) Since compacted zones (dense horizons) have zero porosity, they do not correspond to reservoirs. They may be recognised by their very high resistivity values as well as Neutron, Density and Sonic log values corresponding to matrix characteristics. A clear identification of this type of formation (e.g. compact limestones, sandstones and dolomites, anhydrite, salt, etc.) is generally facilitated by plotting these values onto cross plots.

(3) Within reservoir zones, the caliper log may indicate the presence of mud-cake. This feature is also recognisable on Microlog (i.e. tool providing dual measurements, one mainly influenced by the mud-cake itself and the other by the mud-cake as well as the surrounding formation).

B. Comparison of resistivity curves

Within reservoir zones, the lithology, porosity, shale content, fluid saturation and nature of fluids are involved in almost all types of log; using these results, it is of primary importance to determine the oil-water contact.

In clean water-bearing reservoirs, the depth profiles of $\log_{10} R_t$ and $\log_{10} R_{xo}$ are very nearly parallel, while $R_t/R_{xo} = R_w/R_{mf}$ assuming constant salinity of formations waters.

After superposition of the $\log_{10} R_t$ and $\log_{10} R_{xo}$ curves for the zone thought to contain water, the offset seen between these two curves over the rest of the reservoir is roughly indicative of the water saturation S_w (Fig. 1.55).

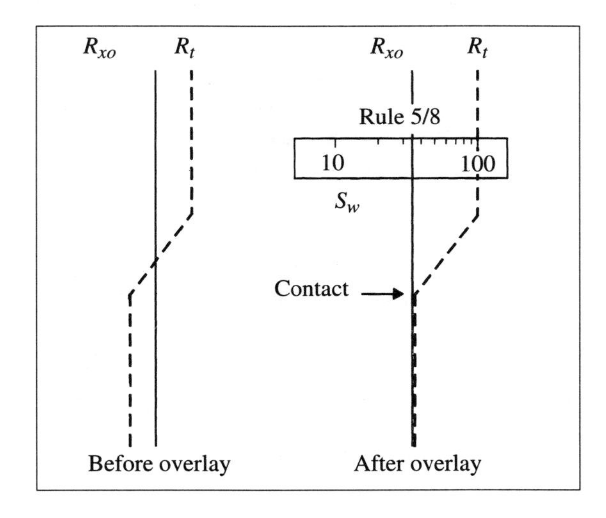

Fig. 1.55 Quick Look method applied to resistivities.

QUICK LOOK

I DEFINITION OF RESERVOIR BOUNDARIES

Elimination of shale layers

- GR
- SP (baseline)
- Caliper (caved sections)
- Φ_N
- Separation $[\rho_b\text{-}\Phi_N] > [\rho_b\text{-}\Phi_N]_{DOL}$
- Lack of fluid invasion, $R_t \simeq R_{xo}$

Elimination of compact layers

- Compact layers
  High R_t and $R_t = R_{xo}$
- Marker beds (salt, anhydrite, etc.) and all types of compact lithology
  ρ_b # ρ_{ma}
  Φ_N # Φ_{Nma}
  Δ_t # Δt_{ma}
- Caliper: bit size (except salt, etc.)

Search for porous and permeable zones

Microlog: positive separation
Caliper: mud-cake

II DETERMINATION OF WATER-OIL CONTACT AND OF WATER SATURATIONS

R_t vs R_{xo} overlay

- Water-bearging zone: low resistivities; R_t curve parallel to R_{xo} over significant interval
- Oil-water contact:

  After overlay of "R_t" and "R_{xo}" profiles

	Rxo / Rt	
HC zone		R_t on right
O/W contact		
Water zone		R_t and R_{xo} overlaying

In the WATER ZONE:	$R_w/R_{mf} = R_t/R_{xo}$.
In the HC ZONE:	R_t, R_{xo} and reading on 5/8 logarithmic scale.

III LITHOLOGY, POROSITY AND HYDROCARBON TYPE

ρ_b vs Φ_N overlay

OIL: weak infuence (predominant effect of lithology)

GAS: gas effect superimposed on lithology

$\rho_b \searrow$ $\Phi_N \searrow$

2 DENSITY 3

45 NEUTRON -15

← Gas effect - - - →

In the WATER ZONE:	LITHOLOGY	Relative positions of ρ_b and Φ_N
	POROSITY	Mean of the ρ_b - Φ_N separation
In the HC ZONE:	FLUID, LITHOLOGY	Relative positions of ρ_b and Φ_N
	POROSITY	Mean or 1/4 of the ρ_b - Φ_N separation

LIMITATIONS OF THE METHOD:

- *Results can be modified by the presence of clay in the reservoir.*
- *Salinity of the formation water should remain constant.*

ENSPM Document.

By assuming $S_{xo} = S_w^{1/5}$ the water saturation S_w can be directly estimated using a logarithmic scale expanded by a factor of 8/5 in comparison with the scale of the resistivity logs. To achieve this, it is simply necessary to overlay the "100" mark of the rule (known as the "5/8 rule") onto the R_t curve, which lies to right of the R_{xo} curve except when perturbed by effects due to the detection system of the tool. The S_w value can then be read off the rule at its intersection with the R_{xo} curve.

It is possible to check whether the reservoirs do not contain formation waters of different salinities by making use of the SP log.

N.B.

It should be noted that oil and gas produce identical responses on resistivity logs, i.e. they both behave as non-conductors, so resistivity measurements are unable to distinguish these two fluids.

C. Comparison of porosity curves

The curves derived from Neutron (porosity Φ_N) and Density (density ρ_b) logs are subsequently used in the determination of lithology, porosity and hydrocarbon type (Fig. 1.57). In this context, the different measurement scales need to be correctly matched against each other (e.g. if the Neutron log is recorded in "limestone matrix" the 2.2 and 2.7 marks on the Density log should be placed opposite the 30 and 0 marks on the Neutron log, respectively).

Within clean water-bearing zones, the position of the two curves is diagnostic of lithology (Fig. 1.56). The value read off the Neutron scale at the mid-point of the separation between the two curves at any given depth corresponds to an approximation of the total porosity of the formation at that depth.

In the hydrocarbon zone, the relative position of the two curves is influenced concurrently by lithology as well as the type and quantity of hydrocarbons. The curve separation is little different in an oil-bearing formation in comparison with a water-bearing formation. The Density and Neutron logs in a gas-bearing formation tend to show decreasing values as the hydrocarbon density falls and hydrocarbon saturation rises.

Porosity is estimated in the same manner as in the water zone. However, in the case of gas, it is preferable to read the value at one quarter of the distance separating the Neutron and Density curves, nearer the Density curve.

N.B.

- *Due to the closely similar behaviour of oil and water as regards porosity measurement, very little difference is observed in the response of Neutron and Density tools; this contrasts strongly with the major differences between gas and liquid.*

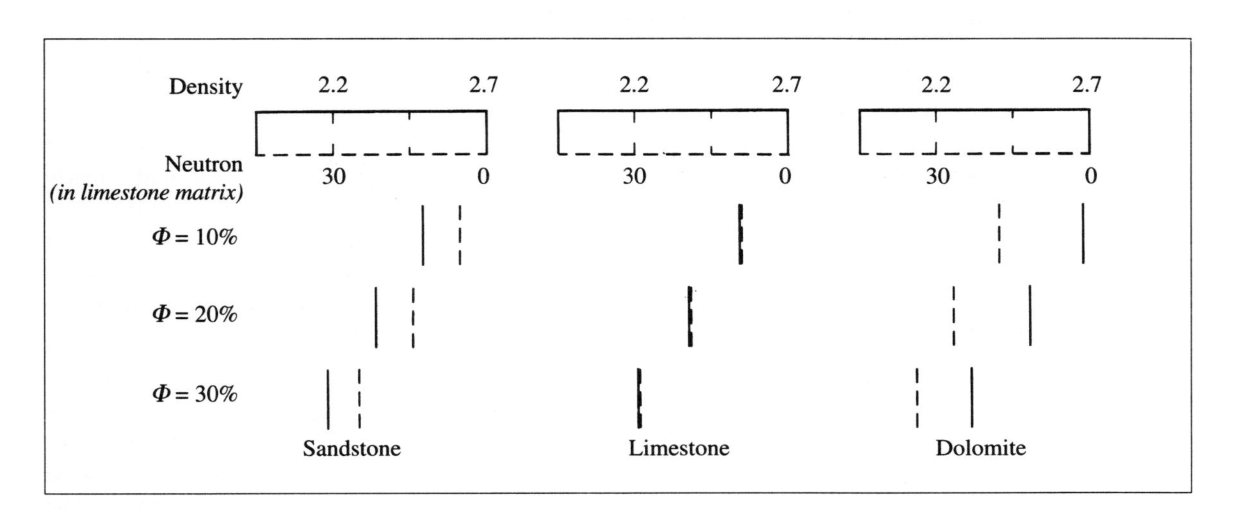

Fig. 1.56 Density and neutron log data used to calculate porosity in different lithologies; the curves are matched against a scale established for limestone formations.

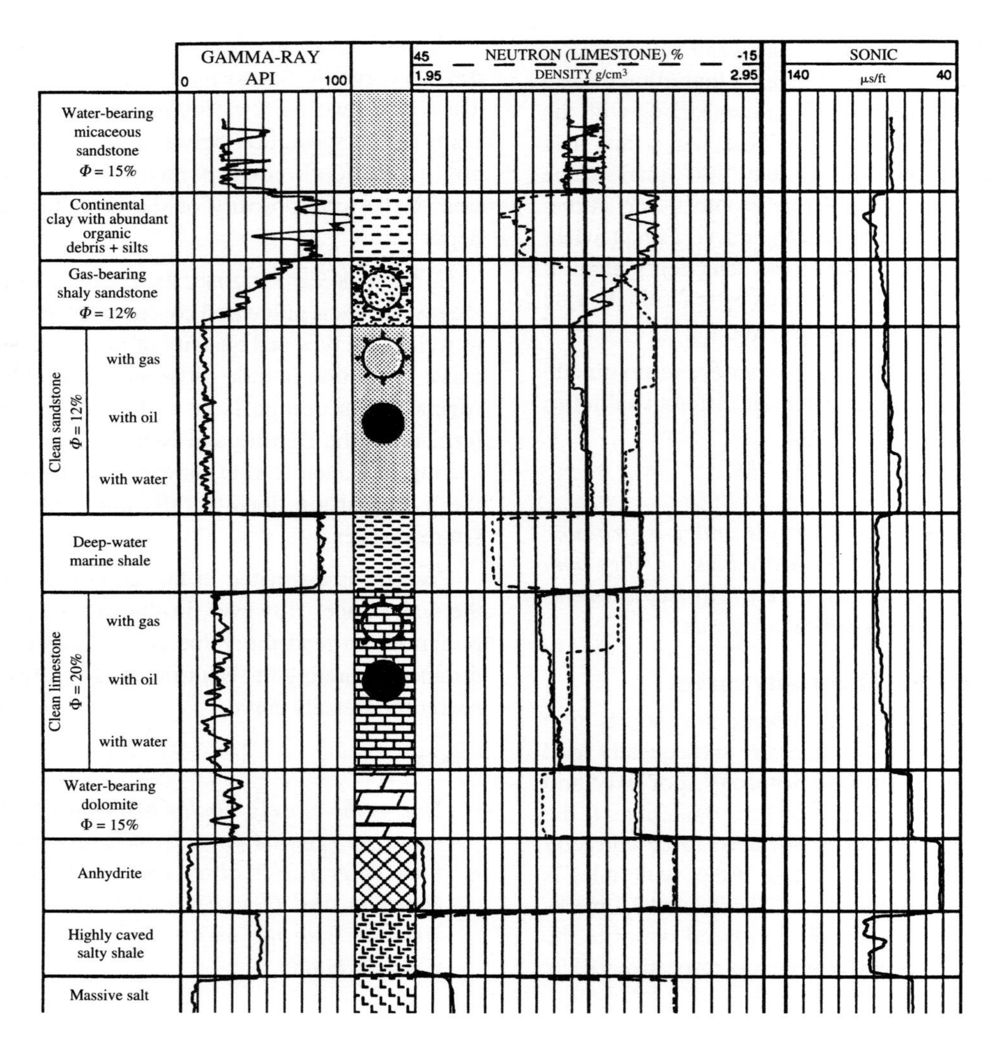

Fig. 1.57 Responses of Gamma-Ray, Neutron, Density and Sonic logging tools (Neutron and Density logs are on scales compatible with limestone lithology). *(Courtesy of ENSPM)*

- *In the case of reservoir rocks containing significant amounts of shale, the interpretation of log data is more problematic and only a quantitative approach is able to provide satisfactory results. Nevertheless, lithology can be assessed by taking into account the shift in Neutron and Density curves towards a position corresponding to pure shale. All other things being equal, this position is intermediate between the clean zone and the end-member shale.*

As an example, a number of conclusions can be drawn from a Quick Look analysis of the well logs presented in Fig. 1.58.

The studied interval (1450-1580 m) corresponds to a reservoir zone containing some shaly, marly and calcareous intercalations. The top of the water zone is situated at 1517 m, while the water saturation decreases upward through the transition zone to a value of 10% in the hydrocarbon zone.

The lithology within the water zone is a calcareous sandstone having a porosity of about 30%. An oil zone about 10 m thick is encountered above the water zone. The gas zone can be clearly picked out above 1502 m. An intermediate zone (1502-1505 m) can be distinguished between the gas and oil zones. The estimated porosity of 30% remains constant throughout the studied interval.

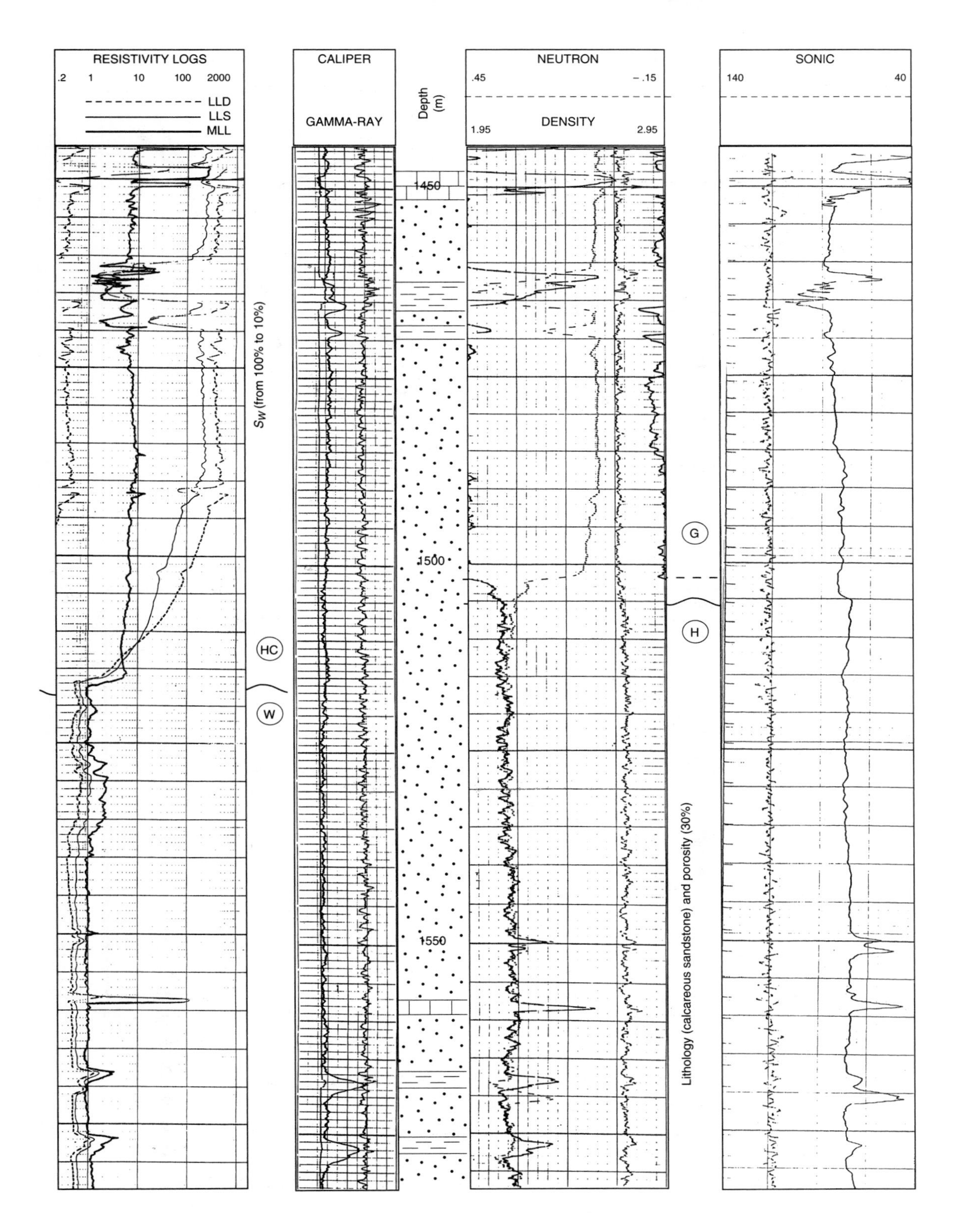

Fig. 1.58 Qualitative interpretation of logs. *(Courtesy of ENSPM)*

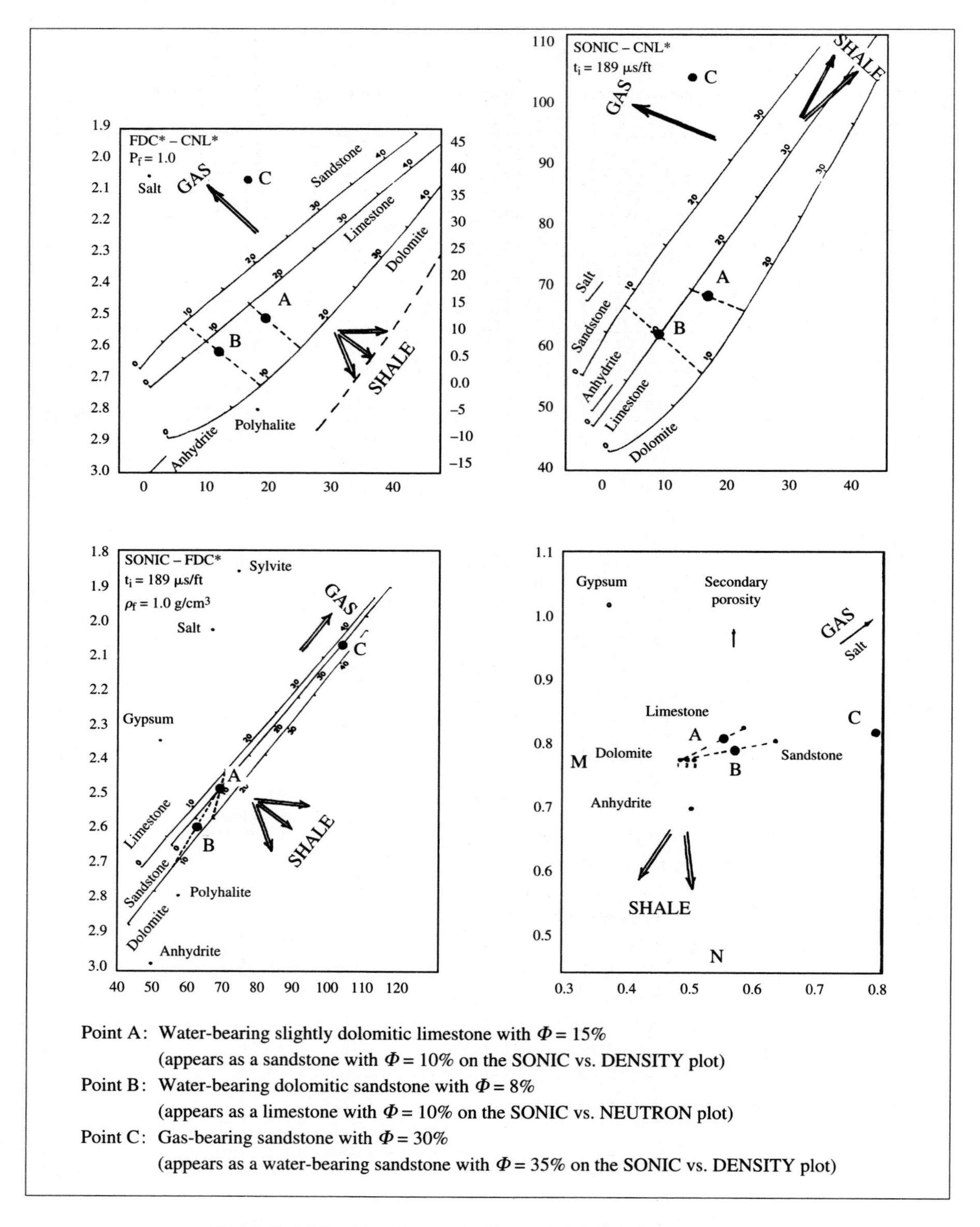

Fig. 1.59 Main types of cross-plot in relation to effects of gas and clay.
(ENSPM document modified from Schlumberger)

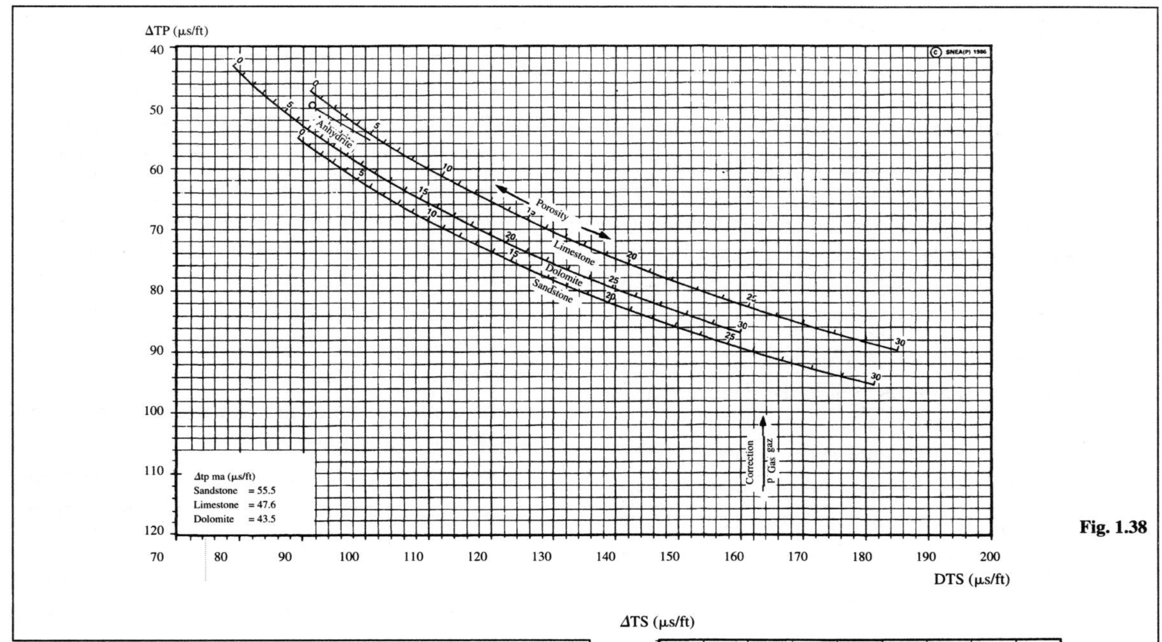

Fig. 1.38

1.3.1.2 Use of cross-plots

Cross-plotting techniques can facilitate the qualitative determination of the required parameters (e.g. lithology, porosity, clay content) by combining the data from Neutron, Density and Sonic logs.

As shown in the example presented in Fig. 1.59, it is generally useful to combine several different types of log — employing several charts at the same time — to detect the effects of gas and shales and thus to avoid errors of interpretation.

The new possibilities offered by tools with full waveform acoustic recording have led to the development of cross-plots which should be used with the same precautions as required for the charts presented in a previous section (Fig. 1.60; *cf.* Section: Acoustic wave velocity and petrophysical properties). It should be kept in mind that the effects of shale, fluids and accessory minerals are also apparent on these new cross-plots. In addition, textural effects have an influence on the measured acoustic parameters, in particular the propagation velocity.

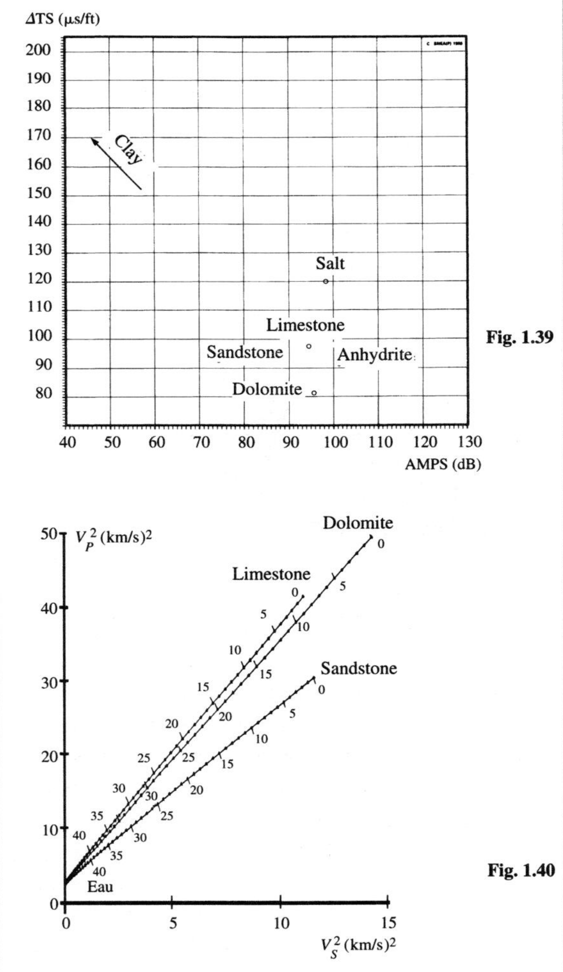

Fig. 1.39

Fig. 1.40

Fig. 1.60 

Cross-plots showing relationship between lithology, seismic slowness and amplitude of *P* and *S* waves.

1.3.2 Quantitative interpretation (*cf.* Fig. 1.61)

A minimum number of measurements are required to carry out the quantitative interpretation of logs with a view to obtaining the porosity (Φ), the water saturation (S_w and S_{xo}, in the uninvaded and the flushed zones, respectively), as well as the characteristics of the matrix and hydrocarbons in the formation:

(1) Three resistivity measurements at different depths of investigation enable the determination of R_t and R_{xo} (resistivities of the uninvaded formation and the flushed zone, respectively).
Approximation to R_t may be obtained directly from deep investigation tools (e.g. Deep Induction log: ILd; Deep Laterolog: LLd) while shallow investigation tools are used for R_{xo} (e.g. Microlaterolog: MLL; ou Micro-SFL, MSFL).

(2) The shale content V_{sh} cannot be precisely determined. It is necessary to calculate a whole series of indicators using different tools, whose measurement are influenced by the amount of clay, in order to obtain a probable value close to the true content. In certain cases, direct use of the measured radioactivity can provide a means of estimating V_{sh}:

$$V_{sh}\,GR = \frac{GR - GR_{min}}{GR_{max} - GR_{min}}$$

where GR_{max} is the maximum radioactivity reading within a pure shale layer and GR_{min} is the minimum radioactivity in a layer that has been checked as being shale-free (i.e. clean).
In fact, this assumes that the measured radioactivity is not due to the presence of certain radioactive minerals found in clastic formations (e.g. heavy minerals, K-feldspars, micas, etc.) or associated with carbonates (e.g. shell debris, radioactive dolomite, limestones containing organic matter). Such occurrences are not so rare as commonly believed.

(3) Resistivity of the formation water R_w can be obtained from salinity measurements carried out during fluid tests (DST, RFT), from calculations based on spontaneous potential or by the combination of resistivity measurements in water-bearing formations.

(4) The porosity Φ, the water saturations S_w and S_{xo}, as well as the type of hydrocarbon and the lithology can be calculated by iteration. Alternatively, they may also be obtained by minimising the discrepancies between computed log values (which are recalculated from an estimated model and the tool response) and the actual measurements from these tools. This procedure makes use of classic equations (Archie, Poupon, etc.) which involve relations between R_w, R_{mf}, R_t, R_{xo}, S_w, S_{xo}, V_{sh}, Φ, ρ_{ma} and ρ_{hc}.
The examination of drill cuttings can also yield indications on lithology.
Formation fluids tests are able to indicate the presence of hydrocarbons and provide information on their characteristics.
Approximations about the nature of the matrix may lead to relatively simple estimates of porosity Φ (see The Quick Look method).
It is noteworthy that as the shale content increases to significant levels, the approximations tend to give results that are farther and farther away from reality.

1.3.3 Other applications of wireline logs

1.3.3.1 Dip determination

A knowledge of the formation dips occurring in wells is one of the key elements enabling correlations between wells. Moreover, it leads to confirmation or rejection of certain tectonic structures suggested by seismic surveys and provides data for seismic processing.

A dipmeter is a logging tool which enables the calculation of the dip angle of various planar surfaces traversed by the well (bed boundaries, faults, etc.); it makes use of least three of the resistivity measurements provided by the logging tool. The autocorrelation of these different measurements reveals the existence of planes whose dips and azimuths can be calculated; dip is the angle formed between the horizontal plane and the line of steepest slope in the plane, while the azimuth is the direction of this line with respect to North.

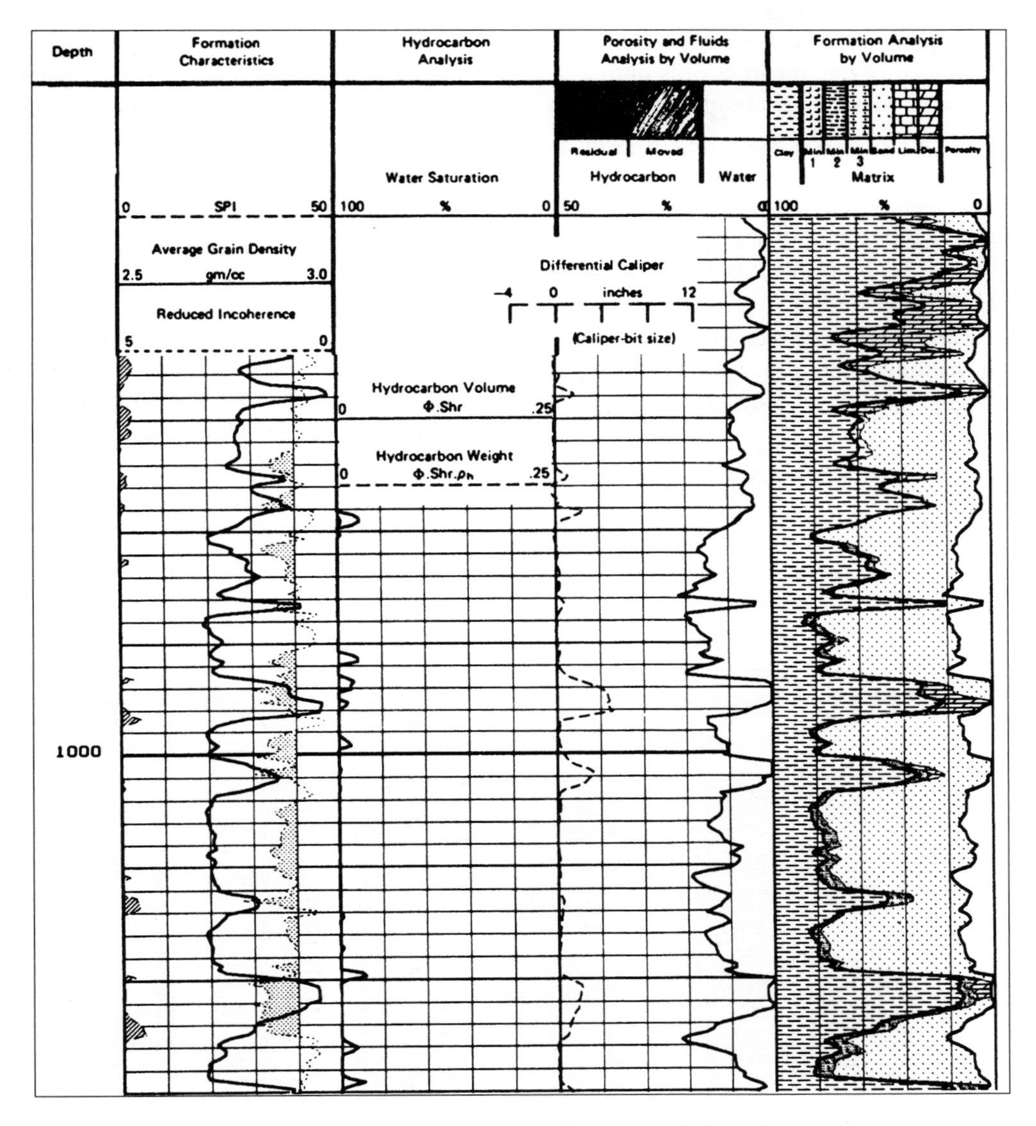

Fig. 1.61 Example of a quantitative interpretation.
(Courtesy of Gaz de France)

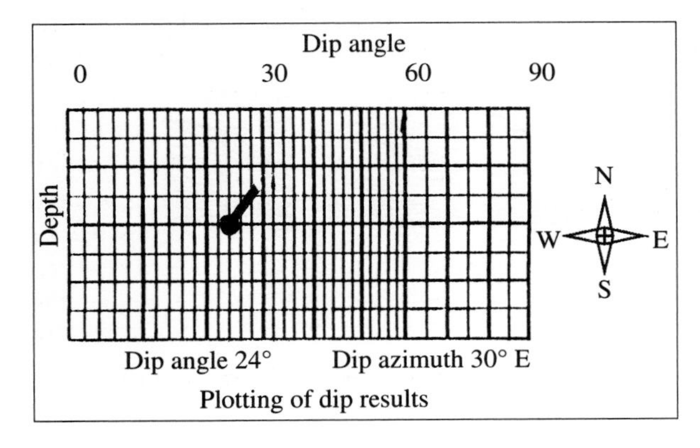

Plotting of dip results

The most common representation consists of an arrow plot with tadpoles indicating the dip of the structural event on a scale from 0 to 90°, and the dip azimuth shown by the orientation of the tadpole (see unnumbered diagram in left column). The angles calculated from the variations in the recorded parameter yield the dips of bedding planes, laminations, faults, etc. (see Figs. 1.63 and 1.64 for examples).

Different ways of presenting dipmeter measurements can be used to plot dip angles directly onto geological and seismic sections. The procedure involves projecting the apparent dips onto a set of vertical planes with different bearings (Fig. 1.62).

This type of plot makes it possible to evaluate the structural dip as a function of depth (see example presented in Fig. 1.66), thus facilitating the structural interpretation by determining the type of folds and their axis, as well as type of faults and their orientation.

Finer scale correlations enable the study of trends in bedding plane dip, which help in the recognition of depositional sequences traversed by the borehole.

The apparent dips on these diagrams are calculated by automatic correlation programs which include algorithms of various different types appropriate to the nature of the interpretation. These algorithms make use of different parameters, some of which are fixed by the program and others which are selected by the operator according to geological information and the inspection of logs, and the type of information sought.

As a consequence, apparent dips should themselves be analysed in the light of the program used, the expected results and the condition of the well.

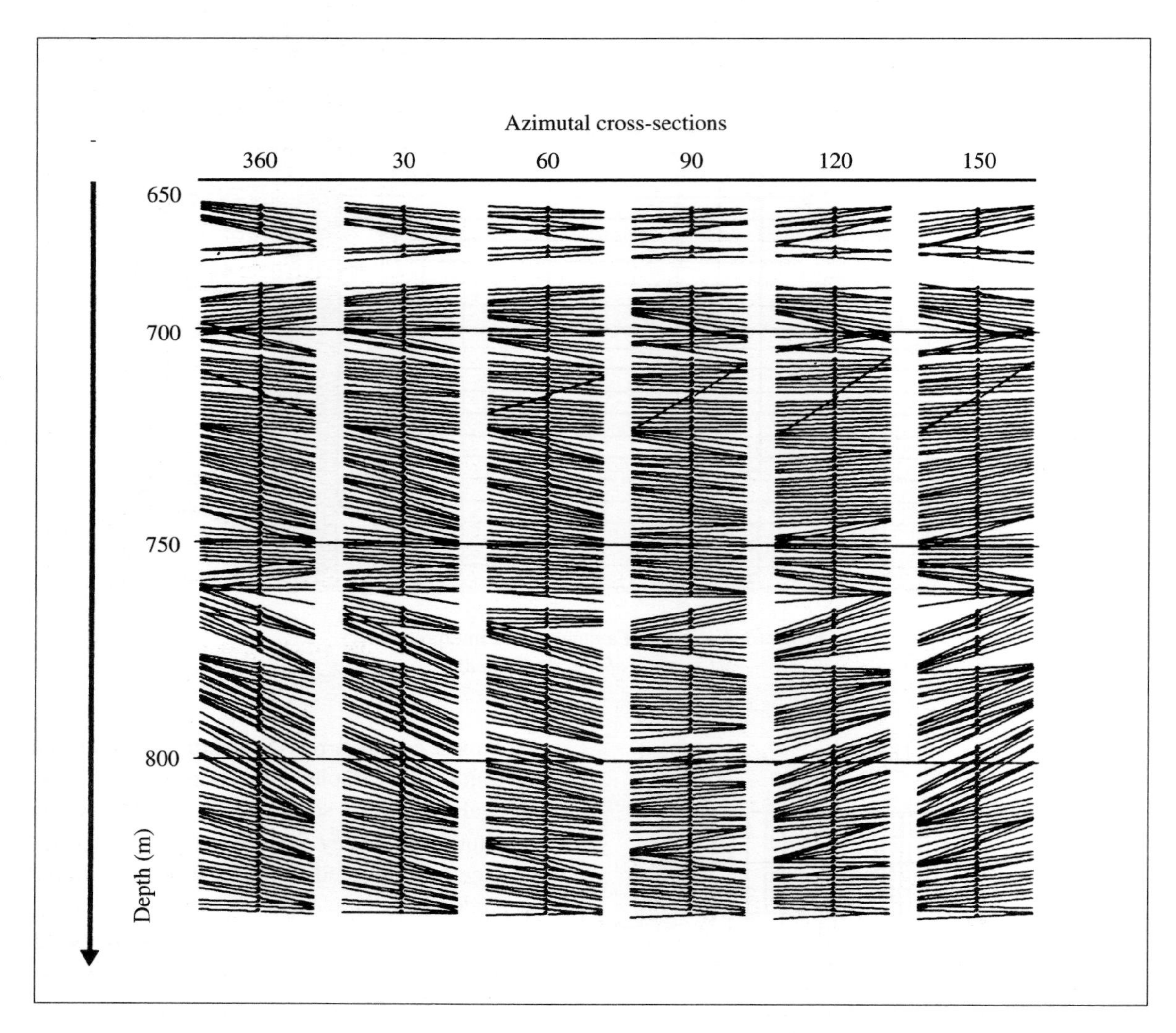

Fig. 1.62 "Stick Plot" showing apparent dips plotted as a function of depth for a set of different bearings. *(Courtesy of Schlumberger)*

The two main types of program used for dip calculation are based on:

(1) Cross-correlations between the resistivity curves recorded by the tool pads (e.g. CLUSTER or MSD methods used by *Schlumberger*); this approach is more generally applied in the study of structural dip and tectonic fabric elements (see Fig. 1.63).

(2) Methods using pattern recognition (e.g. GEODIP and LOCDIP of *Schlumberger*), which are adapted to sedimentological interpretation (see Fig. 1.64).

The two logs presented in Figs. 1.63 and 1.64 come from the same well.

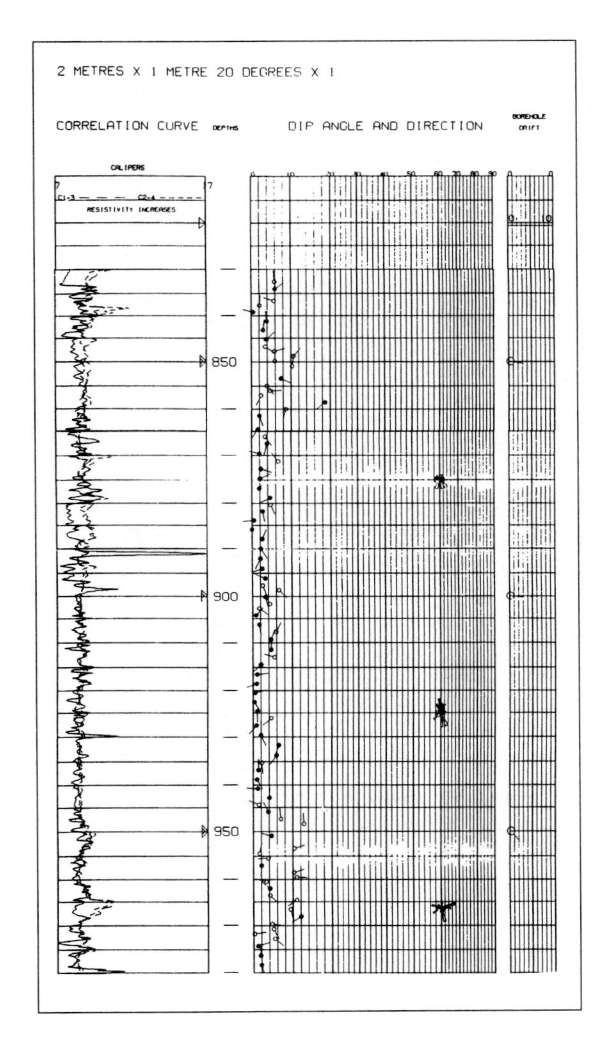

Fig. 1.63 CLUSTER analysis.
(Courtesy of Schlumberger)

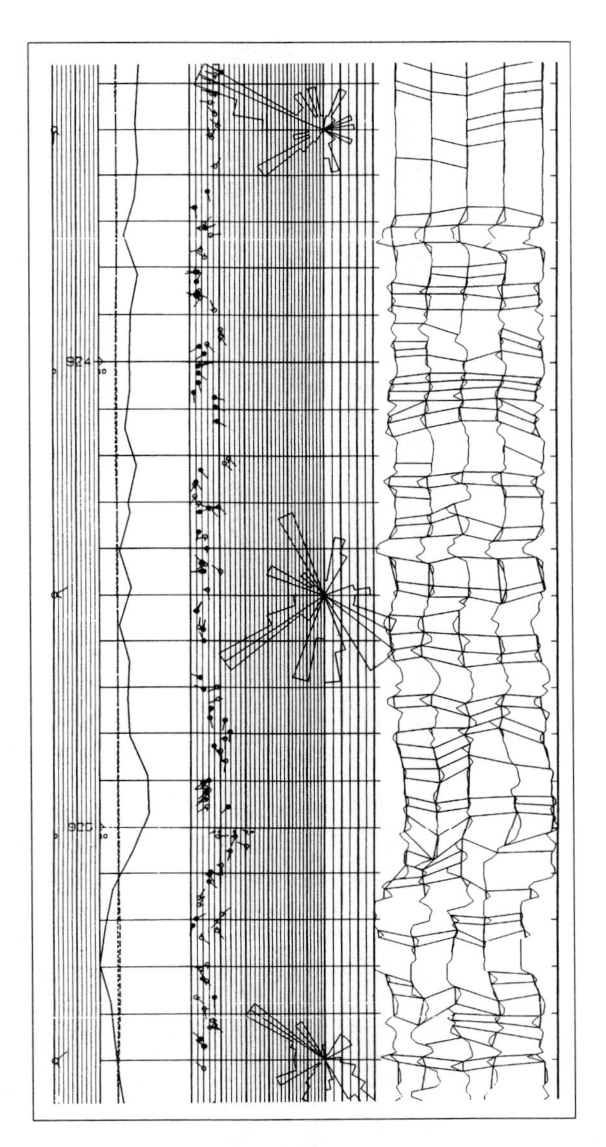

Fig. 1.64 GEODIP analysis.
(Courtesy of Schlumberger)

Certain tools provide a kind of overall image of the borehole wall (e.g. Formation Micro Imaging Tool of *Schlumberger*, BoreHole TeleViewer and Circumferential Acoustic Scanning Tool of *HLS*), which are potentially easier to interpret in terms of the recognition of elements in the well. However, the dip angles extracted by this method have also to be correlated with geological data (Fig. 1.65).

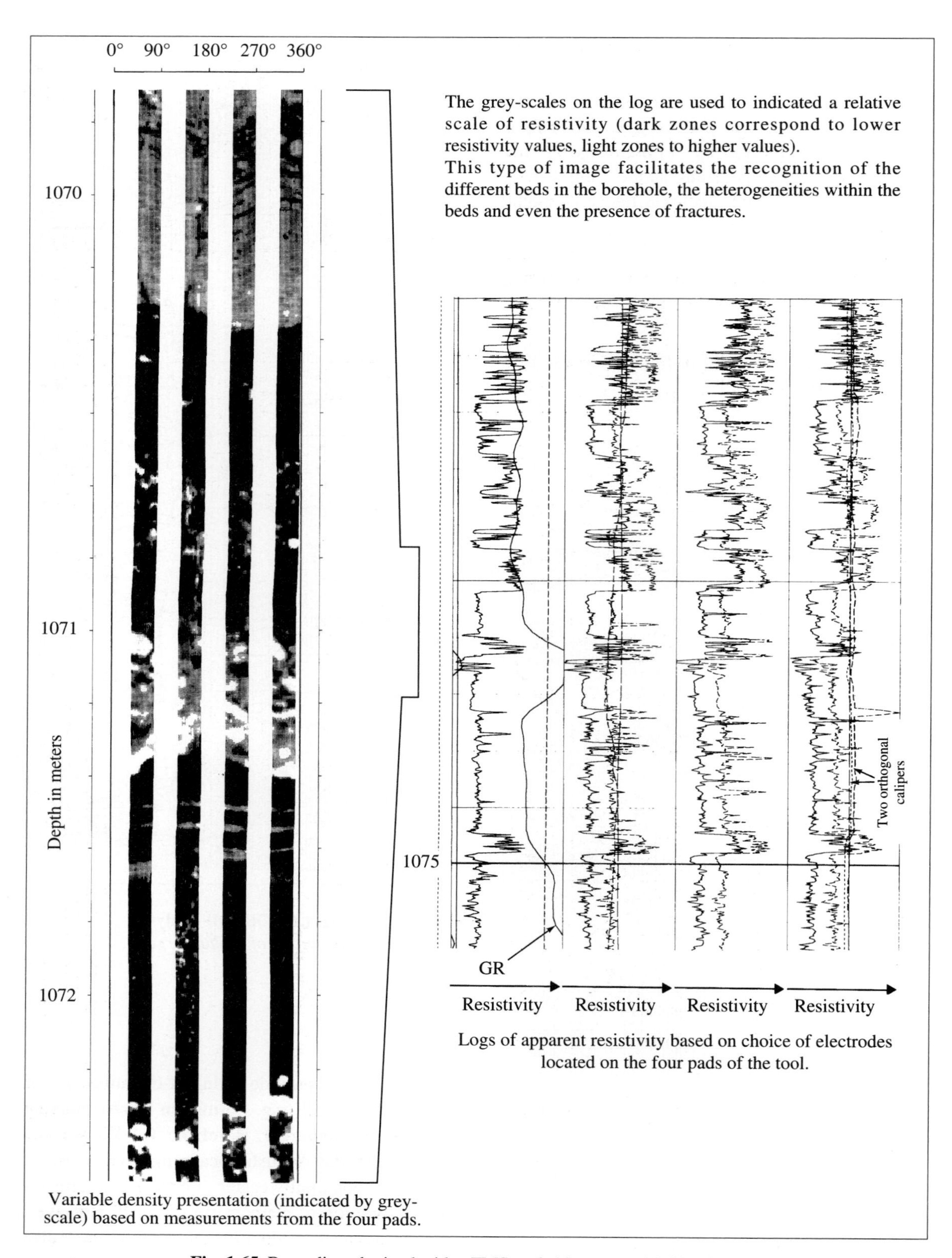

Fig. 1.65 Recording obtained with a FMS tool. (*Courtesy of Schlumberger*)

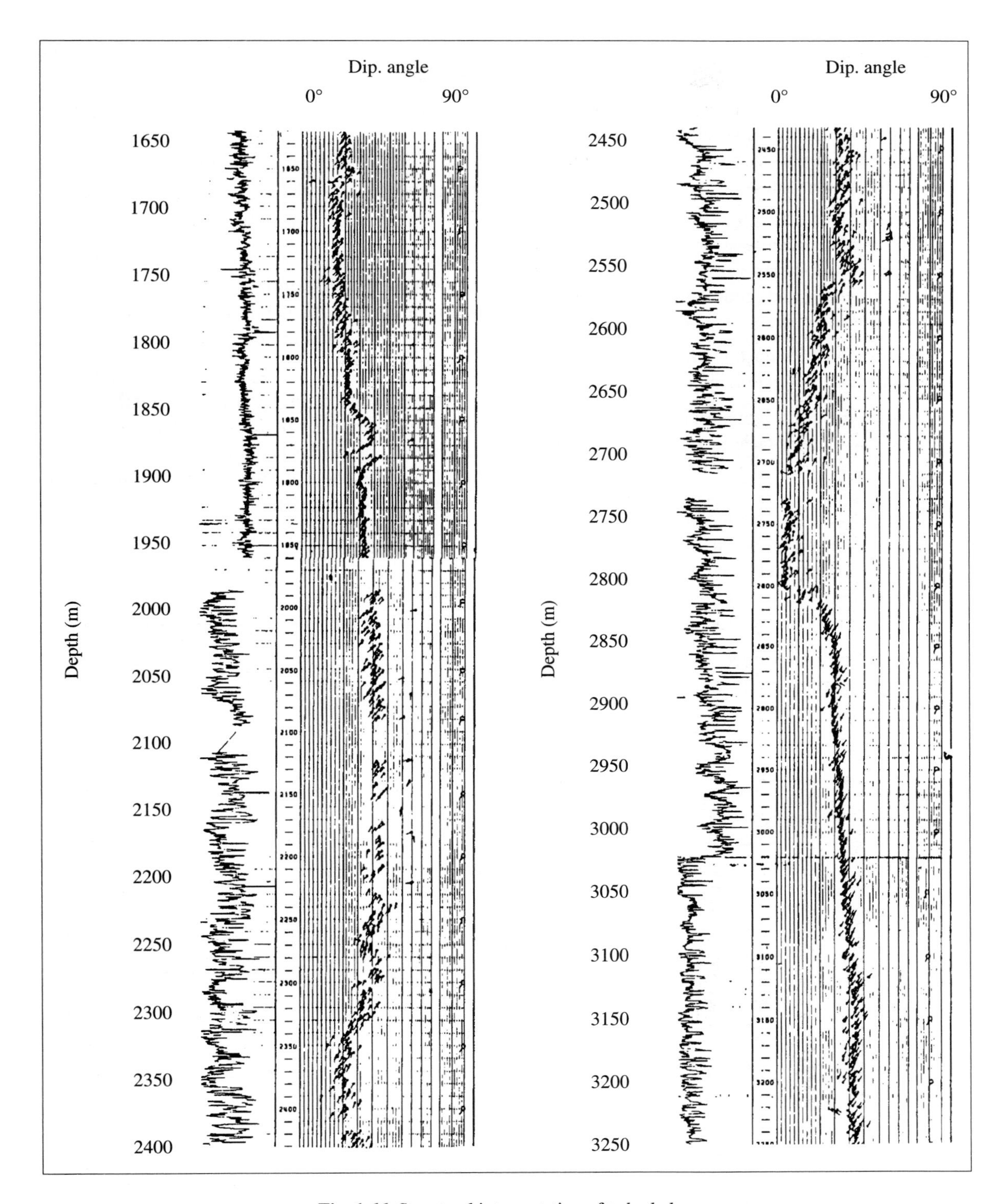

Fig. 1.66 Structural interpretation of tadpole log.
The gradual variation of dip with depth is indicative of folding tectonics
(i.e. passage from one fold flank to another).
Some of the discontinuities may reflect faulted zones (e.g. at a depth of 1880 m).
(Courtesy of Schlumberger)

1.3.3.2 Study of compaction

The compaction of formations as a function of their burial has an effect on porosity and, as a result, also on their content in pore-fluids. Almost all the conventionally used well logging techniques measure this phenomenon.

Compaction effects are particularly apparent in regard to the characteristics of shales. In fact, the resistivity and porosity logs in such formations show a continuous and gradual trend reflecting the decrease in water content with depth.

Various anomalies encountered in these profiles can be related to different geological processes:

(1) Undercompaction is generally linked to rapid burial, or some obstacle to water expulsion from the formation due to the presence of an impermeable caprock. The phenomenon is characterised by a less than normal decrease in porosity with depth (*cf.* Fig. 1.67) resulting in an abnormal compaction trend on the different logs.

(2) Abrupt jumps in porosity may correspond to periods of uplift or erosion of the formation.

These anomalous effects are generally investigated using resistivity and/or sonic curves because these logs are most commonly available over the entire depth of a well.

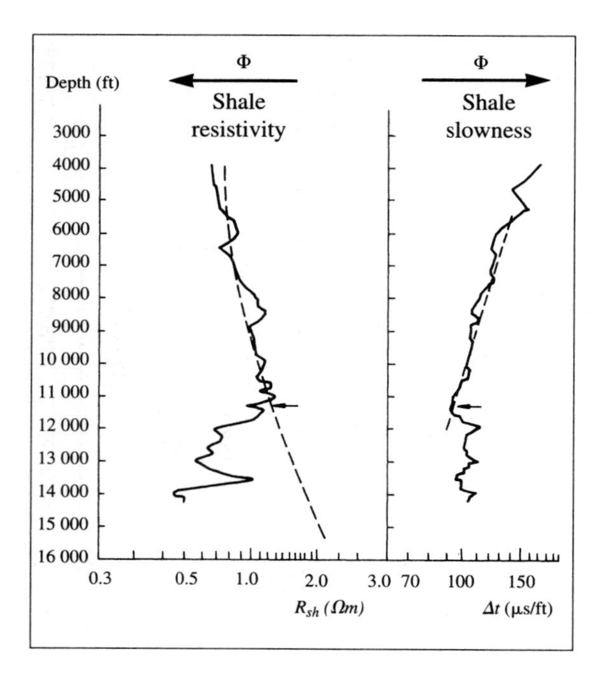

Fig. 1.67 Resistivity and slowness as a function of depth. *(from Kinji Magara, 1978)*

Empirical porosity vs. depth relations have otherwise been established, as illustrated in Fig. 1.68. In the case of shales, the following equation can be written:

$$\Delta t = \Delta t_0 \, e^{-cZ}$$

where the slowness Δt (interval transit-time) and Δt_0 (interval transit-time extrapolated to surface) are in μs/ft, Z is the depth in feet and c is a coefficient representing the slope of the normal porosity vs. depth curve.

Other relationships link the resistivity and acoustic parameters of formations to depth, thus providing one of the methods —among others— used for the construction of synthetic seismic records (see Chapter 4).

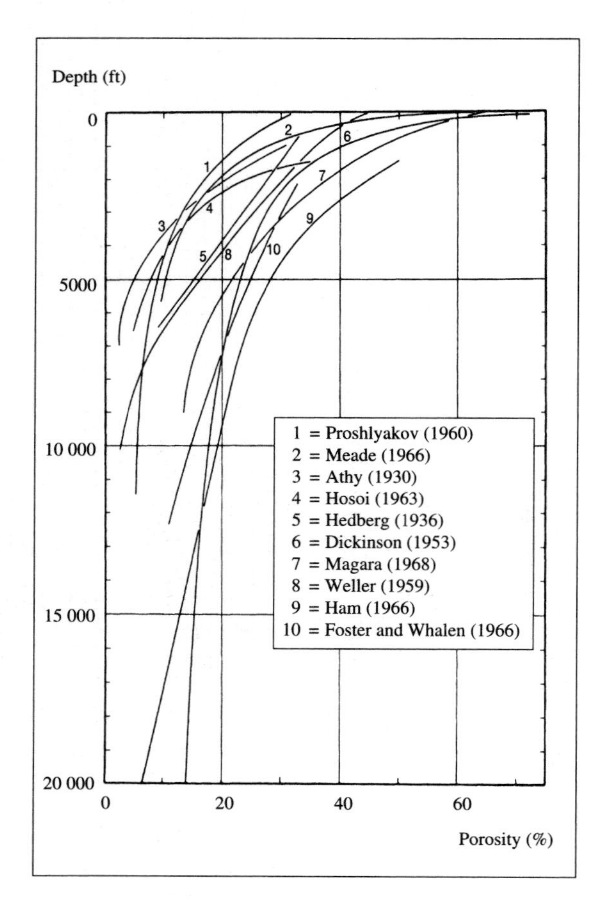

Fig. 1.68 Variation of porosity as a function of depth. *(from Kinji Magara, 1978)*

1.3.3.3 Study of fractured formations

Fractures affect practically all types of logs since they influence the condition of the well (oval hole, rugosity, micro-caved sections, etc.) and cause anomalies in electrical or thermal conductivity (fluid in open fractures, cementation of closed fractures, etc.).

Thus, the identification of fractures (or at least the presence of fractured zones) is based on the identification of anomalies on the various logs (resistivity, porosity, radioactivity, etc.) (Fig. 1.69).

Attempts at assessing the type and degree of fracturing associated with fracture porosity have been carried out using both deep and shallow resistivity measurements (Sibbit and Faivre, 1985), as well as resistivity formation factor.

Although dipmeter surveys facilitate the detection of fractures, and can lead to the determination of their type and orientation, conventional dip calculation programs are unable to correlate the different sets of measurements which are used to reveal the presence of fractures.

The images provided by tools giving an overall image of the borehole (e.g. Formation Micro Imaging Tool of *Schlumberger*, or BHTV) are of considerable help in the identification and characterisation of fracturation.

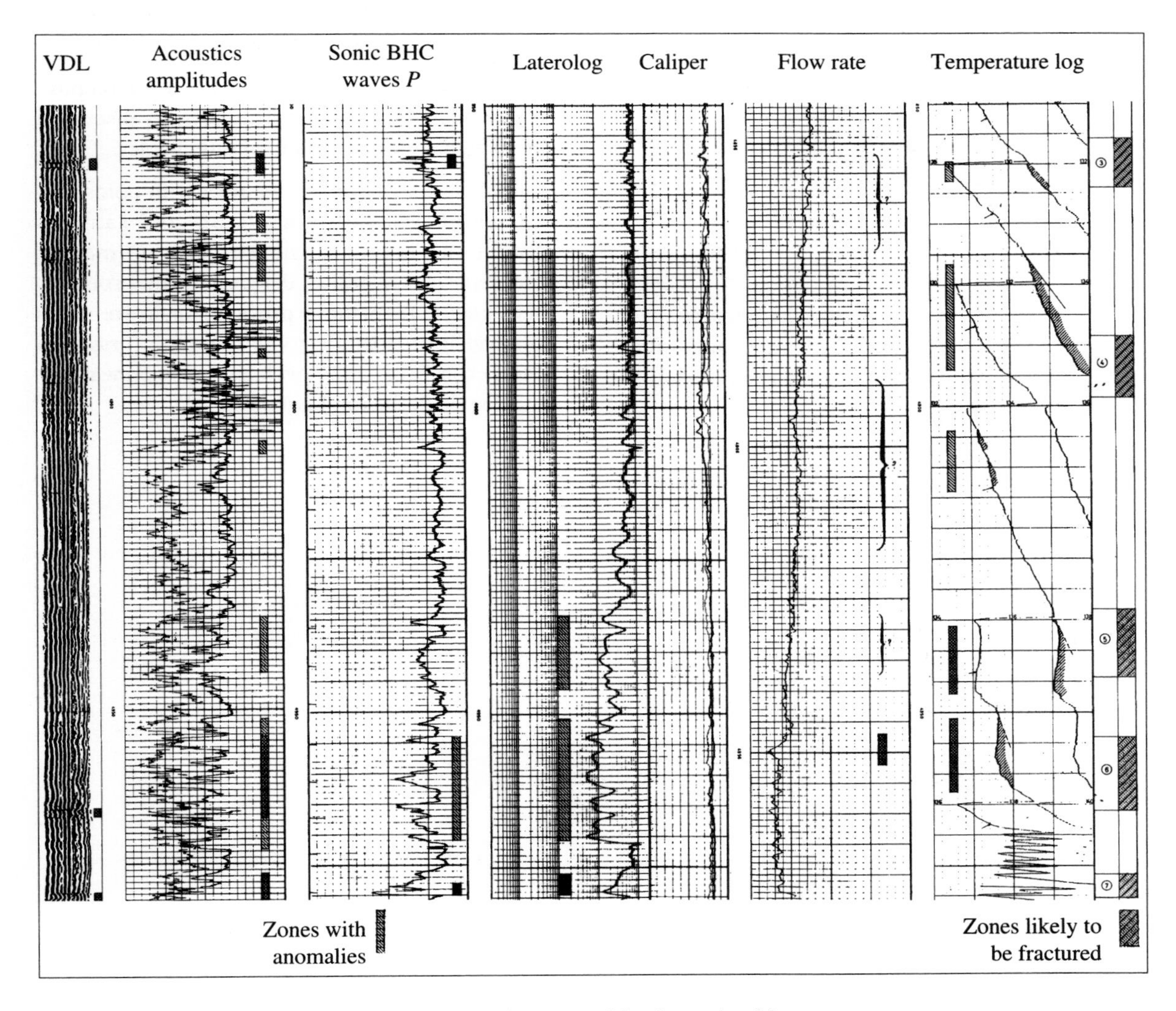

Fig. 1.69 Composite log used for the study of fractures.

The study of acoustic waves would appear to offer an approach to the estimation of permeability in fractured zones (see Section 1.2.1: Acoustic well logging).

1.3.3.4 Permeability

Wireline logging surveys yield practically no information concerning the permeability of formations.

Some empirical relations have been proposed which link the porosity with permeability (measured in millidarcies, mD). Two classic examples are given below:

(1) Wyllie et Rose (1950):

$$K^{1/2} = \frac{c\,\Phi^3}{Sw_{irr}}$$

where c is a function of the hydrocarbon density,

$c = 250$ for oil of density 0.8,

$c = 79$ for gas.

(2) Timur (1968):

$$K = \frac{0.136\,\Phi^{4,4}}{Sw_{irr}^2}$$

However, the over-general character of these relations is usually insufficient, especially since estimation of the irreducible water saturation Sw_{irr} is often problematic.

Analogies between the flow of an electric current and the flow of fluids suggest that tortuosity is indicative of the complexity of the pore-space influencing the transmission of fluids. Tortuosity is conventionally estimated with the parameter m of the formation factor deduced from resistivity logs

$$F = 1/\Phi^m$$

where F is the formation factor, Φ is porosity and m is tortuosity or cementation factor.

Other tools, such Nuclear Magnetic Resonance, which measure the amount of water not bound to the grains of the rock, can provide some indication of the quantity of water that is able to circulate freely. However, this method does not directly yield a value for permeability but may provide Sw_{irr} used in the above equations.

Only acoustic well logging shows some promise of producing results in this respect, as discussed above in the relevant section of this chapter.

At the present time, cross-plots with porosity – permeability measurements on core samples remain the standard method for estimating formation permeability, despite all the problems this entails. Downhole pressure measurements obtained from specific tools (wireline formation testers) can help in the evaluation of permeability on the scale of the formation-interval.

Chapter **2**

SEISMIC WELL SURVEYING

2.1 DATA ACQUISITION

2.1.1 The well velocity survey technique

A well velocity survey is a type of seismic well operation performed to determine the vertical propagation time of a wave emitted at the surface by a seismic source, and then recorded by a geophone. In practice, since the source and the well geophone are not generally situated on the same vertical, the distance separating the verticals which pass through the well geophone and through the source must be taken into account. (Fig. 2.1).

The surface seismic data obtained from seismic reflection survey are set with respect to a reference plane (DP datum plane) and the vertical travel time estimated by the well velocity survey set to the same reference plane. In land seismic surveys, the reference plane is generally chosen at the base of the weathered zone.

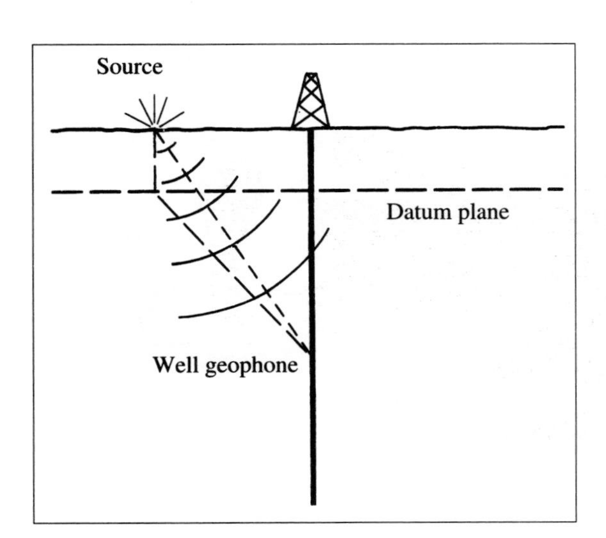

Fig. 2.1 Implementation of seismic well velocity survey.

For well velocity shooting, the assumption can be made that the raypath is vertical as far as the reference plane, but is oblique and rectilinear from the reference plane to the well geophone.

A borehole seismic survey operation is performed using a seismic source, a well geophone, a reference geophone or hydrophone placed near the source and a recording laboratory (Fig. 2.2).

The processing equipment does not require a large number of seismic channels, but must have good recording and precision dynamics as well as a short sampling interval (0.25, 0.5 or 1 ms).

The tool must be small, light and equipped with a good clamping system and a seismic cartridge including either a vertical geophone or a set of three geophones arranged in a triaxial configuration.

The source (Fig. 2.3) used must be repetitive, preferable emitting a short signal with a clearly-defined initial pulse and a wide spectral range: an airgun is generally used in offshore operations. Onshore, the following kinds of source can be used: offshore-type sources in a mud pit, low-charge explosives and dropped weight impulsive sources. The use of vibrator sources is widespread despite their being unfavourable for the picking of first arrivals.

The reference receiver is either a hydrophone for offshore operations, a geophone or mud pit hydrophone for onshore operations.

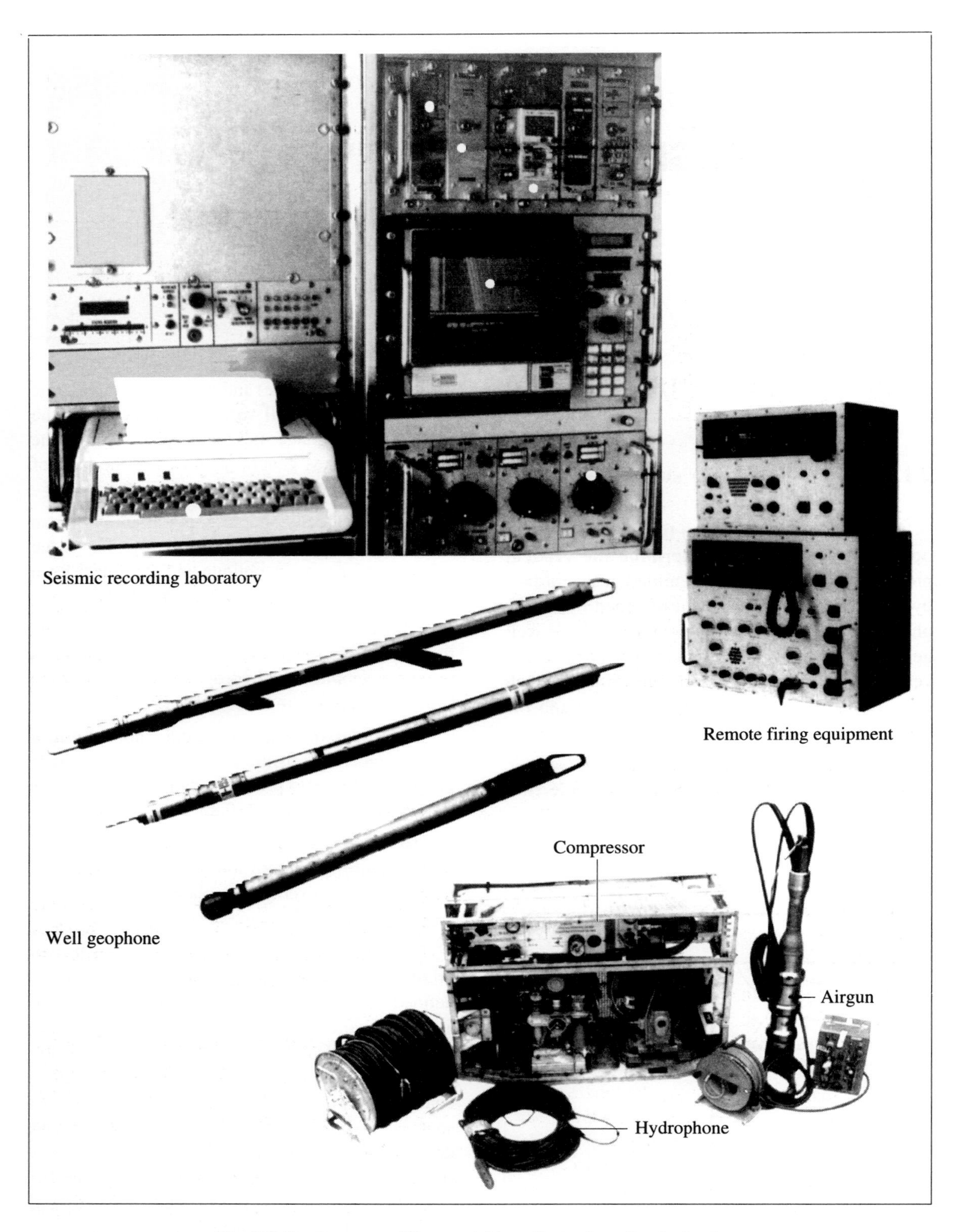

Fig. 2.2 Equipment used for acquisition of seismic well velocity data.
(Courtesy of Schlumberger)

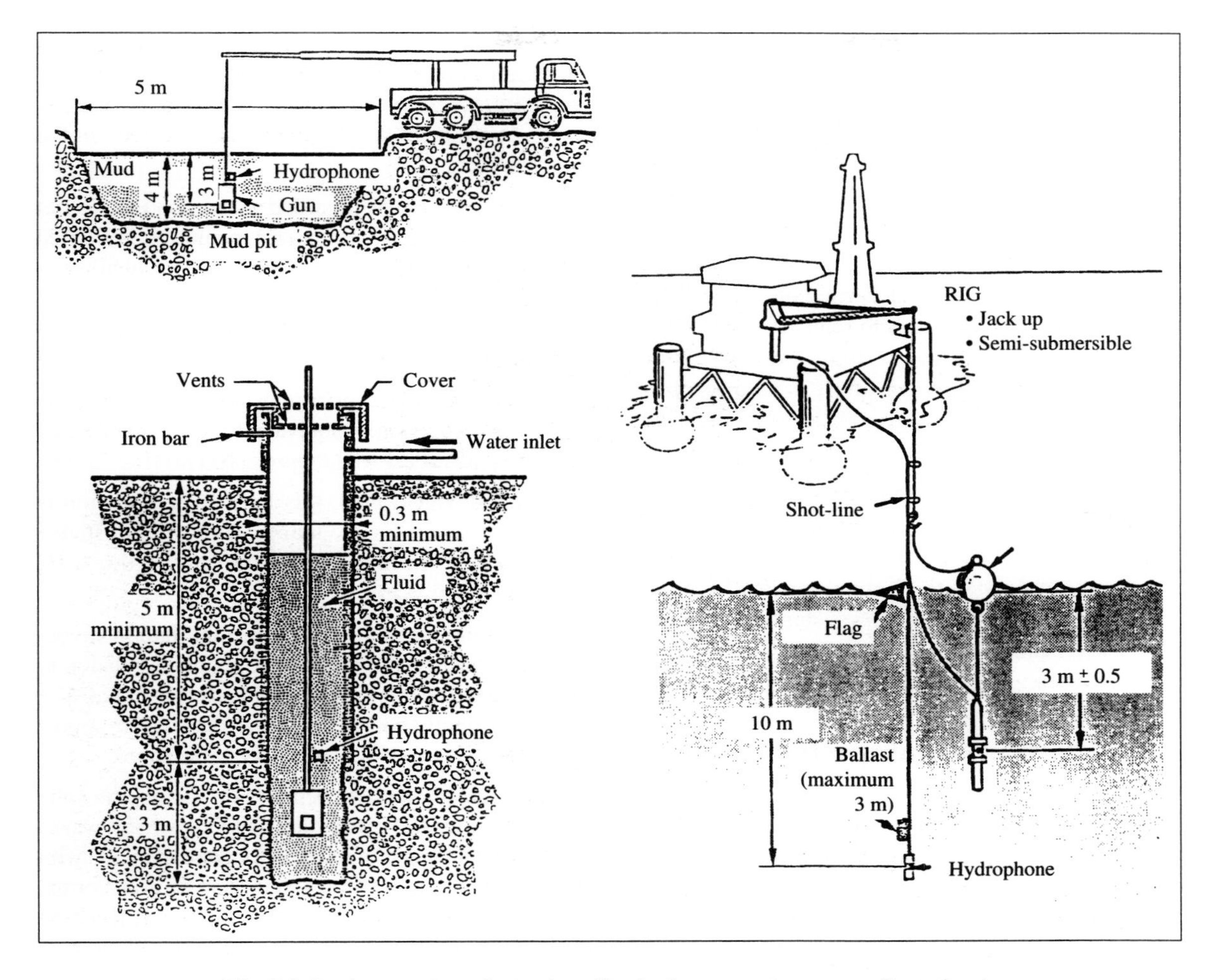

Fig. 2.3 Implementation of seismic well velocity survey (source configurations).
(Courtesy of Schlumberger)

Operation of a seismic well survey

After setting the zero datum at the rotary table or at ground level, a well velocity survey at a given depth comprises the following steps:

(1) Checking the depth of the tool in the well.
The depth is chosen in relation to the position of the velocity and density contrasts obtained on the well logs and according to the quality of the hole. It is important not to position the tool in a zone where caving has occurred. It is essential to have a measuring point at the starting and finishing depths of the sonic log. The intermediate points are chosen close to the markers, preferably situated beneath them so as to avoid interference between the downgoing wave and the upgoing wave reflected off the marker.

(2) Clamping the tool and slackening the cable.
If the tool moves, the position of the measuring point has to be modified. The cable must be slack to avoid cable waves. Clamping is critical to obtain good signal to noise ratio.

(3) Recording of seismic data.
Seismic measurements are made up from the recording of several surface shots using amplification factors and filters at different frequencies as to obtain field records with a signal to noise ratio optimized at the first arrival.

(4) Picking of first arrivals on the well geophone TG and on the reference receiver TR, and verifying the consistency of the arrival times.

(5) Tightening the cable.

(6) Unclamping the tool.

(7) Positioning the tool at the next depth.

The measurement of the first arrival times should be made at several points as the tool is lowered. These points are be used again for checking purposes (depth and repeatability) when the tool is brought back to the surface.

A well velocity survey is carried out to establish the propagation time vs. depth relation $T = F(Z)$. It is also used to perform the sonic log calibration and to obtain sonic and density logs as a function of the vertical two-way travel time through the various formations.

The use of full waveform acoustic logging tools allows the determination of the velocity of compressional waves (P) and shear waves (S), but requires a velocity survey with P and S waves in order to establish the travel time vs. depth relations: $T_P = F_P(Z)$ and $T_S = F_S(Z)$.

2.1.2 Vertical seismic profiles

The operation of vertical seismic profiles (VSP) is very similar to a check shot survey (well velocity survey) at the operational level. A VSP is a downhole seismic operation where a seismic signal emitted at surface is recorded by a geophone situated successively at different depths in the well, with the source kept always on the same vertical as the geophone whatever the depth of the geophone.

If the well is drilled vertically, the source has a fixed position close to the wellhead. If the well is deviated, the source has a variable horizontal position so as to maintain the transmitter and receiver on the same vertical.

The VSP can be regarded as an acoustic log at seismic frequencies. Its lateral resolution is limited to the diameter of Fresnel's first zone, while the lateral investigation is a function of the source offset in relation to the well-head and the structural geometry of the strata.

In the case of a horizontal tabular medium, the lateral investigation is equivalent to the lateral resolution for a vertical well (approx. 100 m) (Fig. 2.4a).

For a deviated well, this lateral investigation is equal to the well's deviation (horizontal distance separating the extreme positions of the well geophone) (Fig. 2.4b).

A possible way of increasing the lateral investigation consists of offsetting the source in relation to the wellhead and downhole geophone (Fig. 2.4c). The lateral investigation in this case is equal to approximately half the horizontal offset of the source.

The choice of offset depends on the depth of the target. It is limited by the fact that the incident waves must have an angle of incidence of less than 30° with the markers to satisfy the assumption of near-normal incidence for the calculation of the reflection coefficient and to avoid guided and refracted modes.

In practice, the offset D must be less than three quarters of the depth of the principal geological target H ($D \leq 3/4\, H$).

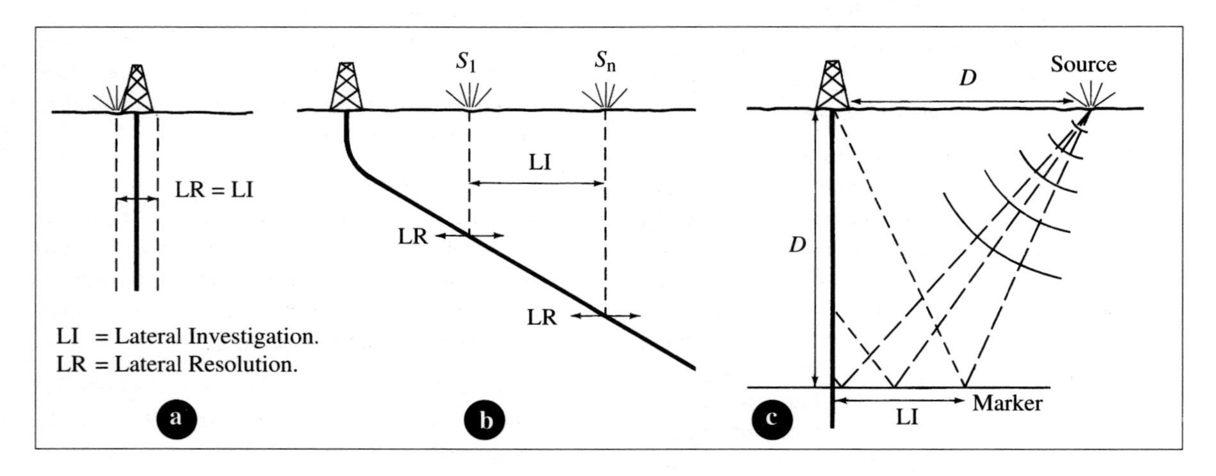

Fig. 2.4 Lateral range of investigation (LI) and lateral resolution (LR) in a vertical seismic profile.

VSP operations

The procedure for a VSP operation is identical to that carried out in well velocity surveying. However, the depth sampling interval is set at closer and more regular intervals. The maximum spacing between the successive depths depends on the minimum velocity (V_{min}) of the formation and the maximum frequency (F_{max}) that must be recorded in order to respect the Z sampling theorem (Shannon, 1949) needed to avoid aliasing and ensure high-quality data processing.

The maximum sampling ΔZ_{max} (two samples in one wavelenght) is given by the relationship:

$$\Delta Z_{max} = V_{min} / 2F_{max}$$

For example, if : $V_{min} = 1\,500$ m/s,
and : $F_{max} = 150$ Hz,
then : $Z_{max} = 5$ m.

The VSP recording is composed of upgoing and downgoing body waves of the P and/or S type, as well as guided interface modes linked to the well and the well fluid. The guided modes, usually termed tube waves, are dispersive waves of the Stoneley type.

The upgoing body waves are primary or multiple reflected waves. Only the primary reflected waves intersect the first arrivals. The downgoing body waves comprise waves emitted by the source forming the direct arrivals, and all the multiple events created by seismic markers situated above the well geophone.

Figure 2.5 shows a compressional wave VSP with a complex set of tube waves labelled TW1 to TW6.

A simple way (although not always practical) of attenuating the tube waves created by surface noise generated by the seismic source consists of lowering the column of mud in the well and/or deviating the source in relation to the wellhead.

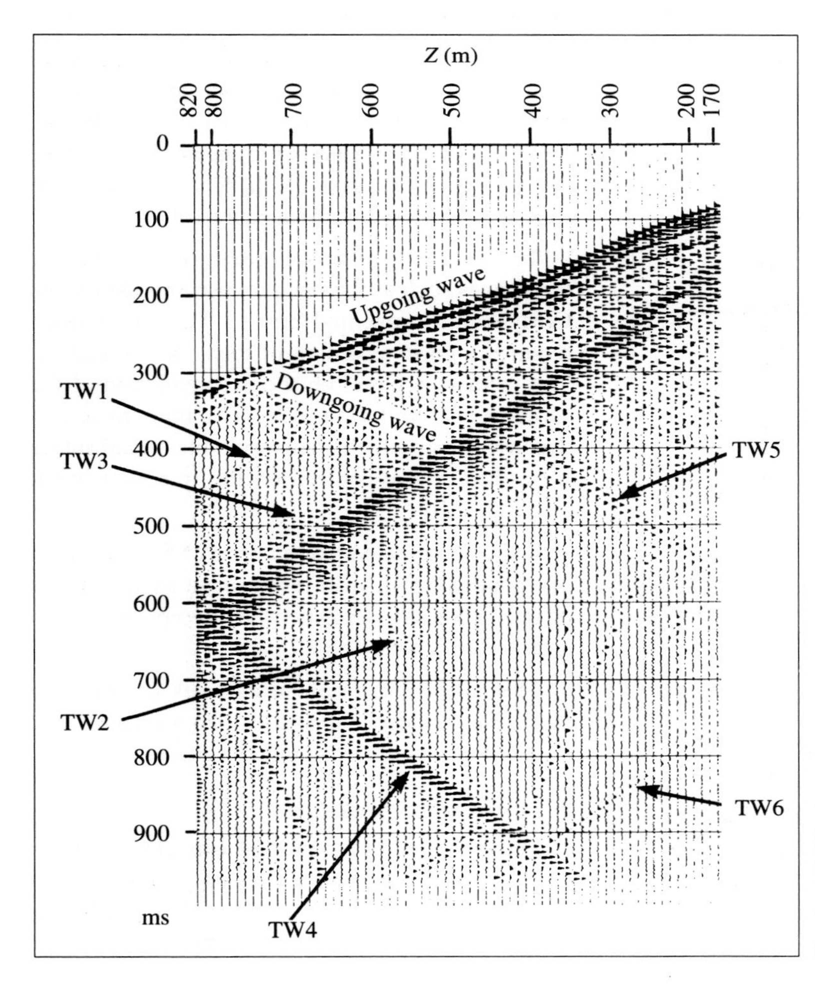

Fig. 2.5
Example of VSP recording.
(TW : tube waves).
TW1, TW3 and TW6 are downgoing;
TW2, TW4 and TW5 are upgoing.
(Courtesy of Gaz de France and IFP)

2.2 PROCESSING OF WELL VELOCITY SURVEY AND VERTICAL SEISMIC PROFILING DATA

2.2.1 Processing of well velocity survey data

a. Estimation of first arrival times

This is a delicate operation to achieve. In fact, an incorrect estimate of the position of the shot points at the surface can yield erroneous propagation velocities. Particularly in the case of an offshore VSP, the position of the source can vary in relation to the reference point. Variations may occur in both horizontal and vertical planes if the source is lowered hurriedly. This type of error can be avoided by monitoring the operation and noting every factor that might give rise to such errors, as well as by checking the arrival times at the reference receiver (usually situated close to the wellhead) (Fig. 2.6).

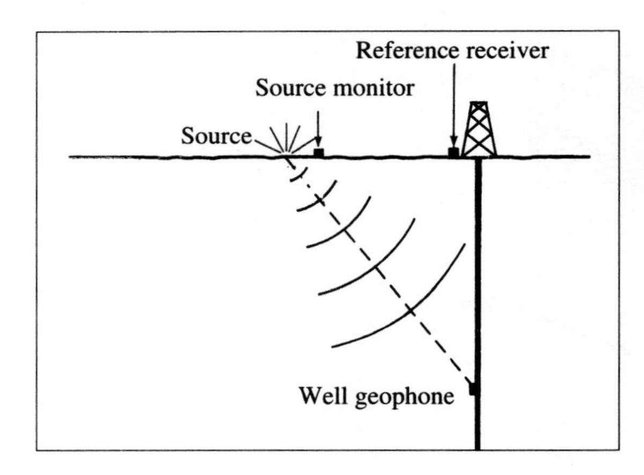

Fig. 2.6 Implementation of well velocity survey data using reference receiver.

Likewise, the discrepancies introduced by electronic systems must be taken into account.

During wave propagation between the surface and the receiving geophone, the signal is perturbed by certain phenomena such as attenuation, dispersion, diffractions and interference by thin layers and multiples.

A good recovery of the signal therefore requires the use of amplification factors and filters which may introduce a slight delay into the recording of arrival times.

The records are composed of several traces, one of them is the time break trace, the others are the received signal recorded by the borehole geophone with different filters and amplification listed on an operator's report.

Finally, the different picking techniques employed can be a source of discrepancy between the time measurements.

In a well velocity survey operation, the picking is done by hand and either the first peak or the first break is sought. Looking for an extremum is sensitive to the distortion of the signal and determination of the first break of the signal is sensitive to the signal-to-noise ratio. In a VSP, the data are recorded on magnetic tape so that picking can be refined by means of numerical techniques.

Although a careful and consistent utilization of the recordings will avoid the major errors, it cannot prevent all errors (sometimes reaching 5 ms). In fact, Dillon and Collyer (1985) have shown that, even the signal-to-noise ratio is good, the identification of the first arrival is by no means unequivocal.

This has led to the development of specific techniques such as the one proposed by Dillon and Collyer and presented below. Their technique requires a good recording of the source signature. The method is based on the deconvolution of the source signature followed by a "minimum phase" deconvolution which transforms the signal into a zero-phase pulse centred on the true first break (see Fig. 2.7).

It should be noted that an error of 5 ms between two points 1000 m apart results in an approximation of only 1.5 μs/ft, while between two points 200 m apart the error would be 7.7 μs/ft.

This simple example illustrates the need to make use of closely spaced calibration points and to interpolate between significant extrema.

b. Correction procedures

The measured times must be corrected to obtain vertical times between the datum plane and the position of well geophone, as shown in Fig. 2.8. The corrections are performed in two steps.

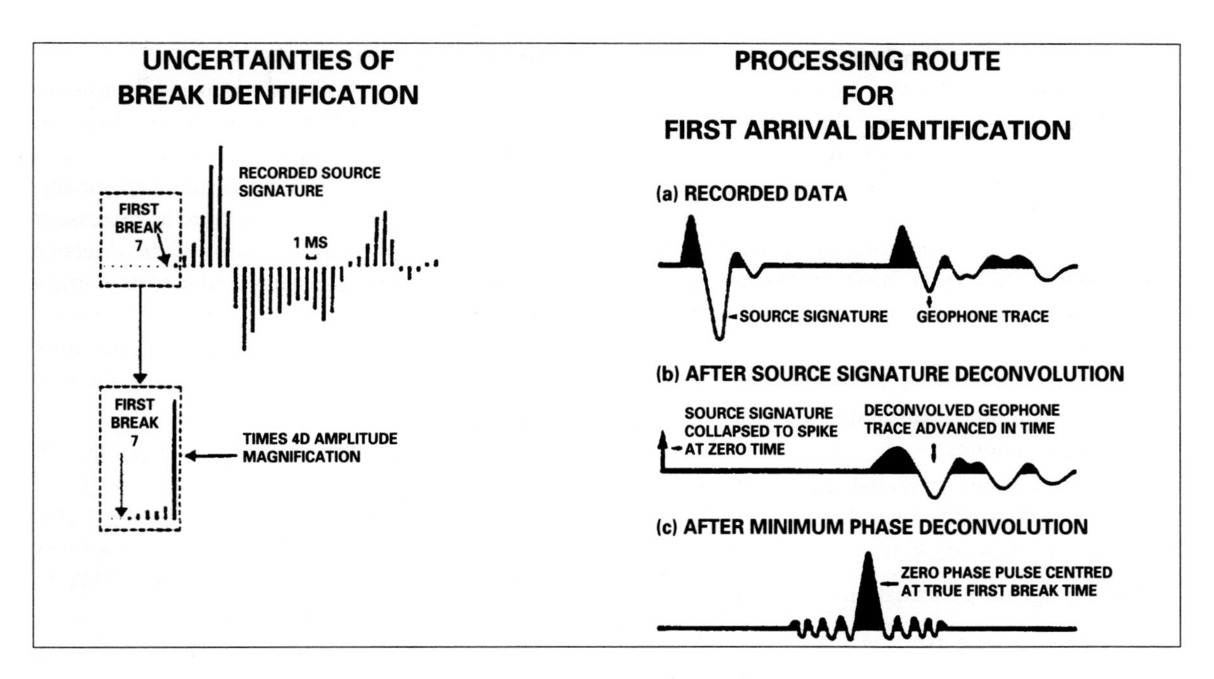

Fig. 2.7 Uncertainty in identifying true first arrivals. First arrival picking with source signature recording.
(Dillon and Collyer, 1985)

The first step is carried out to obtain the rectilinear slant times between the datum plane (DP) (vertically beneath the source) and the position of the well geophone.

The second step is carried out to obtain the vertical travel times between the DP and the well geophone, taking into account the horizontal distance D between the vertical passing through the source position and the vertical passing through the well geophone.

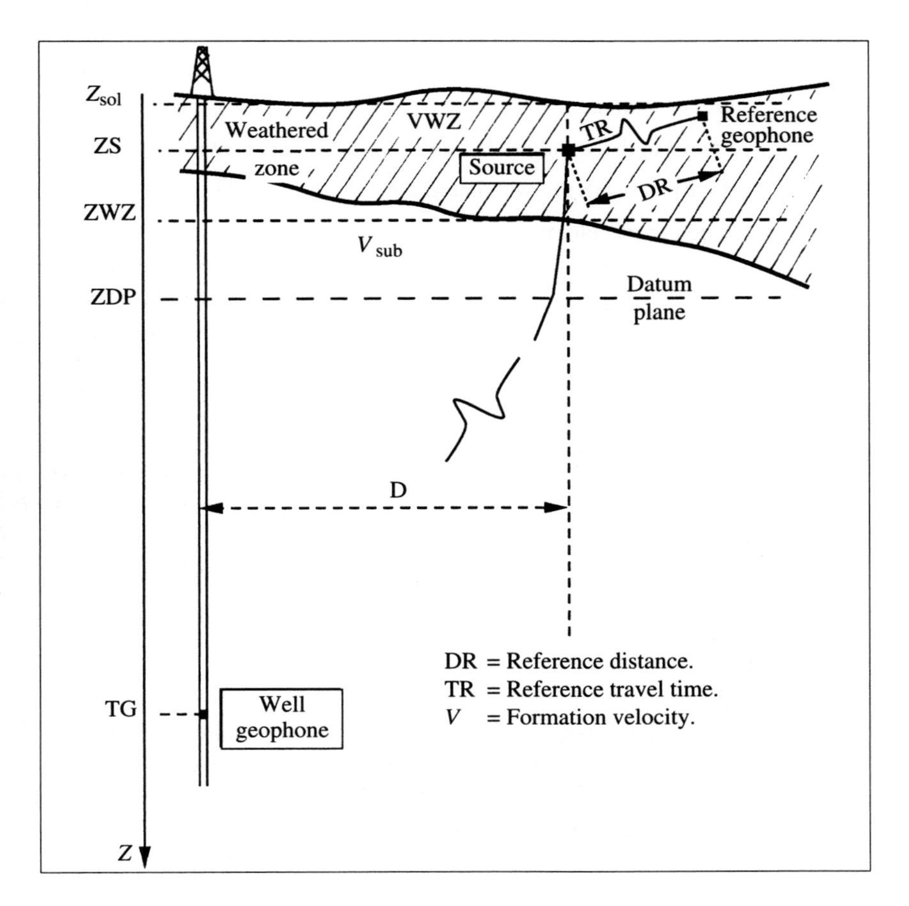

Fig. 2.8 ▶
Well velocity surveying.

- **First step: correction of slant time**

The correction of slant time includes correction of the weathered zone and source signature fluctuation. The weathered zone correction is similar to that practised in surface seismic surveying, i.e. referring the measured time to the datum plane and taking account of weathered zone parameters. The signature fluctuation correction is made to take account of time break (TB) fluctuations and variations in the source positions, i.e:

$$TGC = TG + CWZ + CSIG$$

TG slant time measured at the well geophone;

TGC slant time corrected at the well geophone;

CWZ weathered zone correction;

CSIG signature fluctuation correction.

- Weathered zone correction:

$$CWZ = -\frac{ZWZ - ZS}{VC} + \frac{ZWZ - ZDP}{V_{sub}}$$

ZDP depth of the Datum Plane DP;

ZWZ depth of weathered zone;

ZS depth of source;

VWZ velocity in the weathered zone;

V_{sub} velocity beneath the weathered zone;

VC correction of velocity;

$ZS \geqslant ZWZ \quad VC = V_{sub}$

$ZS < ZWZ \quad VC = VWZ$

- Signature fluctuation correction:

(1) Offshore and onshore using offshore sources in mud pit:

$$CSIG = -TR + DR/VF$$

DR distance from source to reference hydrophone;

TR time measured at the reference hydrophone;

VF velocity of propagation in the fluid (water or mud).

(2) Onshore:

$$CSIG = -TR + TRM$$

TR time measured at the reference geophone;

TRM mean time measured at the reference geophone.

- **Second step: slant correction**

This correction is made by estimating the average velocity from the DP to the well geophone. It assumes a rectilinear raypath between the DP and the well geophone. A layered structure with variable indices of refraction generally leads to an increase in the distance travelled by the wave, if the direction of propagation is not perpendicular to the stratification. The seismic travel time is therefore different from that measured by a logging tool, where the raypath is a straight line from the shot point to the receiver.

A rigourously corrected time can only be calculated using a model which allows the angle of incidence of the wave to make contact with the markers. Such an approach is difficult to achieve and is insufficient to obtain the total traveltime of the wave in the vicinity of the well.

The vertical time is calculated using the following relations:

$$VW(ZG) = \frac{\sqrt{D^2 + (ZG - ZDP)^2}}{TGC(ZG)}$$

$$TVC(ZG) = \frac{ZG - ZDP}{VM(ZG)}$$

ZG depth of well geophone;

D horizontal distance between source and well geophone;

TGC(ZG) corrected slant time at well geophone;

ZDP depth of Datum Plane (DP);

TVC(ZG) vertical time between well geophone and DP.

If the number of measurement points in the well is large enough, the corrected vertical times can be used to estimate the interval velocities and the root-mean-square (RMS) velocities, which is close to those obtained by velocity analysis in surface seismic reflection.

These velocities are calculated using the following relations:

Average velocity VM(ZG)

Interval velocity $VTRAN_{(i,\,i+1)}$

$$\Delta TVC_{(i,\,i+1)} = TVC(ZG_{i+1}) - TVC(ZG_i)$$

$$\Delta ZG_{(i,\,i+1)} = ZG_{i+1} - ZG$$

$$VTRAN_{(i,\,i+1)} = \Delta ZG_{(i,\,i+1)} / \Delta TVC_{(i,\,i+1)}$$

Average root-mean-square velocity $VRMS_{(i)}$:

$$VRMS_{(1)} = VM(ZG_i) = VTAN_{(0,\,1)}$$

$$VRMS^2_{(i)} = \frac{\left[\sum_{j=1}^{i} VTRAN^2_{(j-1,\,j)}\ \Delta TVC_{(j-1,\,j)}\right]}{TVC\,(ZG_i)}$$

Figure 2.9 summarizes the formulae used for corrections and conversion to vertical times in different configurations (offshore and onshore), assuming different hypotheses for the weathered zone and the DP position. An example of well velocity surveying (operator listings, recordings, presentation of results) is given in Figs. 2.10-2.12.

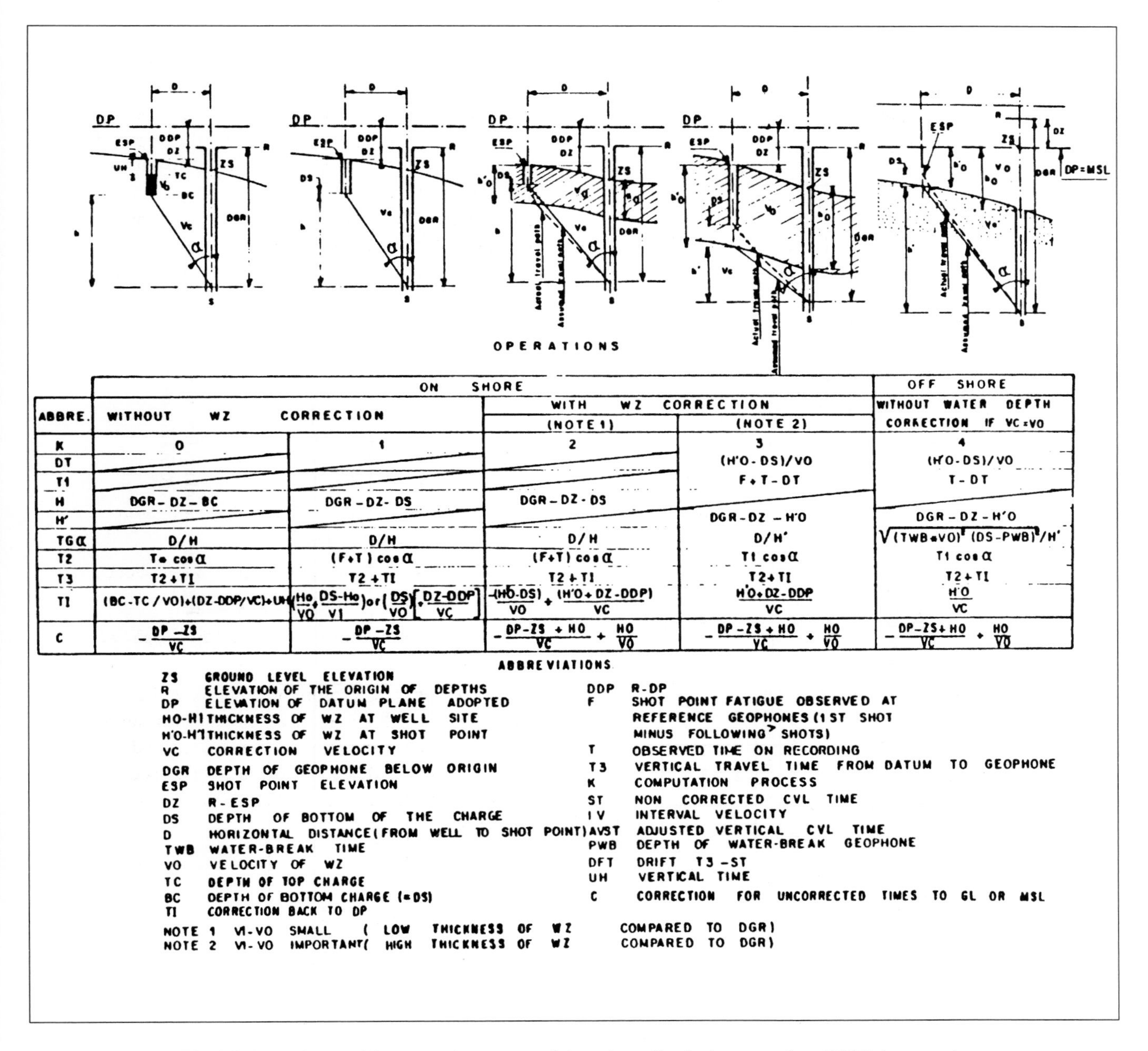

ABBRE.	ON SHORE: WITHOUT WZ CORRECTION		ON SHORE: WITH WZ CORRECTION (NOTE 1)	ON SHORE: WITH WZ CORRECTION (NOTE 2)	OFF SHORE: WITHOUT WATER DEPTH CORRECTION IF VC=VO
K	0	1	2	3	4
DT				(H'O-DS)/VO	(H'O-DS)/VO
T1				F+T-DT	T-DT
H	DGR-DZ-BC	DGR-DZ-DS	DGR-DZ-DS		
H'				DGR-DZ-H'O	DGR-DZ-H'O
TGα	D/H	D/H	D/H	D/H'	$\sqrt{(TWB \cdot VO)^2 (DS-PWB)^2}/H'$
T2	T cos α	(F+T) cos α	(F+T) cos α	T1 cos α	T1 cos α
T3	T2+TI	T2+TI	T2+TI	T2+TI	T2+TI
TI	(BC-TC/VO)+(DZ-DDP/VC)+UH	$\left(\frac{Ho}{VO}+\frac{DS-Ho}{V1}\right)$ or $\left(\frac{DS}{VO}\right)\left[+\frac{DZ-DDP}{VC}\right]$	$-\frac{(H'O-DS)}{VO}+\frac{(H'O+DZ-DDP)}{VC}$	$\frac{H'O+DZ-DDP}{VC}$	$\frac{H'O}{VC}$
C	$-\frac{DP-ZS}{VC}$	$-\frac{DP-ZS}{VC}$	$-\frac{DP-ZS+HO}{VC}+\frac{HO}{VO}$	$-\frac{DP-ZS+HO}{VC}+\frac{HO}{VO}$	$-\frac{DP-ZS+HO}{VC}+\frac{HO}{VO}$

Fig. 2.9 Formulae used for correcting measured times in well velocity surveying. (*CGG document*)

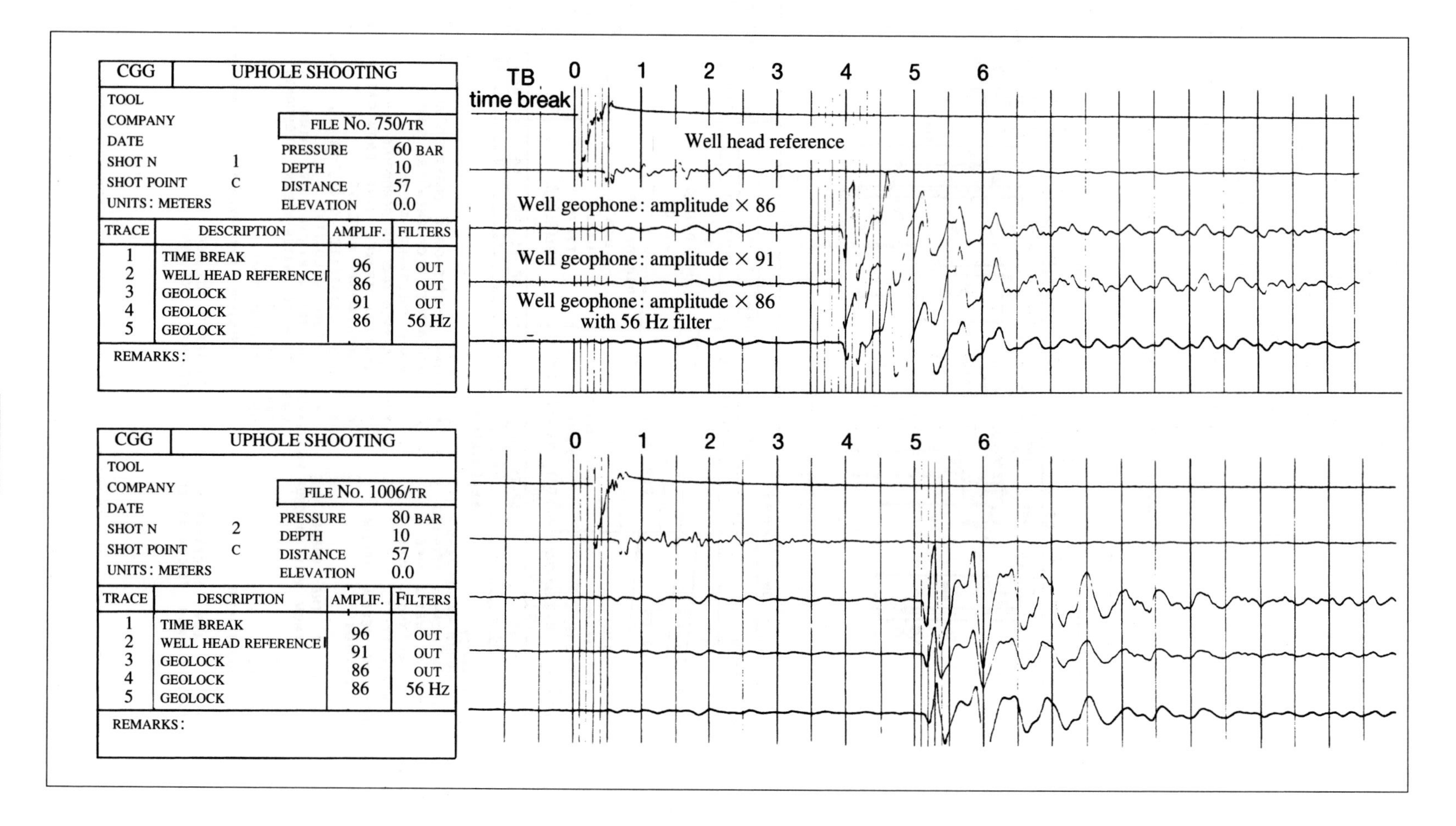

Fig. 2.10 Recordings from well velocity shooting.
(CGG document)

OPERATOR'S LISTING

JOB CGG			
COMPANY			
OPERATOR		SEISMO-TOOL:	GEOLOCK
ORIGIN OF MEASUREMENTS:	TR	LABO:	A5 676 X
DEPTH UNITS		SOURCE:	AIR GUN
PRESSURE UNIT:	BAR		
DATE			

TIR	HEURE	P/R	N PT	ZPT	D	PC	CHARGE	DEFT	GAINS					OBSERVATIONS
1	2.00	750	C	0.0	57	10	60		96	86	91	86		
2	2.20	1006	C	0.0	57	10	80		96	91	86	86		
3	2.39	1204	C	0.0	57	10	100		96	91	86	86		
4	3.15	1390	C	0.0	57	10	120		96	91	86	86		
5	3.29	1447	C	0.0	57	10	140		96	91	86	86		
6	3.90	1610	C	0.0	57	10	140		96	96	91	91		FILM BRUTTE
7	4.05	1728	C	0.0	57	10	140		96	96	91	91		
8	4.29	1900	C	0.0	57	10	140		96	101	96	96		
9	4.45	2100	C	0.0	57	10	140		96	106	96	96		
10	4.55	2225	C	0.0	57	10	140		96	106	101	101		
11	4.96	2229	C	0.0	57	10	140		96	106	101	101		
12	5.05	2175	C	0.0	57	10	140		96	106	101	101		
13	5.15	2000	C	0.0	57	10	140		96	106	101	101		
14	5.25	1800	C	0.0	57	10	140		96	106	96	96		
15	5.35	1600	C	0.0	57	10	140		96	101	96	96		
16	5.45	1500	C	0.0	57	10	140		96	101	96	96		
17	6.05	308	C	0.0	57	10	60		96	91	86	86		

P/R *Rotary table elevation*
NPT *Source type*
ZPT *Shot point depth*
D *Offset of shot point (from well)*
PC *Shot depth*
Charge *Impulse pressure (bars)*

Fig. 2.11 Well velocity survey listing.
(CGG document)

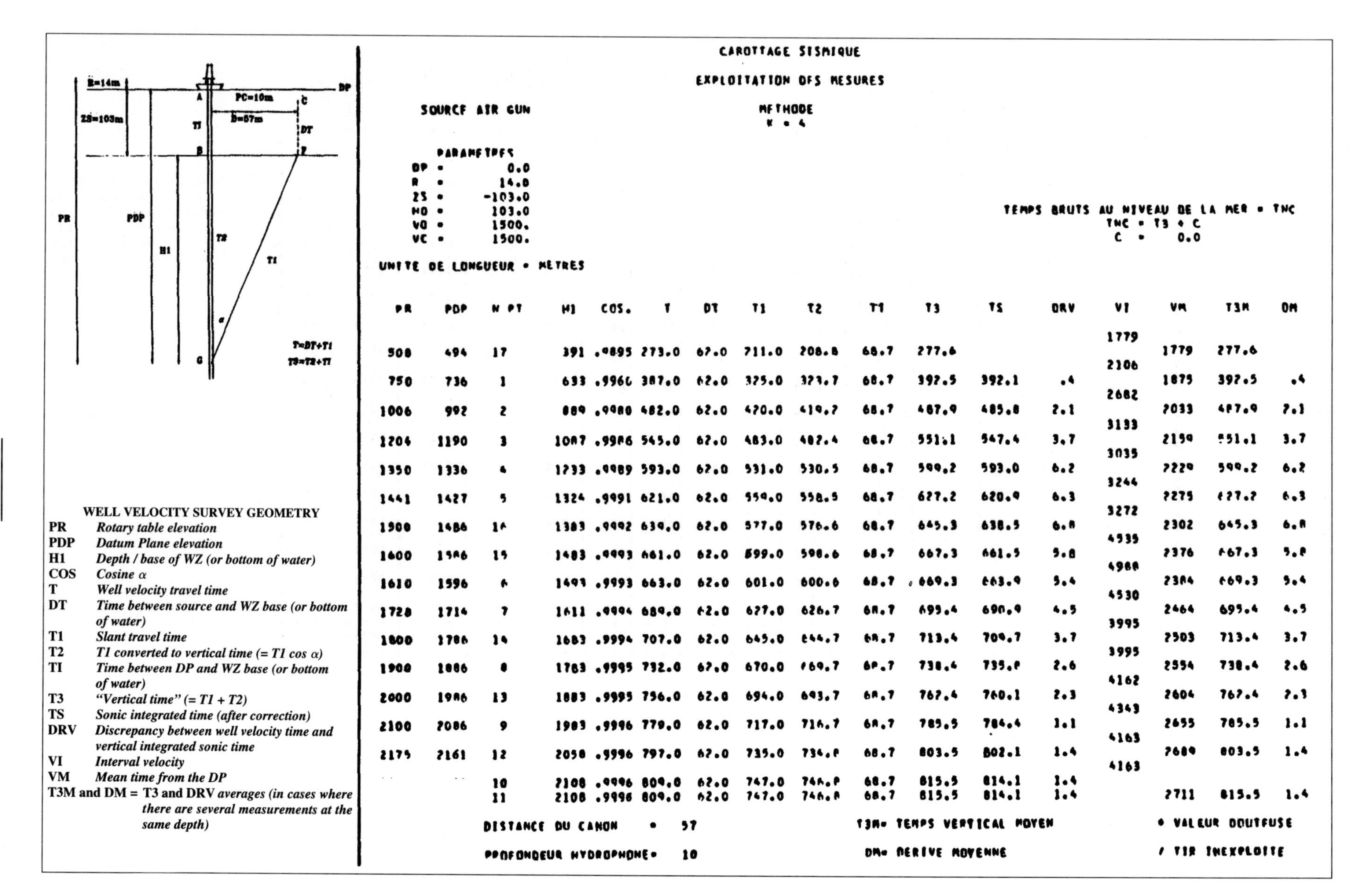

CAROTTAGE SISMIQUE

EXPLOITATION DES MESURES

SOURCE AIR GUN

METHODE
K = 4

PARAMETRES
DP = 0.0
R = 14.0
ZS = -103.0
HO = 103.0
VO = 1500.
VC = 1500.

TEMPS BRUTS AU NIVEAU DE LA MER = TNC
TNC = T3 + C
C = 0.0

UNITE DE LONGUEUR = METRES

PR	PDP	N PT	H1	COS.	T	DT	T1	T2	TI	T3	TS	DRV	VI	VM	T3M	DM
													1779			
500	494	17	391	.9895	273.0	62.0	211.0	208.8	68.7	277.6				1779	277.6	
													2106			
750	736	1	633	.9960	387.0	62.0	325.0	323.7	68.7	392.5	392.1	.4		1875	392.5	.4
													2682			
1006	992	2	889	.9980	482.0	62.0	420.0	419.2	68.7	487.9	485.8	2.1		2033	487.9	2.1
													3133			
1204	1190	3	1087	.9986	545.0	62.0	483.0	482.4	68.7	551.1	547.4	3.7		2159	551.1	3.7
													3035			
1350	1336	4	1233	.9989	593.0	62.0	531.0	530.5	68.7	599.2	593.0	6.2		2229	599.2	6.2
													3244			
1441	1427	5	1324	.9991	621.0	62.0	559.0	558.5	68.7	627.2	620.9	6.3		2275	627.2	6.3
													3272			
1500	1486	16	1383	.9992	639.0	62.0	577.0	576.6	68.7	645.3	638.5	6.8		2302	645.3	6.8
													4535			
1600	1586	15	1483	.9993	661.0	62.0	599.0	598.6	68.7	667.3	661.5	5.8		2376	667.3	5.8
													4988			
1610	1596	6	1493	.9993	663.0	62.0	601.0	600.6	68.7	669.3	663.9	5.4		2384	669.3	5.4
													4530			
1728	1714	7	1611	.9994	689.0	62.0	627.0	626.7	68.7	695.4	690.9	4.5		2464	695.4	4.5
													3995			
1800	1786	14	1683	.9994	707.0	62.0	645.0	644.7	68.7	713.4	709.7	3.7		2503	713.4	3.7
													3995			
1900	1886	8	1783	.9995	732.0	62.0	670.0	669.7	68.7	738.4	735.8	2.6		2554	738.4	2.6
													4162			
2000	1986	13	1883	.9995	756.0	62.0	694.0	693.7	68.7	762.4	760.1	2.3		2604	762.4	2.3
													4343			
2100	2086	9	1983	.9996	779.0	62.0	717.0	716.7	68.7	785.5	784.4	1.1		2655	785.5	1.1
													4163			
2175	2161	12	2058	.9996	797.0	62.0	735.0	734.8	68.7	803.5	802.1	1.4		2689	803.5	1.4
													4163			
		10	2108	.9996	809.0	62.0	747.0	746.8	68.7	815.5	814.1	1.4				
		11	2108	.9996	809.0	62.0	747.0	746.8	68.7	815.5	814.1	1.4		2711	815.5	1.4

DISTANCE DU CANON = 57

PROFONDEUR HYDROPHONE = 10

T3M = TEMPS VERTICAL MOYEN

DM = DERIVE MOYENNE

* VALEUR DOUTEUSE

/ TIR INEXPLOITE

Fig. 2.12 Presentation of well velocity survey results.
(CGG document)

2.2.2 Data processing of vertical seismic profiling

The conventional processing of VSP data gives a stacked trace comparable to a synthetic seismic record without multiples, or a set of seismic traces comprising a high-resolution seismic section in the immediate vicinity of the well.

Whatever the acquisition geometry, the processing of a VSP survey can be subdivided into different stages.

The first step of processing consists of:

(1) demultiplexing the data,

(2) correlation, if the seismic source is a vibrator,

(3) correction for the signature fluctuation effect,

(4) correction for tool rotation and well deviation (3 axes borehole geophone are required),

(5) elimination of poor quality recordings,

(6) stacking of recordings made at the same geophone position,

(7) corrections for spherical divergence and absorption,

(8) separation into components in the case of triaxial well tool.

The second processing step includes the picking of first arrival times, using the same techniques as those used in well velocity surveying in order to establish a time vs. depth relation $T = F(Z)$ and a velocity model at the well.

The third processing step consists of separating compressional from shear waves, and upgoing from downgoing wavefields, using velocity filters and polarization filters.

Shear waves have lower velocities than compressional waves, exhibiting a particle-movement direction (vibration) at right angles to the propagation direction. Compressional waves have a vibration direction parallel to the propagation direction. The shear waves are of two kinds: *SH* (vibration perpendicular to the propagation plane) and *SV* (vertical vibration in the plane of propagation). The downgoing waves are characterised by positive apparent velocities $(\Delta z/\Delta t)$ and the upgoing waves by negative apparent velocities.

The separation of the two wavefields can be performed by the application of apparent velocity filters in either the space-time domain or the frequency domain. In the frequency domain, upgoing and downgoing waves are divided into sets with negative and positive wavenumbers (k'). A simple way of extracting them is to retain in the (f, k) plane only those energies that are found in f the positive or negative halves of the wavenumber field. In the space-time domain, separation of the two wavefields can be achieved by the application of filters based on the average or anti-average principle (Lamer 1982, Coppens 1982 and Hardage 1985). This type of algorithm extracts the desired signal by subtracting a noise model that has been estimated as accurately as possible. In the case of VSP, the noise model actually corresponds to the downgoing wavefield and the desired signal is the upgoing wavefield.

One way of estimating the downgoing wavefield is to apply a static time shift to all the VSP traces, using a value which is equal to the first arrival time but with a change in sign. Then, a filter can be used to recover only the infinite apparent velocities (e.g. by means of compositing or median filtering). Numerous specific algorithms have been developed to separate the upgoing and downgoing waves, including trace pair filtering (point-to-point predictive filter) (Mari et al., 1986 and 1989) and various multi-channel algorithms, of which one of the best known is from to Seeman and Horowicz (1983).

After wave separation, the choice of processing differs according to the acquisition geometry, well profile and geological structure.

- If the source and receiver can be considered as being on the same perpendicular as the reflectors (the simplest case is that of a vertical well in horizontal strata with the source situated close to the wellhead) the processing steps are as follows:

(1) Deconvolution of the upgoing by the downgoing waves. Application of a deconvolution operator at each geophone position allows the removal of both source signal and downgoing multiples.

(2) Flattening of the deconvolved upgoing waves is carried out at each depth point by the application of a static correction equal to the first arrival time measured at the geophone position under

consideration. This operation renders the VSP recording comparable in time (two-way travel time) to a recording obtained by surface seismic reflection.

(3) Obtaining a VSP stacked trace. The deconvolved and flattened upgoing waves are stacked in a corridor which is placed immediately after the first arrival (Fig. 2.13). This restricted vertical summation known as a corridor stack gives a trace in the seismic frequency bandwidth without any assumption about the source signature. After deconvolution, the seismic signal is a zero-phase signal.

The VSP stacked trace is comparable to a synthetic seismic record obtained from sonic and density log data. A stacked trace obtained in this way may contain upgoing multiples. To remove the effects of multiples, a narrow stacking corridor is chosen in order to accept only the reflected signal received just after the first arrival.

Thus, the corridor stack is analogous to a synthetic seismic record — without multiples — in the frequency band of the received signal. In this way, it is comparable to a surface seismic CDP stacked trace.

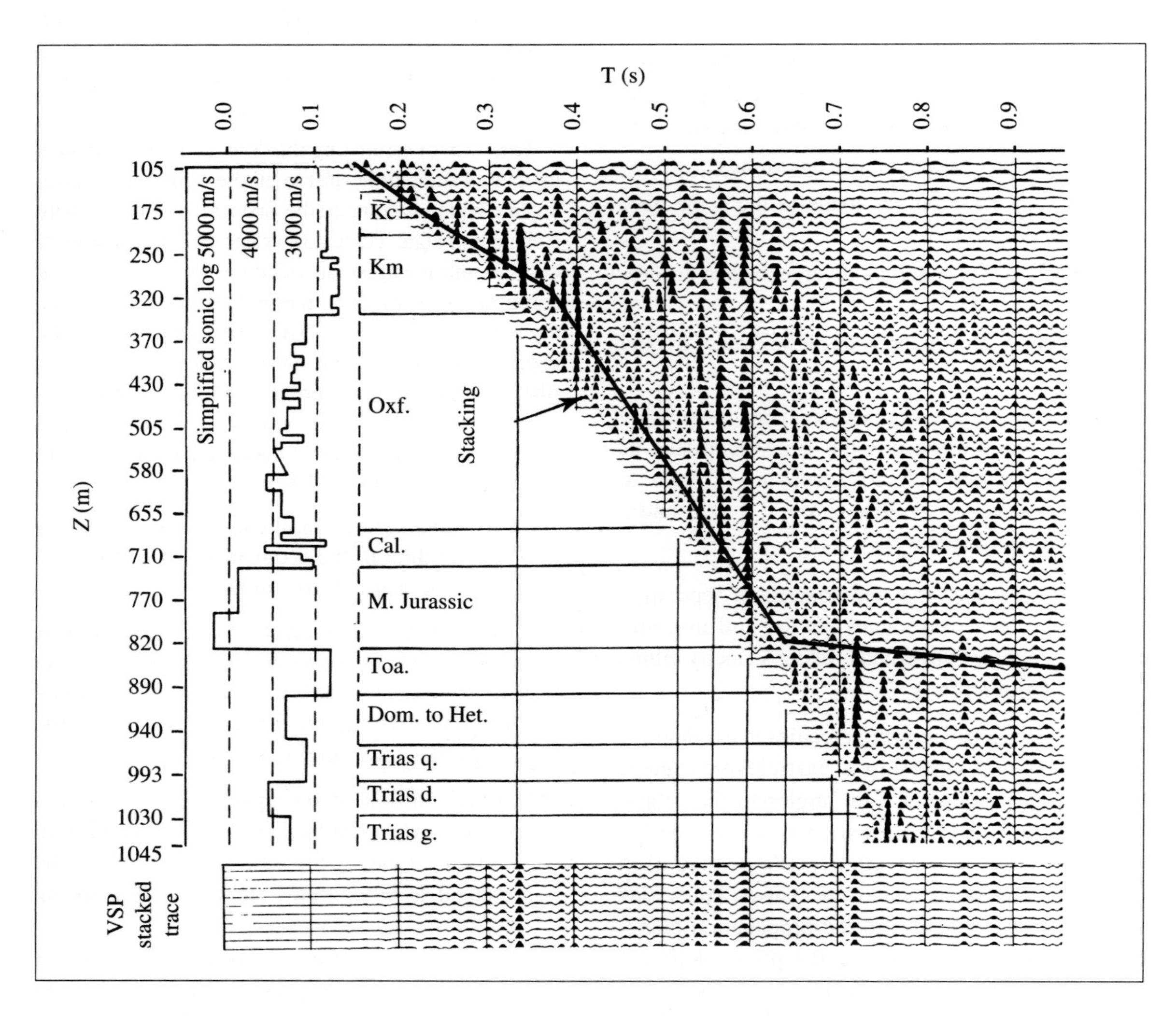

Fig. 2.13 Stacking corridor applied to processed VSP data, with resulting stacked trace.
(Mari and Coppens, 1987)

• If the source and receiver cannot be considered as being on the same perpendicular as the reflectors (the simplest case would be a vertical well in a horizontally layered medium where the source is offset from the wellhead), the data processing is as follows:

(1) Deconvolution of the upgoing waves. The deconvolution operator is unique. Since it is extracted from traces recorded at the bottom of the well, a source signature is not required.

(2) Moveout correction of the deconvolved upgoing waves. These corrections are carried out by introducing a velocity model derived from the first arrival times, or a model based on ray tracing techniques designed to take account of the acquisition geometry.

(3) Flattening of deconvolved upgoing waves after move-out correction. This is performed by the application of static corrections at each geophone position. The static correction corresponds to the first arrival time reduced to the vertical.

(4) Migration. The method most commonly used in VSP is the one proposed by Wyatt and Wyatt (1982).

The VSP seismic section obtained after migration is directly comparable with a surface seismic reflection section. The migrated VSP section has a lateral investigation ranging from tens to hundreds of meters. However this type of section may contain upgoing and downgoing multiples. Fig. 2.14 shows the variation of lateral investigation as a function of the position of the well geophone, the source and the seismic marker.

Figure 2.15 shows an example of processing of VSP data recorded between 1045 and 105 m, the source being slightly offsetted (30 m) from the wellhead. The spacing between successive geophone positions varied from 3 to 23 m. The figure presents a VSP after editing, showing both downgoing and upgoing waves. Upgoing waves were deconvolved

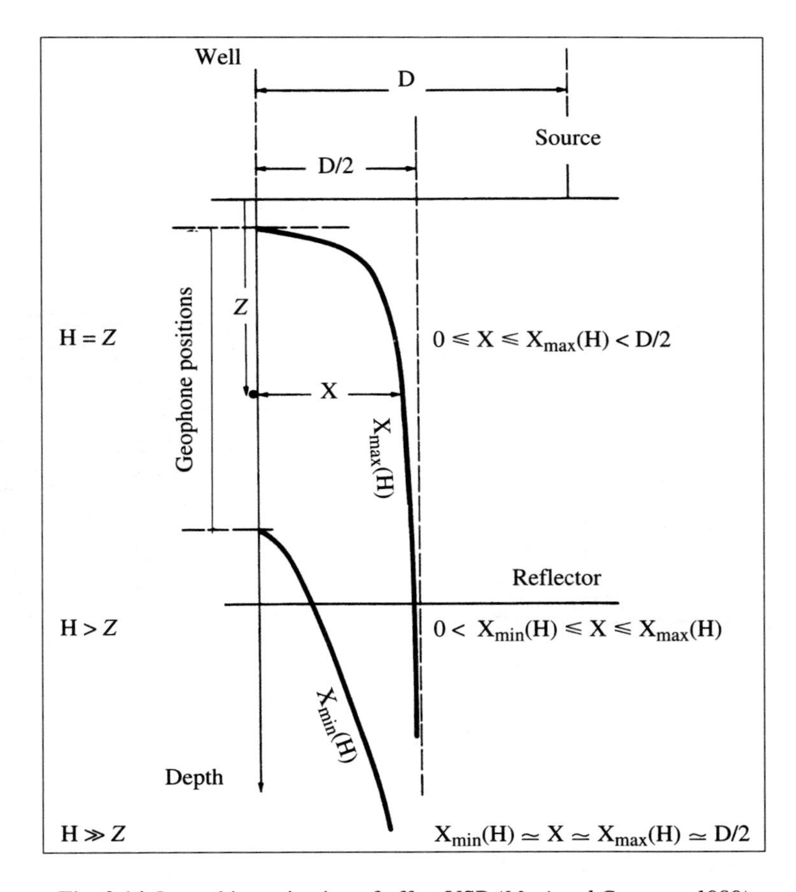

Fig. 2.14 Lateral investigation of offset VSP.*(Mari and Coppens, 1989)*

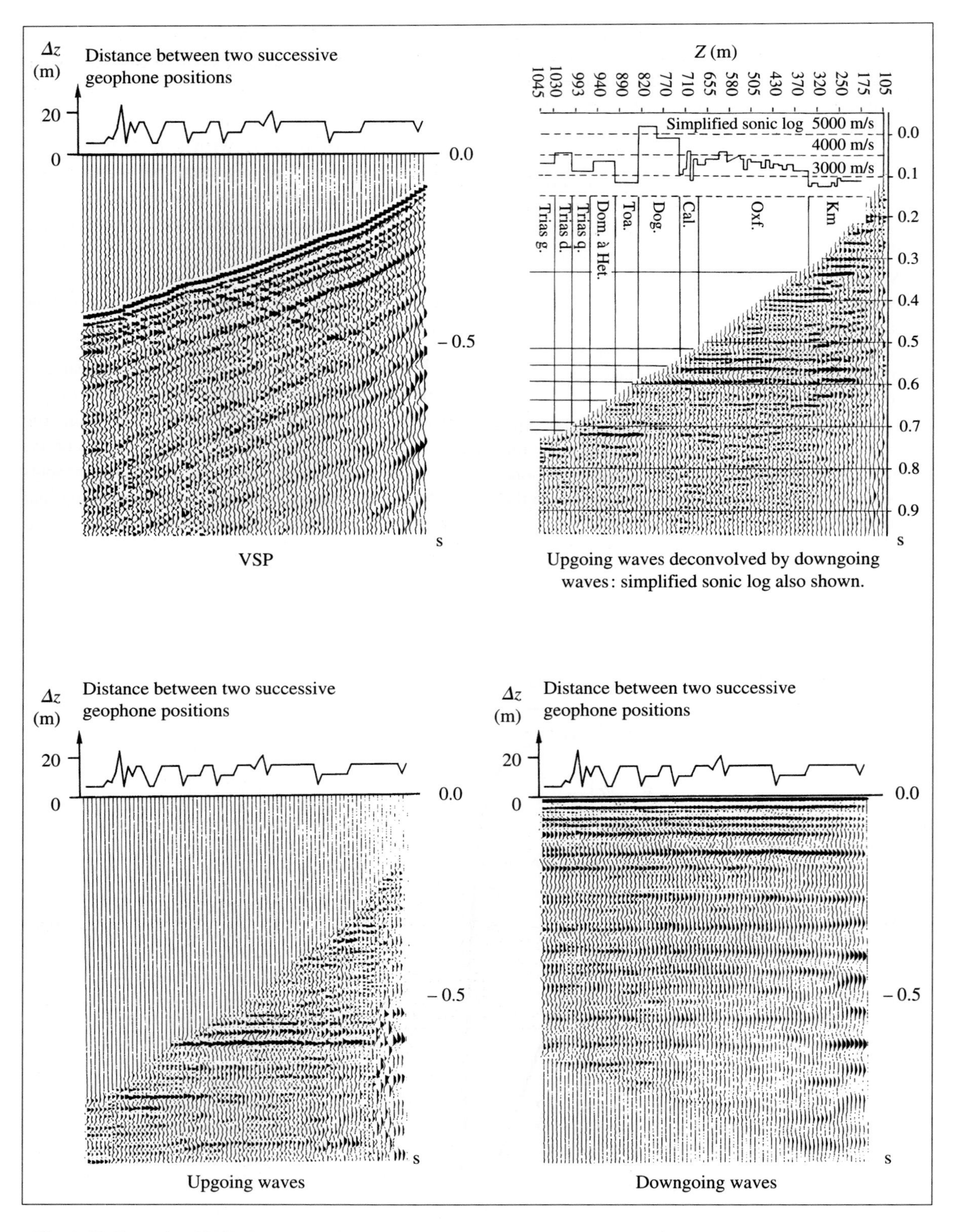

Fig. 2.15 Example of VSP data processing. Time (s) vs. depth (m) sections, with Δz indicating distance between two successive geophone positions. *(Gaz de France-IFP document)*

using the downgoing waves and correlated with a simplified sonic log. The geological depth calibration (*Z*) given above the VSP was established from downhole logging data.

After deconvolution of the upgoing waves by the downgoing waves, the VSP trace obtained in the stacking corridor is utilized to match the seismic surface survey with the downhole data as illustrated in Fig. 2.16. The fit is obtained by using a deconvolution technique applied to surface seismics (see Section 4.3.1: Stratigraphic deconvolution in Chapter Four: Synthetic seismic records) which takes account of the source signature while attenuating the effect of multiples.

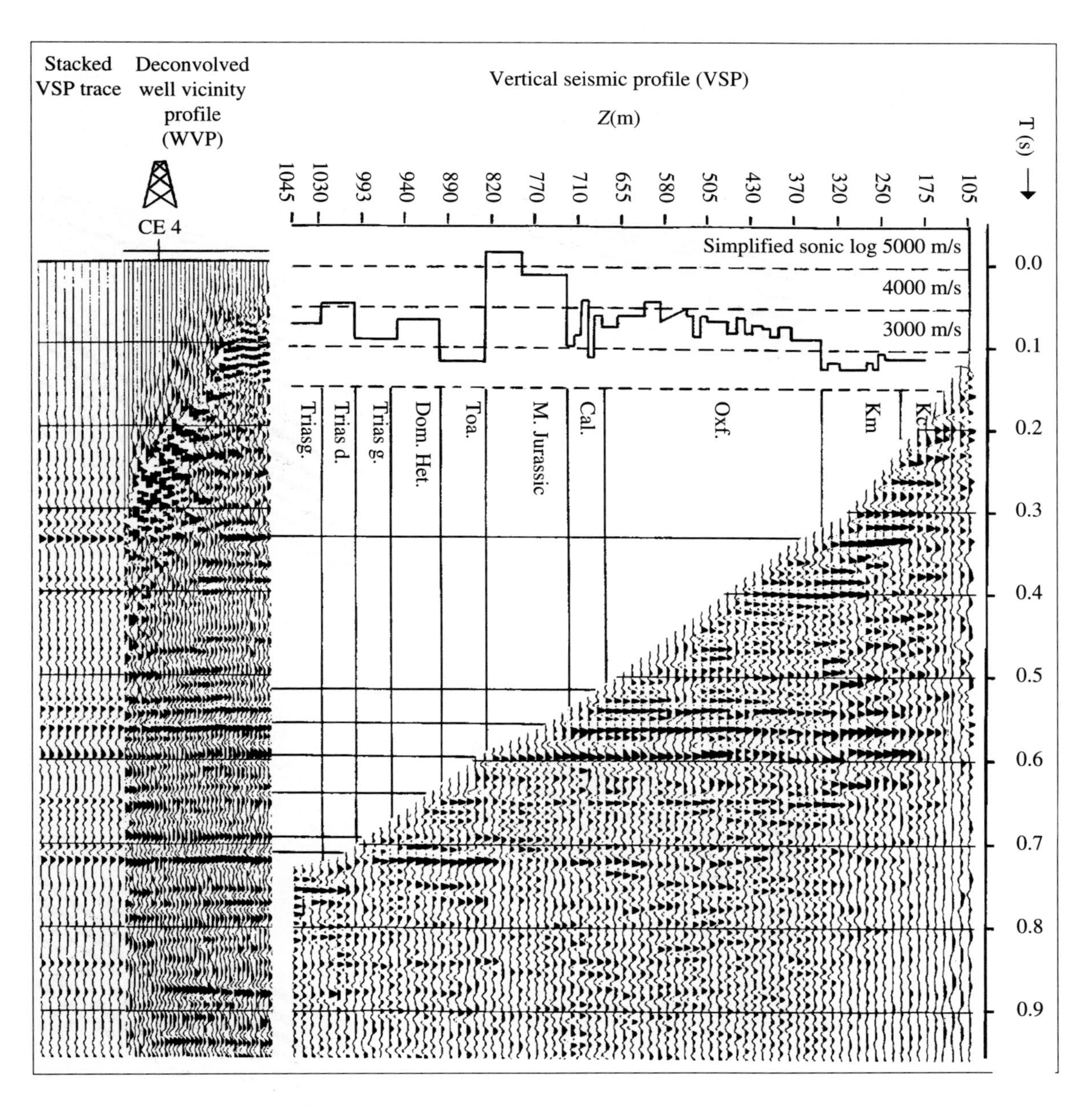

Fig. 2.16 Matching of surface seismic survey with VSP data.
(After Mari et al., 1987)

Figures 2.17a and b show the data processing of an offset VSP. In this acquisition, a weight dropper impulse source of the Soursile type (registered trademark of *IFP*) was offset 250 m from the well head. The tool is a three component borehole geophone (SPH 300, registered trademark of *ARTEP*).

The section (Fig. 2.17a) represents the data obtained on the vertical component *Z* of the well geophone. Those obtained on the horizontal component *X* are oriented in the plane passing through the source and the well. A strong field of downgoing *SV* waves can be seen on the horizontal component at 0.6 s and 1 s for geophone depths of 930 m and 1600 m.

In the example shown in Fig. 2.17b, the use of polarization filtering has led to a separation of the *P* and *SV* waves (vibration direction included in the source-well plane) and to obtain VSP sections in *P* and *SV* waves (Fig. 2.17b).

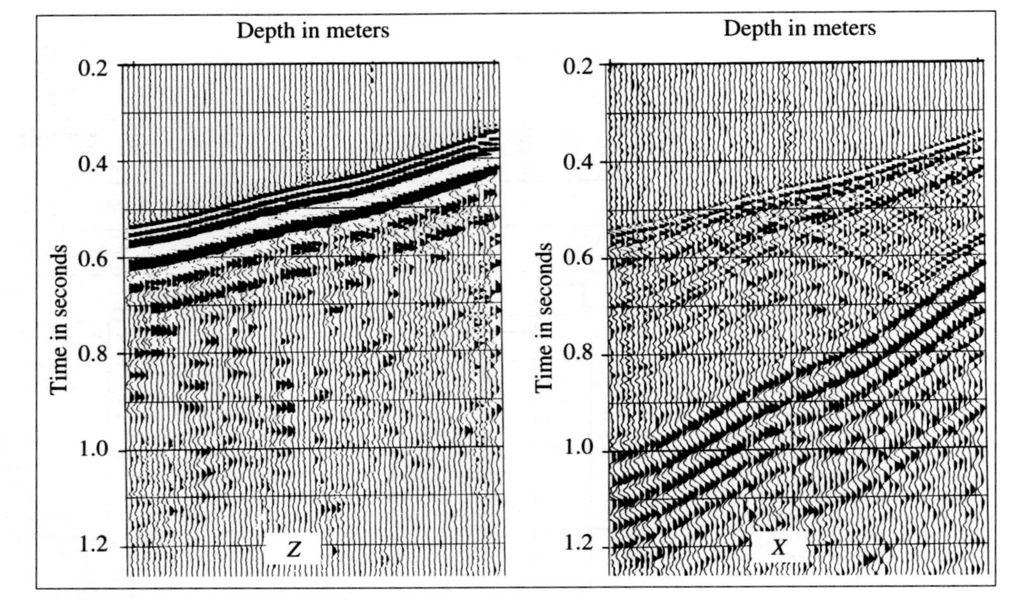

Fig. 2.17a ▶ VSP: horizontal (*X*) and vertical (*Z*) components of the well geophone recording. *(Mari and Coppens, 1989)*

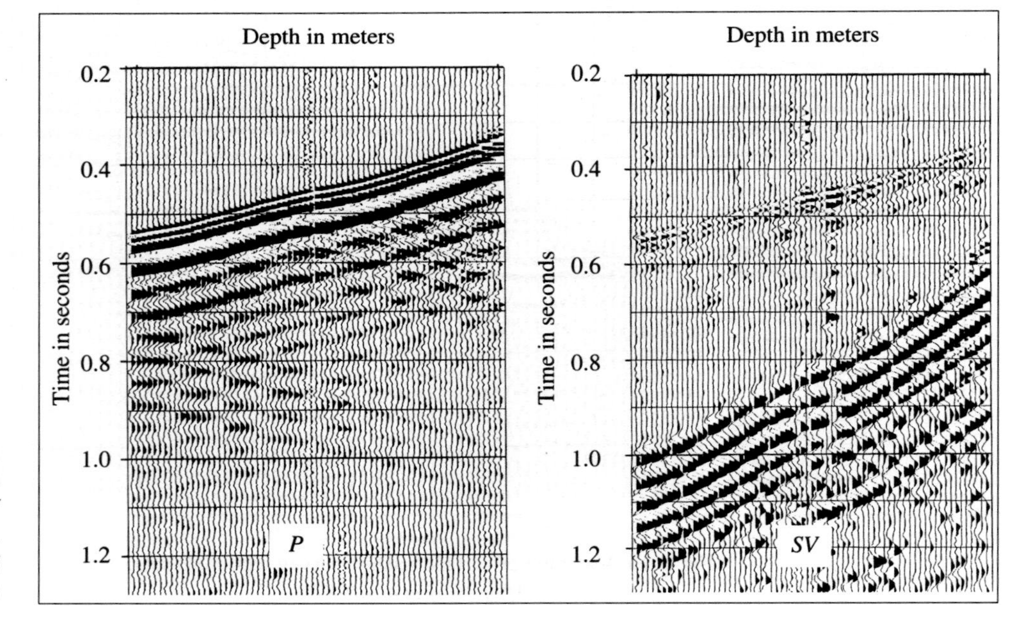

Fig. 2.17b ▶ VSP: separation of *P* and *SV* waves. *(Mari and Coppens, 1989)*

The presence of residual compressional waves may be noted on the *SV* wave section.

The separate processing of the VSP data in terms of *P* and *SV* wavefields makes it possible to obtain migrated VSP seismic sections (Fig. 2.18b).

The *SV* wave migrated section is shown with a time scale half of that used for the *P* wave section (200 ms in *S* = 100 ms in *P*), corresponding to a V_P/V_S ratio of 2. Correlation of the two sections cannot be achieved by eye on the basis of seismic features. Instead, the depths and times of primary reflections have been identified in the upgoing *P* and *SV* wavefields, after their separation and before migration (Fig. 2.18a), using the time-depth relations $T_P = F_P(Z)$ and $T_S = F_S(Z)$ obtained by picking first arrivals.

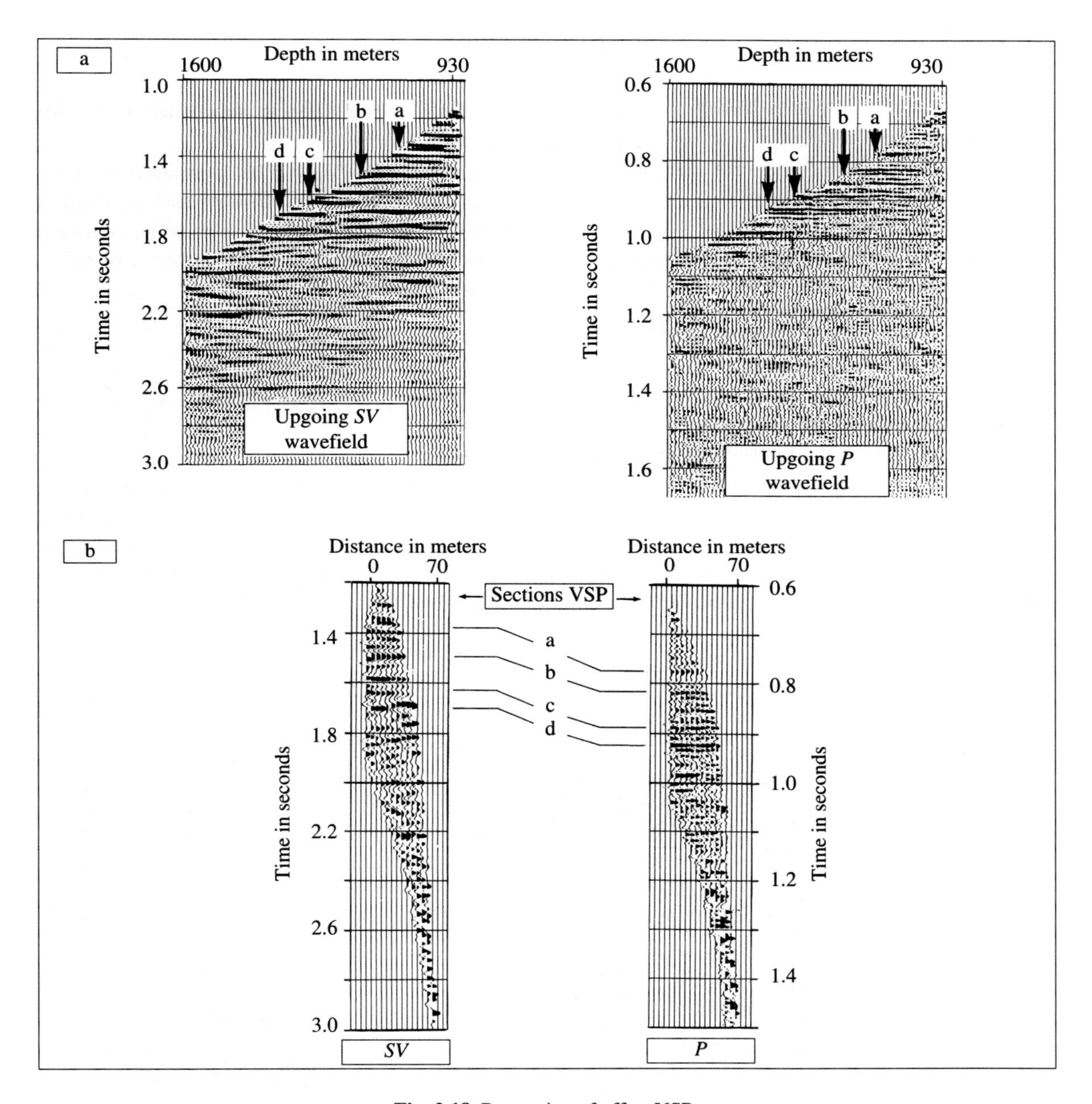

Fig. 2.18 Processing of offset VSP:
a) extraction of upgoing *SV* and *P* waves, b) VSP section based on migrated *P* and *SV* waves.
(Mari and Coppens, 1989)

2.2.3 Some applications of VSP

The principal applications of vertical seismic profiling are:

(1) time-depth calibration of seismic reflection surveys,
(2) measurement of P and/or S wave velocities,
(3) prediction of undercompacted zones,
(4) identification of primary and multiple reflections,
(5) estimation of the dip of reflectors,
(6) localization of fault planes,
(7) correlation between P and S reflected arrivals,
(8) obtaining detailed seismic data close to the well,
(9) prediction of reflectors below the bottom of the well,
(10) identification of fractured and highly permeable zones,
(11) measurement of anisotropy.

Examples of the use of VSP in the prediction of deeper reflectors or undercompacted zones and measurement of reflector dips are given in Chapter Six.

VSP also provides a means of obtaining certain lithological parameters, of which the most commonly sought are acoustic impedance, attenuation and the V_P / V_S ratio (yielding Poisson's ratio). It also enables the identification of highly permeable and fractured zones.

a. Determination of acoustic impedance

A log of impedance $\rho_i\ V_i$ (product of density and velocity) can be estimated from the reflectivity function r_i, using the relation:

$$\rho_i\, V_i = \rho_1\, V_1 \prod_{j=2}^{i} (1 + r_{j-1}) / (1 - r_{j-1})$$

where $\rho_1 V_1$ is the acoustic impedance of the first medium.

The impedance log can be constrained by imposing certain acoustic impedance values $\rho_P V_P$ (thus $\rho_P V_P / \rho_1 V_1$ = constant). These points of constraint imply complementary relations in the set of reflection coefficients:

$$(1 + r_{p-1}) / (1 - r_{p-1}) = \frac{\rho_p\, V_p}{\left(\rho_1\, V_1 \prod_{j=2}^{p-1} (1 + r_{j-1}) / (1 - r_{j-1}) \right)}$$

The VSP stacked trace can be considered as a function of reflectivity and then used to calculate the acoustic impedance log. In this instance, the high-frequency component of the impedance log is obtained, as illustrated in Fig. 2.19.

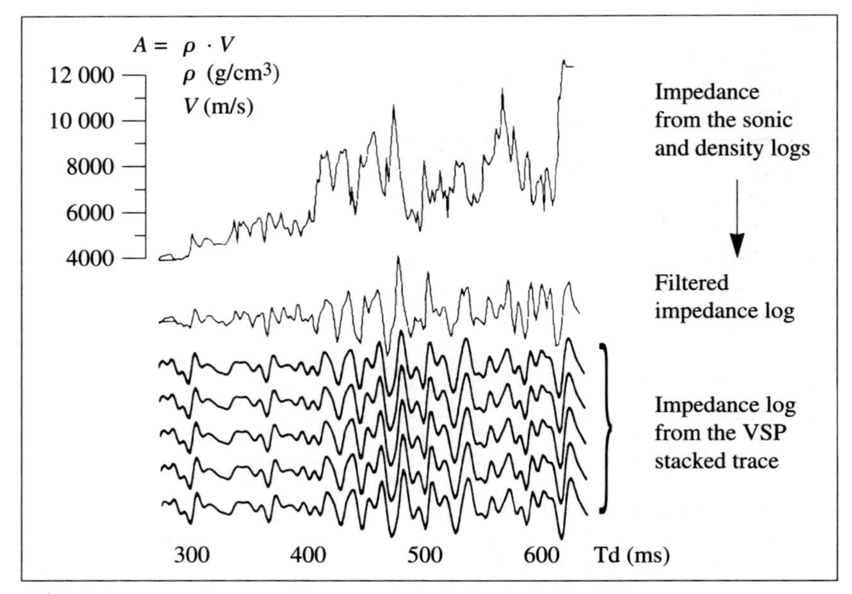

Fig. 2.19
Comparison between impedance logs derived from VSP and from logs (sonic-density).
(Tariel and Michon, 1982)

Figure 2.19 shows a comparison between the raw and low-cut frequency filtered impedance logs (derived from sonic and density logging) and the impedance log obtained by integration of the VSP stacked trace.

The acoustic impedance log can also be estimated using inversion techniques (Grivelet, 1985; Macé and Lailly, 1986), using a slope algorithm which determines the impedance distribution needed to minimize the least-square deviations between the calculated and observed seismic records. The observed results may correspond to the upgoing wavefield or the raw VSP data. Since seismic data has a limited frequency content, the solution produced by inversion is not unique. To escape from this drawback, it may be desirable to use an impedance model — constrained or partially constant — with imposed values at certain geophone positions in the well.

Figure 2.20 presents the results obtained by the Grivelet method (1985) applied to a real example with and without constraints. In this approach, the observed seismic record is the upgoing wavefield obtained by a conventional wave-separation process. The points of constraint, which are marked by arrows, allow for an acoustic impedance calibrated in absolute values.

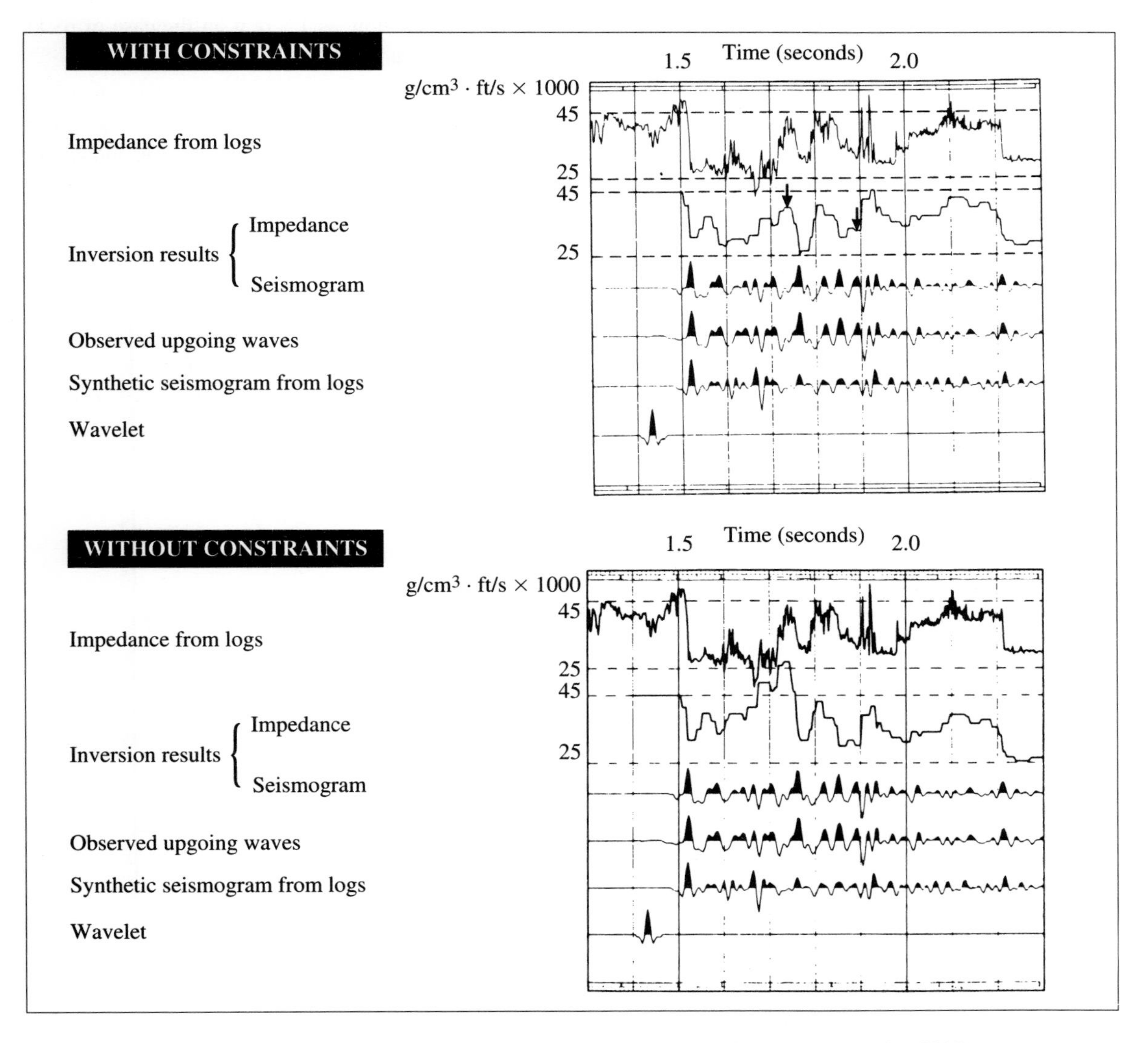

Fig. 2.20 Estimation of acoustic impedance from data inversion. *(After Grivelet, 1985)*

b. Poisson's ratio in VSP using P and S waves

Since the use of tools with triaxial geophones is becoming more and more widespread, VSP's are now a common means of analysing the propagation of the different kinds of waves, i.e. head waves (*P* or *S* type) and Stoneley waves, better known as tube waves, an example of which is discussed above.

The velocity of *S* waves depends solely on the shear modulus μ and the density ρ:

$$V_S = \sqrt{\frac{\mu}{\rho}}$$

The velocity of P waves depends on the density ρ, the bulk modulus *k* and the shear modulus μ:

$$V_P = \sqrt{\frac{K + \left(\frac{4}{3}\right)\mu}{\rho}}$$

The comparison of *P* wave seismic sections and *S* wave seismic sections can provide indications concerning the elastic coefficients of rocks as well as their lithological and petrophysical characteristics.

The ratio between the longitudinal and transverse wave velocities (V_P/V_S) leads to a direct calculation of Poisson's ratio (Fig. 2.21):

$$\sigma = \frac{1}{2}\,\frac{\left(\frac{V_P}{V_S}\right)^2 - 2}{\left(\frac{V_P}{V_S}\right)^2 - 1}$$

The V_P/V_S ratio varies theoretically between $\sqrt{2}$ and infinity in an isotropic medium, which corresponds to a Poisson's coefficient between 0 and 0.5. In an anisotropic medium, V_P/V_S can be less than $\sqrt{2}$ and falls as low as 1.3 (e.g. in the case of rocks saturated in gas).

Gregory (1976) has shown that the ratio between longitudinal and transverse velocities V_P/V_S can indicate whether the formations are consolidated or poorly consolidated and whether there is hydrocarbon impregnation.

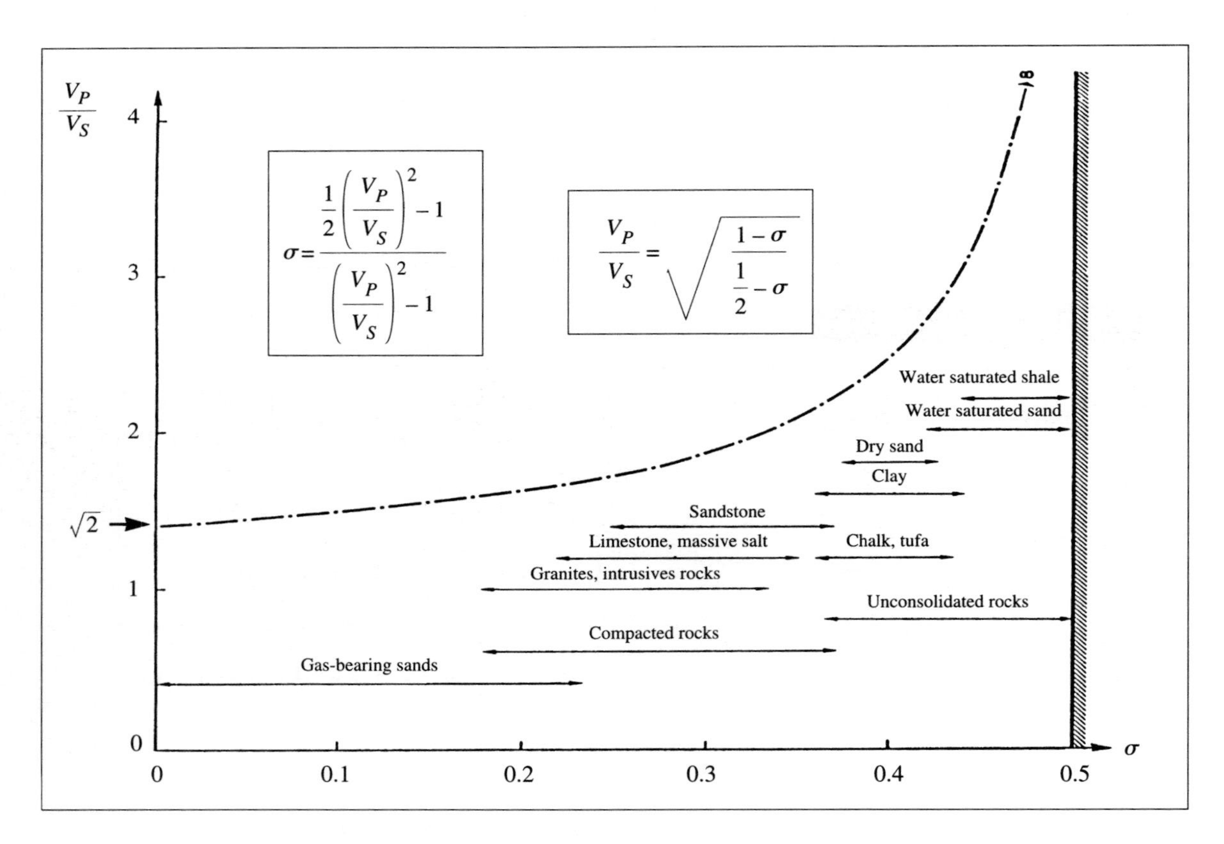

Fig. 2.21 Longitudinal/transverse velocity ratio as a function of Poisson's ratio σ. *(IFP document)*

Precise measurements carried out by Vasiliev and Gurvich (1962) indicate that for compact rocks, σ is generally in the range between 0.20 and 0.35 (V_P/V_S = 1.6-2.0), and for unconsolidated rocks between 0.35 and 0.48 (V_P/V_S = 2-5). Domenico (1977) has shown, in the Ottawa sands, that σ is around 0.10-0.15 for sand impregnated with gas and around 0.40 for sand impregnated with water.

Broadly speaking, values of $V_P/V_S > 2$ ($\sigma > 1/3$) are often characteristic of unconsolidated formations saturated in water, while values of $V_P/V_S < 2$ ($\sigma < 1/3$) indicate well-consolidated formations or sands impregnated with hydrocarbons.

Since gases are more compressible than liquids, the velocity of P waves in a gas-saturated rock is generally distinctly lower than it would be if the rock were water saturated (Domenico, 1976; Gregory, 1976). Moreover, under the same conditions, the velocity of SH waves hardly varies, the very slight variation being due to the density difference between the two fluids (Ensley, 1984).

As regards the VSP, the V_P/V_S ratio is directly obtained from the picking of first arrivals on VSP traces recorded at the same geophone positions for P and S waves.

The time-depth relationship thus obtained linking P and S waves is the only reliable means of identifying the reflections which correspond to the same geological horizons on the P and S wave sections. This leads to a calculation of the V_P/V_S variation profile, which is able to yield the lithological information mentioned above.

In practice, however, the precision of the lithological information thus obtained is limited by the vertical resolution of the surface survey. In the case of the VSP, more detailed lithological information can be expected, although limited by the vertical resolution of the VSP.

Figure 2.22 presents a VSP recorded in three components, obtained with the use of an oblique falling weight, after separation of the compressional waves (vertical component Z) and the shear waves (horizontal components $H1$ and $H2$). In this example, the source was a horizontal hammer (Marthor registered trade mark *IFP*) striking an inclined target.

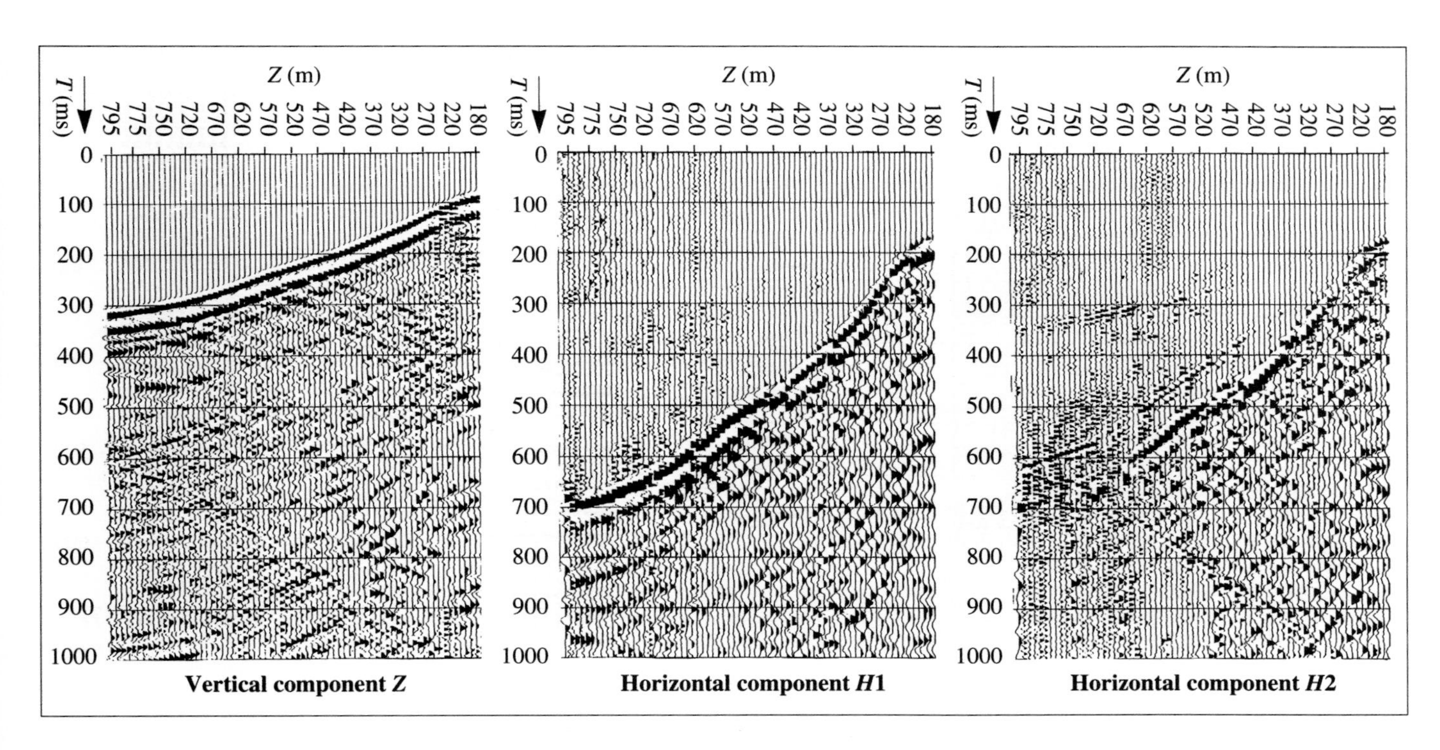

Fig. 2.22 VSP obtained with a hammer striking an inclined target.
(Mari and Coppens, 1989)

Figure 2.23 shows the P and S waves VSP stacked traces, as well as the time-depth correlation between P reflections and S reflections, a calibration with the acoustic impedance log derived from the sonic and density logs together with its match to a lithological log. The γ_T values represent the travel time ratio $\Delta Tp/\Delta Ts$ (which is equivalent to the V_S/V_P ratio) between seismic horizons recognised as being at the same depth.

It should be pointed out that correlations based on the characteristics of P and S waves are only valid for seismic horizons associated with strong reflection coefficients.

c. *Attenuation in VSP*

The decrease in amplitude of a seismic wave propagating underground has different origins, notably: spherical divergence, transmission, conversion, diffraction, the effect of internal multiples and the attenuation quantified by a quality factor Q which characterises the energy dissipation of the medium. The Q factor decreases as the attenuation increases.

Numerous measurements in the laboratory and on the ground have shown that P wave attenuation is at its highest in partially saturated sandstones. In porous sandy units, the attenuation is higher than in the surrounding shale. Attenuation is linked to the degree of compaction of sediments, the presence of gas, the degree of fluid saturation and the type of fluid present.

For an evaluation of the characteristic attenuation of a given lithology, the saturation in fluids or the overall "physical state" of the rock, an attempt should be made to compensate the amplitudes as much as possible for phenomena other than attenuation. However, this is a very delicate operation.

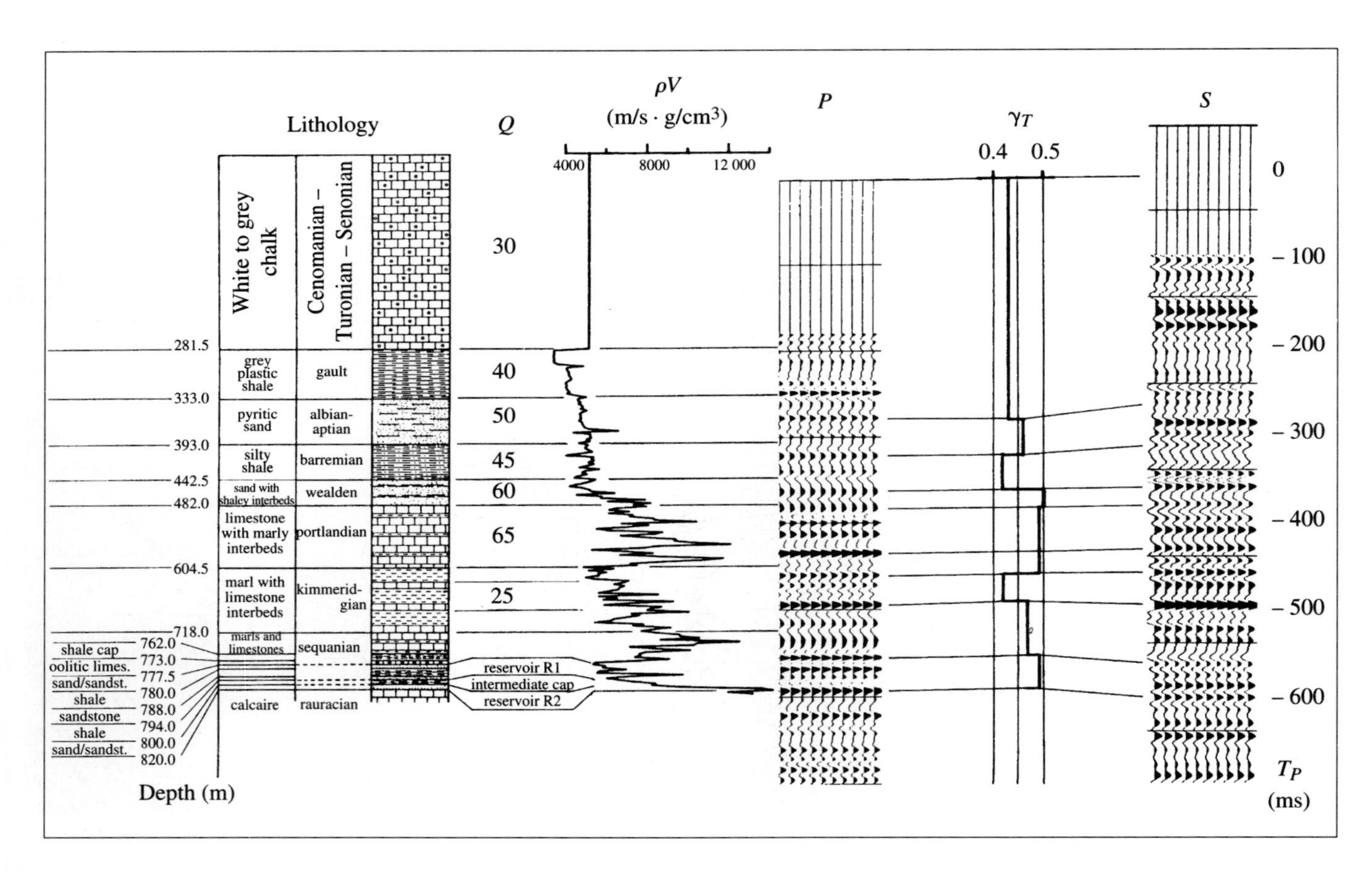

Fig. 2.23 Correlations between downhole datasets (VSP, logs and lithology) derived from the processed VSP data presented in Fig. 2.22.
(Mari and Coppens, 1989)

VSP's provide a type of data that is suitable for quantifying the "macro" attenuation of geological units taken on the scale of the seismic wavelength involved.

The study of attenuation is generally carried out on the first arrival of the downgoing wavefield, after compensation for spherical divergence and calibration of the signal emitted by the source. The most widely used techniques are the spectral ratio method (Hauge, 1981 ; Kan et al., 1981 ; Stainsby, 1985) and the rise-time method for VSP's recorded with impulsive sources (Kjartansson, 1979).

As an example, Fig. 2.23 presents the attenuation values determined by the compressional wave rise-time method. We note a good correlation of the γ_T coefficient with the quality factor Q (there is a positive correlation between γ_T and the Q coefficient, which is a characteristic feature of geological units with low attenuation and vice versa).

Figure 2.24 shows a pseudo-attenuation log (Mari, 1989) obtained by the rise-time method using VSP data (Fig. 2.25) from the SR 180 well (see following section). The velocity and attenuation logs obtained from the VSP data are compared with the sonic, gamma ray (GR) and lithological logs at the well. It can be seen that, where the attenuation and velocity logs are anti-correlated

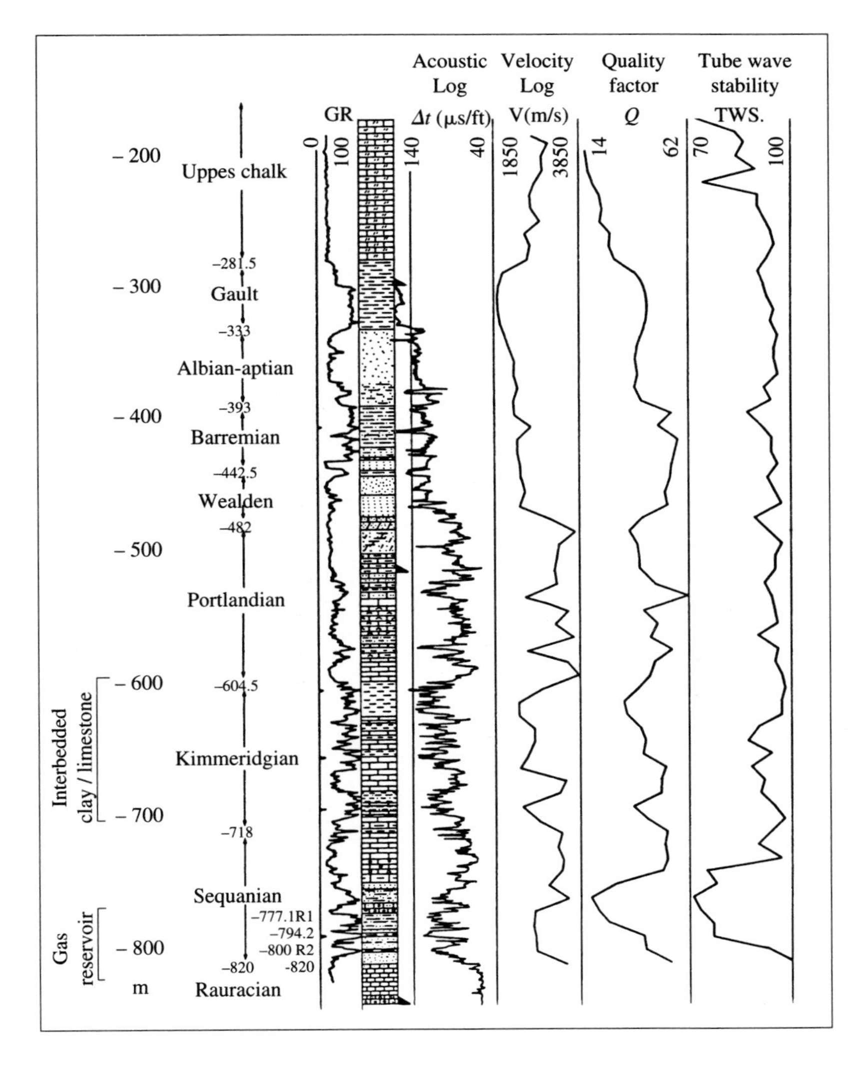

Fig. 2.24 Lithology and attenuation (Q). *(Mari, 1989)*

(above 600 m), the gamma ray (GR) values are low or almost constant. On the other hand, in the clayey limestones of the Kimmeridgian (600-700 m), where the attenuation and velocity logs are slightly correlated, the gamma ray (GR) curve shows strong variations. This suggests that the attenuation of the compressional waves is linked to the occurrence of thin beds of limestone and clay. The strong attenuation observed between 777 and 820 m is associated with the presence of a reservoir zone partially saturated with gas.

d. Identification of highly permeable and fractured zones

Low frequency Stoneley waves, usually called tube waves, can be used to identify fractured and permeable zones.

Tube waves are created by particle vibration in the mud column. The ground roll generated by the seismic source excites the mud column when it reachs the well and gives rise to tube waves. Once initiated, the tube wave is propagated upwards and

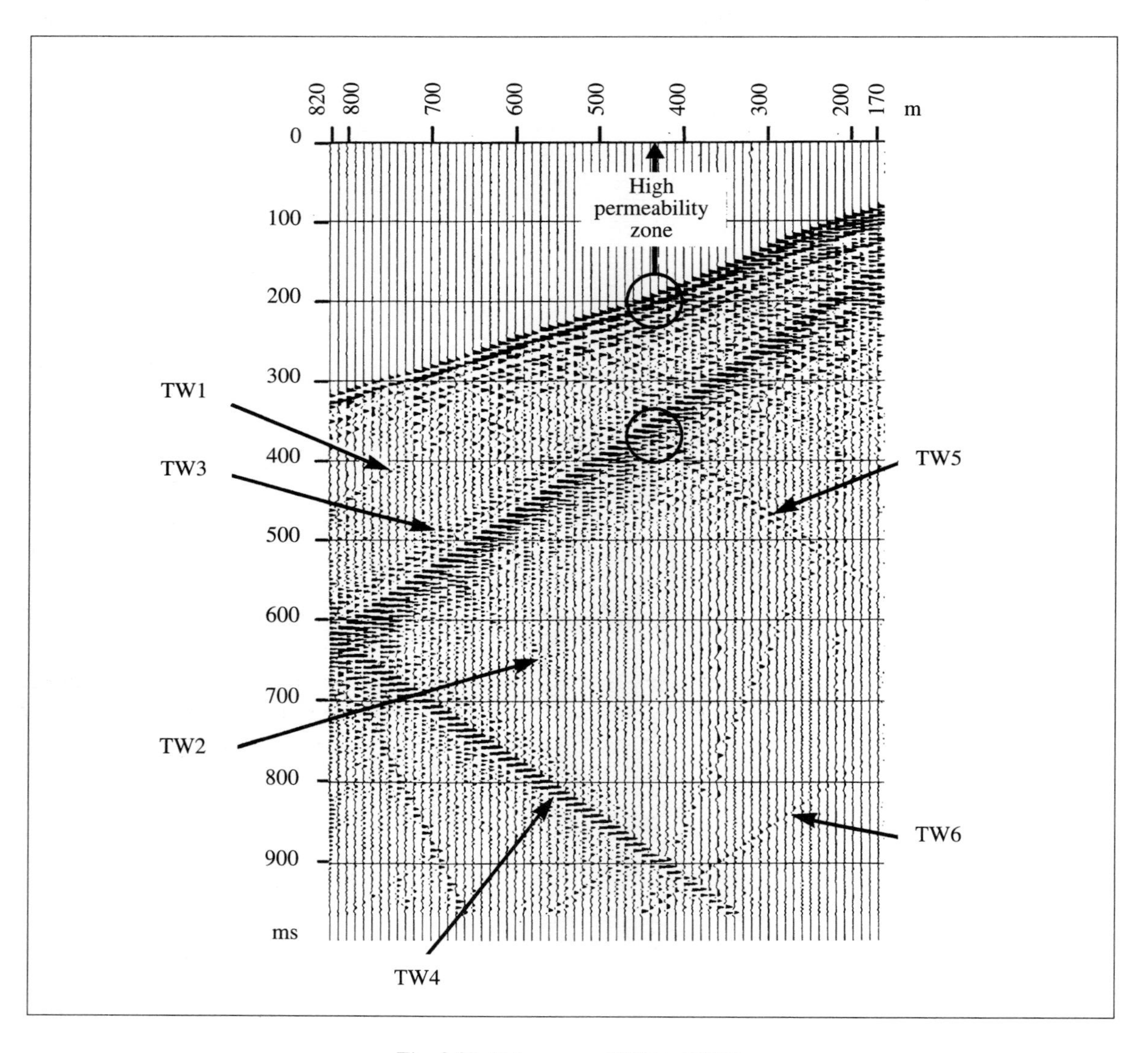

Fig. 2.25 Tube waves (TW1 to TW6).
(Courtesy of Gaz de France-IFP)

downwards in the well and generates secondary tube waves each time there is a significant acoustic impedance contrast. Tube waves can also be created by compressional waves when these latter encounter a strong impedance contrast in the well or when they cross a permeable or fractured zone.

The VSP shown in Fig. 2.25 is very rich in tube waves (labelled TW1 to TW6).

The seismic source generates surface waves (ground roll) which give rise to the TW3 tube waves. They are reflected at a depth of 440 m, which corresponds here to the top of a porous and permeable sandy formation. The reflected tube wave is labelled TW5. In the same formation, the downgoing compressional wave generates a downgoing tube wave labelled TW1, clearly showing the presence of a high permeability zone at a depth of 440 m.

Similar phenomena can occur in the presence of fractured zones.

The variation of phase velocity and attenuation as a function of the frequency of the Stoneley waves (low-frequency, 10-200 Hz) is dependent on the petrophysical parameters of the medium (notably the porosity and permeability of the formation, the viscosity and bulk modulus of the pore fluid). Hence, it is possible to obtain an estimate of the permeability by dispersion analysis, using a dispersion model proposed by Biot (1956) and Rosenbaum (1974) applied to high-frequency (a few kHz) acoustic well logging.

Since the Stoneley waves are interface waves, the information thus obtained refers to a zone close to the borehole wall.

The study of fractured zones situated from the well cannot be undertaken by analysing the interface waves. Other techniques then must be used, such as microseismics, which aims to localize the sources of microseismicity (Becquey, 1989).

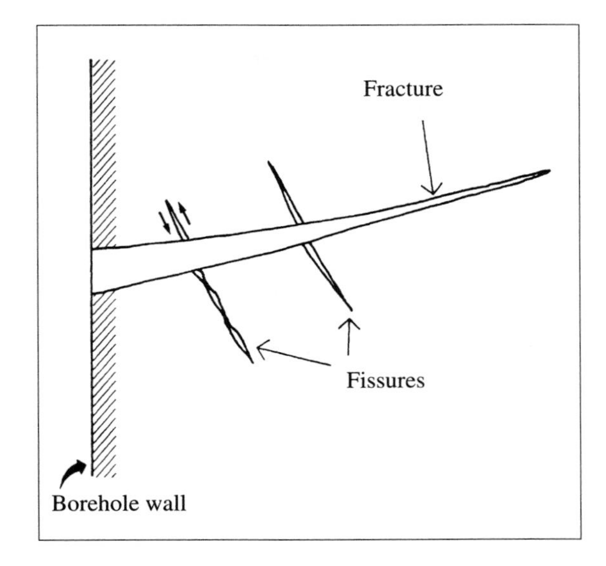

Fig. 2.26 Sources of well microseismicity.
(Becquey, 1989)

The source of microseismicity (Fig. 2.26) may be linked either to fissure propagation, as in the case of fracturing, or the relative movement of the fissure borders. The fluid, which is either in place or injected, is free to move between the borders of the fissure. Alternatively, an irregular flow of fluid may occur across closed passages in the fissures.

In the first case, the polarization of the emitted P wave gives the direction of the source (deeper) or shallower than the geophone depth) and the difference in arrival time of the P and S waves can indicate the distance.

In the second case, the lips of the fissure slide in relation to each other and generate a shear wave polarized in the direction of the fissure.

Chapter **3**

SEISMIC AND WELL LOGGING SURVEY CALIBRATION

The comparison of the travel times of an acoustic wave between two depth points, as obtained by two different methods (well velocity survey (VSP or check shots) and acoustic logging), can only be made if the conditions of propagation are relatively similar, i.e. the same type of wave in the same geological medium following the same path.

Thus, comparisons can be made between the vertically corrected time (TVC) obtained from a well velocity survey (set to a reference depth, usually the Datum Plane DP) and the integrated sonic transit time (TTI'), which itself has been corrected to vertical time (true vertical time) and corrected to the DP.

However, it can be seen that residual discrepancies remain (D = TVC – TTI') which vary in magnitude with depth (Fig. 3.2). These deviations (or mismatches) can easily reach 10 ms, and sometimes much more according to acquisition conditions and the nature of the formations. The deviation generally increases with depth, although sometimes it decreases, particularly in surface formations.

The calibration of sonic logs to seismic survey data consists of limiting the mismatches observed at the well velocity measurement depths. This is done by modifying the sonic log at given depth (generally on a thick depth interval) so as to esta-

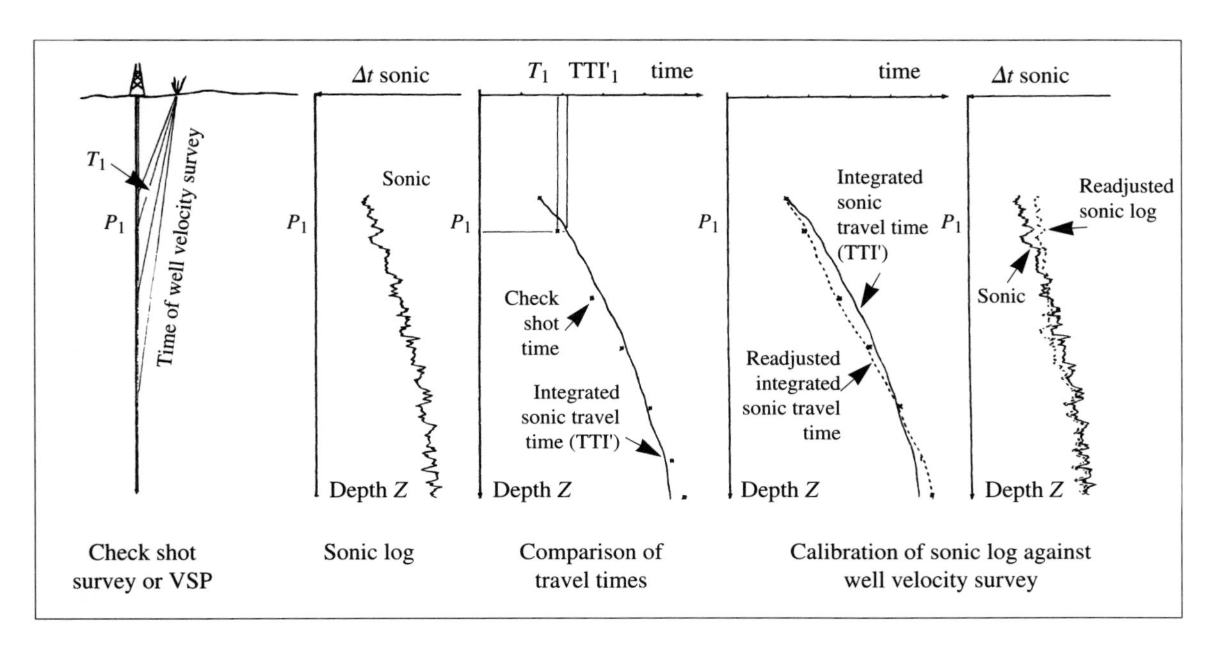

Fig. 3.1 Calibrating the sonic log to well velocity survey (or VSP).

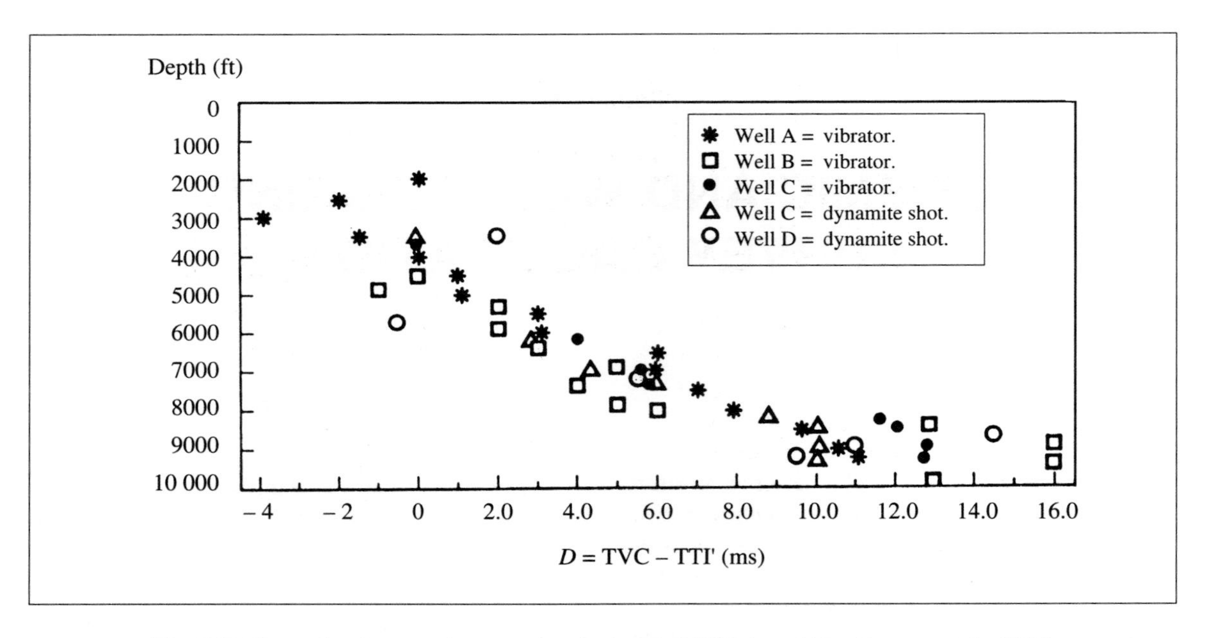

Fig. 3.2 Mismatches between integrated sonic time and VSP time. *(After Stewart et al., 1984)*

Figure 3.2 presents the results of a study carried out in the Anadarko basin: the discrepancies between the integrated sonic time and the VSP travel time can reach 15 ms in this particular example.

blish a $T = F(Z)$ relation based on an integration of this recalibrated log. Although the time-depth relation is compatible with the seismic data, it is based on a higher sampling rate comparable to that of the acoustic log.

Once this new relation is defined, it is easy to convert depth logs to time logs —on the seismic scale— in order to ensure fully consistent use of the different data.

In order to more easily compare the log data with the seismic data, a synthetic record is computed (Fig. 3.3) using reflection coefficients obtained from the time logs. The impulse log thus obtained is convolved with signals typical of or close to those acquired by the seismic survey (see Chapter 4 on synthetic seismograms). However, there is a prior need to make the log and seismic data compatible by correcting the raw values and adjusting them to raypaths that are as similar as possible. In this way it is assumed that acoustic and seismic waves are propagating through the same geological media.

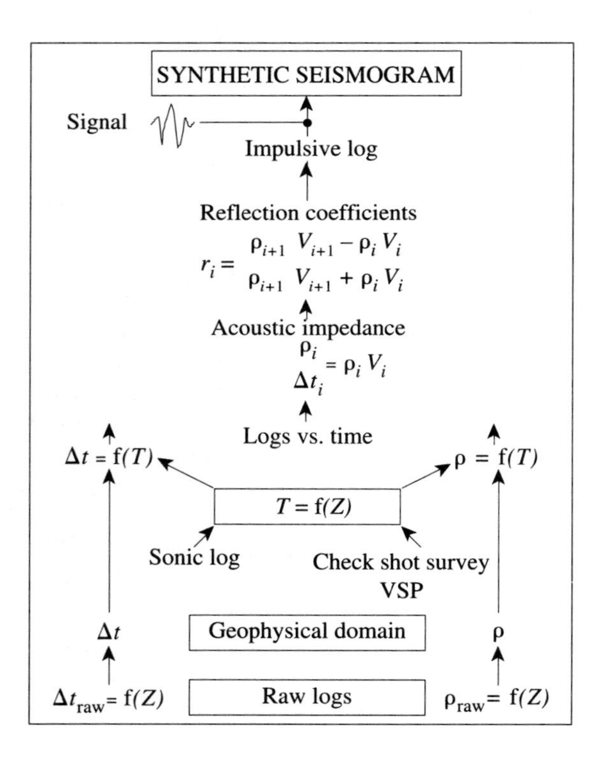

Fig. 3.3 Production of synthetic seismogram.

3.1 DISCREPANCIES BETWEEN SONIC LOG AND SEISMIC VELOCITIES

3.1.1 Preliminary remarks on geological media

Well velocity survey and well logging operations are not carried out in the same geological media. The comparison of times obtained from these two operations is carried out by adjusting the times to a similar trajectory (normal incidence wave propagation). To do this, different environmental corrections must be made using geometrical data (source location, borehole trajectory, ...).

A. Location of measurements

The spatial location of each downhole geophone station can be obtained from continuous directional surveys such as those made during dipmeter or gyroscope logging, or from single or multishot directional surveys carried out by the drillers. The different types of measurement techniques can result in accuracies as poor as 10-15 m at 3000 m wells at 45° deviation.

B. Influence of surrounding media on measurements (Fig. 3.4)

a. Uninvaded and flushed zones

Acoustic and density logs have a shallow depth of investigation (less than 1 m), analysing a rock-volume which is limited to the immediate surroundings of the borehole in the flushed zone where mud has driven out the formation water and part of the hydrocarbons. This flushed zone is generally not very extensive (a few decimeters at the most). The seismic wave emitted at the surface is mainly propagated through the uninvaded formation before reaching the receiver via the flushed zone. The velocity and density characteristics of the geological media depend on the fluids (mud, formation water and hydrocarbons) and hydrocarbon saturation. These fluid characteristics are therefore in relation with the geological media investigated by the survey. Corrections for the sonic and density values can be envisaged through a complete logging interpretation. After correction, the acoustic and density logs enable the estimation of porosities, as well as water saturation and nature of

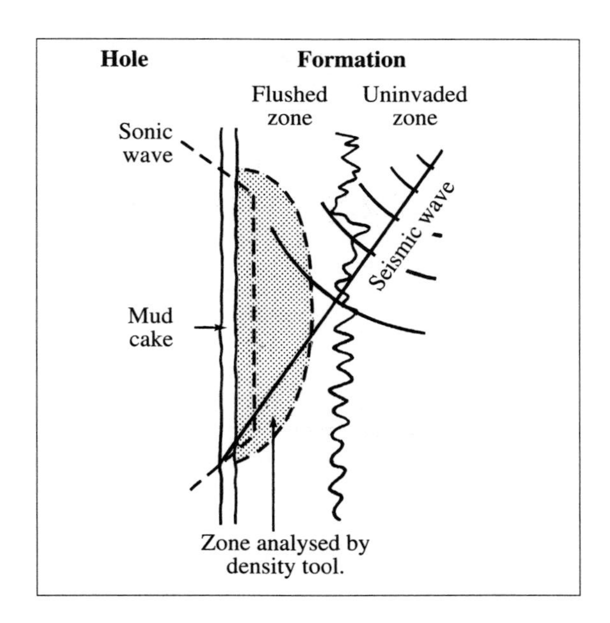

Fig. 3.4 Propagation media for seismic and sonic waves.

the fluids (formation water and hydrocarbon), if they are not already known (see Interpretation of recorded logs, p. 60).

b. Lateral facies transitions

The seismic wave path can be at distances relatively far from the well depending on the well/shot point configuration. Lateral facies variations are accompanied by changes in the elasticity and lithology and should be taken into account when comparisons in travel times are being made, due to the existence of lateral variations in velocity.

c. Dip in seismically fast formations

In the case of so-called "seismically fast" formations, the formation dip can lead to wave propagation by the fastest trajectory (Fig. 3.5a).

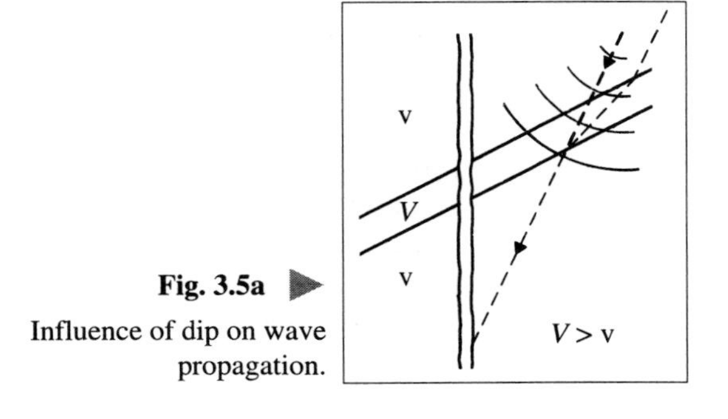

Fig. 3.5a Influence of dip on wave propagation.

The sonic and seismic waves are affected differently according to the position of the well and seismic shot points in relation to the dip.

d. Anisotropy

Wave propagation is influenced by the anisotropic character of the formations (e.g.: velocities normal to the bedding can be lower than longitudinal velocities).

e. Raypath

Integration of the sonic well log does not necessarily give a travel time corresponding to the path taken by the seismic wave (Fig. 3.5b). Although the correction for well deviation (conversion to vertical times) minimises this error, it does not take into account all the effects mentioned above and does not enable the recovery of time discrepancies.

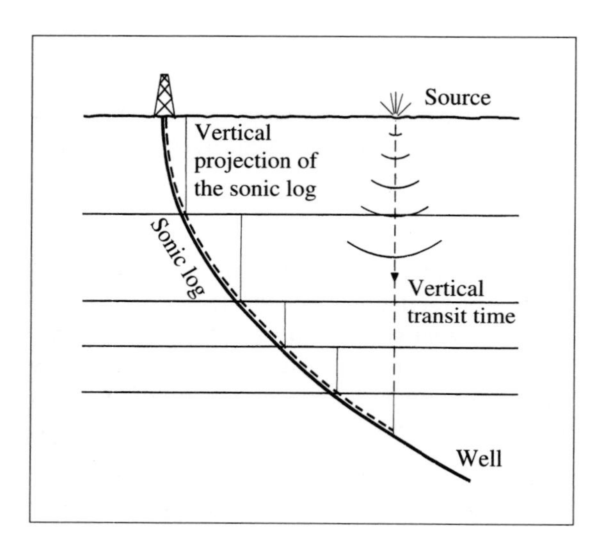

Fig. 3.5b Influence of well deviation on the estimation of vertical times.

Only sophisticated simulations using all available information (seismic sections, logs and cores) are able to reduce the errors incurred in this way. However, this type of costly approach is not generally adopted, at least for a simple sonic calibration; moreover, the precision of the applied corrections is probably illusory considering the relative lack of knowledge of the geological media (with the exception of flushing and well deviation corrections).

3.1.2 Causes of discrepancies between seismic and well logging travel times

A. Measurement of acoustic wave propagation time

In the case of well seismic surveys, the measurement of travel time is direct (from a source at the surface to a receiver in the well). On the sonic log, however, it is calculated by the integration of Δt; under these conditions, the errors on Δt accumulate progressively over the depth of the log.

B. Velocity of wave propagation

The low frequencies seismic waves propagate slower than the high frequency acoustic waves (dispersion effects).

Velocity is in fact a function of frequency f (10-20 kHz in sonic logging, 10-150 Hz in seismics) as well as the attenuation Q of the media traversed by the waves, i.e.:

$$\frac{V(\omega_1)}{V(\omega_2)} = 1 + \frac{1}{\pi Q} \text{Ln}\left(\frac{\omega_1}{\omega_2}\right) \quad \text{where } \omega = 2\,\pi f$$

For media exhibiting very little attenuation (carbonates, ...), for example, the error is about 1.5%, for $Q = 100$.

For attenuating media (shaley, ...), for example, the error is about 5%, for $Q = 30$.

C. Noise and attenuation (using sonic log with bias system)

Slowness measurement using the bias system is affected by noise and attenuation. Large discrepancies do not arise as long as the anomalies remain of limited occurrence and are only infrequently repeated. But if these phenomena persist, the measured log can exhibit serious discrepancies in the integrated time values (*cf.* Section 1.2.1.2: Acoustic well logging: data checking and corrections). Generally speaking, a qualitative interpretation of the dataset may indicate the presence of such phenomena.

A full waveform recording enables preprocessing of the data aimed at increasing the signal-to-noise ratio, thus eliminating the impulse noise and improving the quality of slowness logs.

D. Influence of media traversed by log

a. Raypaths

In VSP, the seismic wave is emitted from a shot point which does not lie vertically above the reception point, whereas in sonic well logging, the wave is propagated along the hole axis.

The conversion of the measured seismic time to vertical time is merely an approximate correction since the raypath is not always a straight line.

In the case of deviated wells, replotting of the sonic time on a vertical axis only provides an approximate slowness log. In fact, this correction assumes a horizontal homogeneity of the formation along the entire deviation of the well.

The borehole environment can lead to a slowing down of the wave emitted by the sonic tool (caving and state of the well wall).

b. Media

The propagation of the acoustic wave depends on the arrangement of the grains in the rock and on the type, size and distribution of the pores, in other words, on the elastic characteristics of the rock and the fluid it contains. The wave frequency has an influence on this propagation, so differences in travel times between a high-frequency "sonic" wave and a low-frequency "seismic" wave can be linked with two frequently encountered phenomena (*cf.* Section 1.2.1: Acoustic well logging):

(1) poor cohesion between the grains,

(2) vugs.

Studies using all the available information (logging surveys, mud logs, cuttings, cores, geology, etc.) allow the observation of inconsistencies between the sonic data and other porosity logs. These inconsistencies can be related to problems arising from vugs or cohesion of the rock (i.e. study of "abnormal trends" on different logs linked to porosity: Neutron, Density, Sonic, Resistivity logs, etc.).

c. Invasion

The acoustic wave emitted by the sonic tool propagates along the hole, while the wave emitted at the surface travels mainly in the uninvaded formations and traverses the flushed zone before reaching the receiver. As a result, there is a difference between the travel times. This difference depends on the fluids (mud, formation water, hydrocarbons) and the saturation hydrocarbon in these two zones.

A complete log interpretation allowing an estimation of the porosities, as well as the water saturation and nature of the fluids (formation water and hydrocarbons) if these are not known can be performed for log correction. Nevertheless, calculation of these corrections is probematic due to considerably more complex relations than the simple formulae —usually of a linear type— that are conventionally used (e.g. Wyllie, etc.) (see Section 1.2.1: Acoustic well logging).

3.2 ADJUSTMENT OF SONIC INTEGRATED TRANSIT TIME USING SEISMIC TRAVEL TIME

As mentioned in the chapter on acoustic well logging, the integration of acoustic propagation time Δt measured by the sonic tool over the entire recording-depth interval (after correction for the media) yields the integrated transit time TTI.

In order to obtain an integrated time at each depth of the sonic log, using a common reference depth with respect to the well velocity survey, it is necessary to:

(1) add a T_0 time corresponding to the transit time from the common reference depth to the beginning of the sonic log,

(2) convert the times to a vertical scale in the case of a deviated well (in the same manner as for check shot travel times).

The integrated time TTI' obtained in this way is directly comparable with the well velocity corrected vertical time (TVC) (see Section 2.2 on well velocity data processing). This procedure also provides the discrepancy D between these two times (Fig. 3.6).

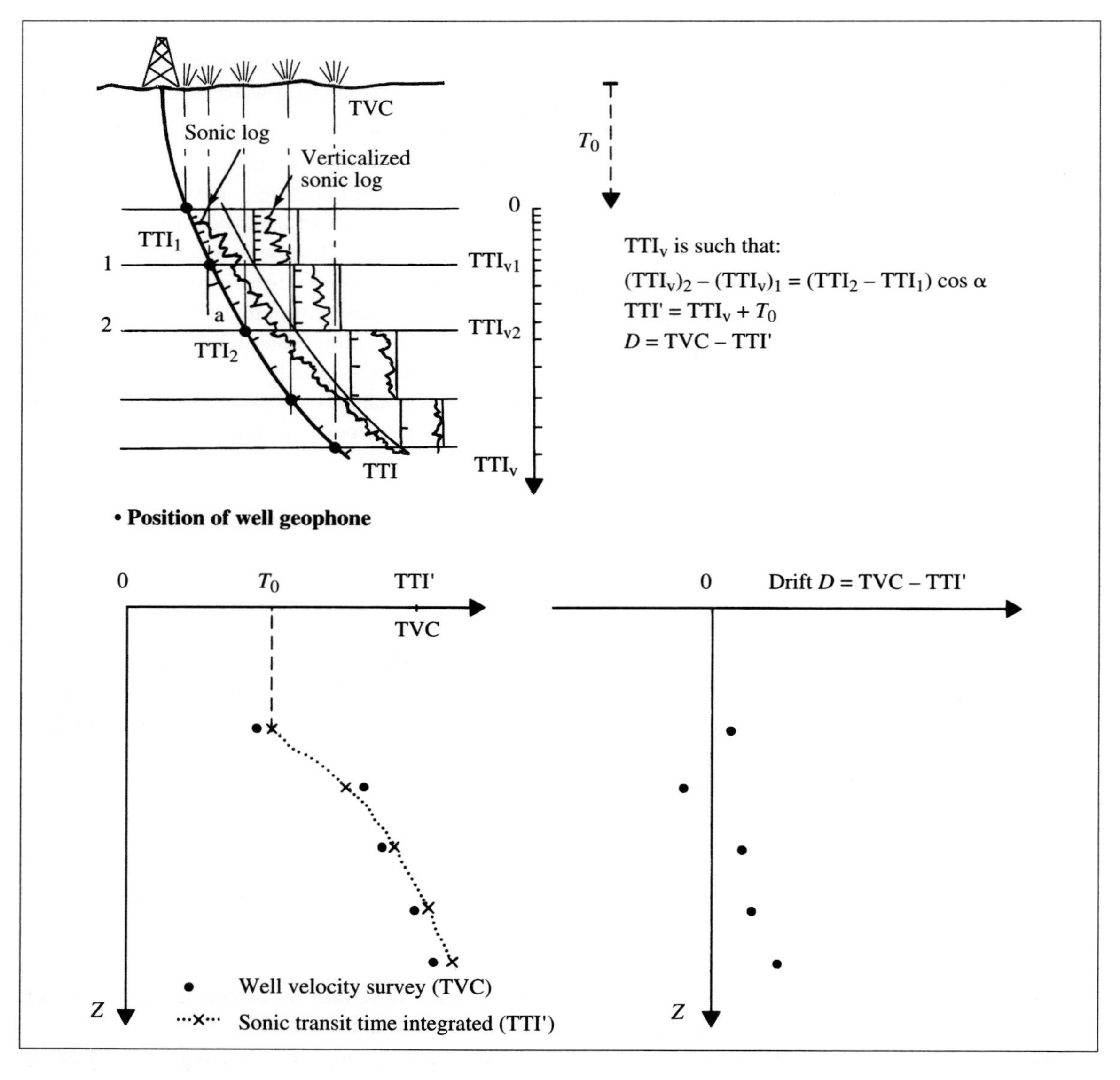

Fig. 3.6 Study of discrepancies between sonic integrated times and times obtained from well velocity surveying.

3.2.1 Drift calculation between seismic time (TVC) and integrated sonic time (TTI)

As stated above, the integrated sonic transit time TTI' is computed from a Δt integration and adding a T_0 value onto the travel times measured from the DP reference surface down to the top of the zone where the sonic log. The values corresponding to the difference between TVC and TTI' are the drift values D expressed as a depth function.

N.B.

As mentioned above, well velocity times must be corrected for the raypath followed. In order to obtain vertical times, corrections should be made for the tool position in relation to the shots. A slanting rectilinear raypath is generally assumed (Fig. 3.7). The times are normalised to a reference depth (DP) using measurements which involve wathered zone (WZ) or by considering the wave velocity in the water column in offshore case.

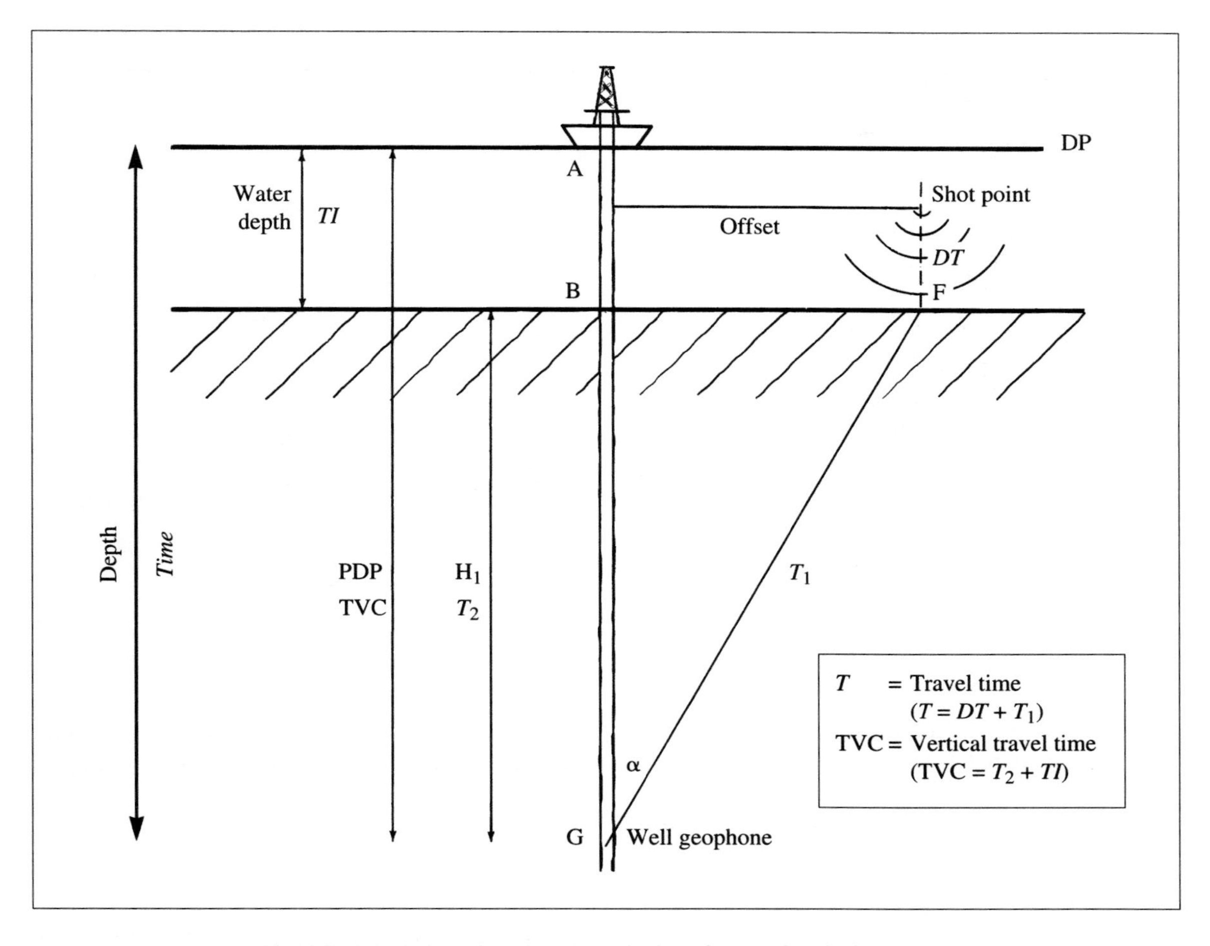

Fig. 3.7 Calculation of vertical time obtained from well velocity survey.

In most cases, the reference depth for the log is set at the rig floor, the rotary table or the kelly bushing, and sometimes at ground- or at sea-level. The reference depth for the well logs must be the same as the reference depth used in the well velocity survey.

3.2.2 Drift curve

As the drift values can be affected by different problems in relation with sonic log or with check shots, the sonic log will be adjusted using a drift curve.

This curve is traced from drift points (representing the drifts values, i.e., transit time) set to the reference depth (Datum Plane, DP), either by using of straight line segments (for linear adjustment) or by higher degree curves.

The value of T_0 added to the sonic TTI' is chosen to adjust the TTI' on the check shot survey time (suitable corrected) for example at the top of the sonic log.

For linear adjustment, the changes of slopes in the drift curve are best placed at formation boundaries (variations visible on sonic, density and other wireline logs indicating a change in the lithological character or nature of the fluids). As the drift curve is used to correct the Δt prior to the creation of a synthetic section, artificial reflections may be created by abrupt and unjustified changes to the Δt log. These points, also called "knees", should not be spaced too closely to each other (i.e. separated by at least 200 m).

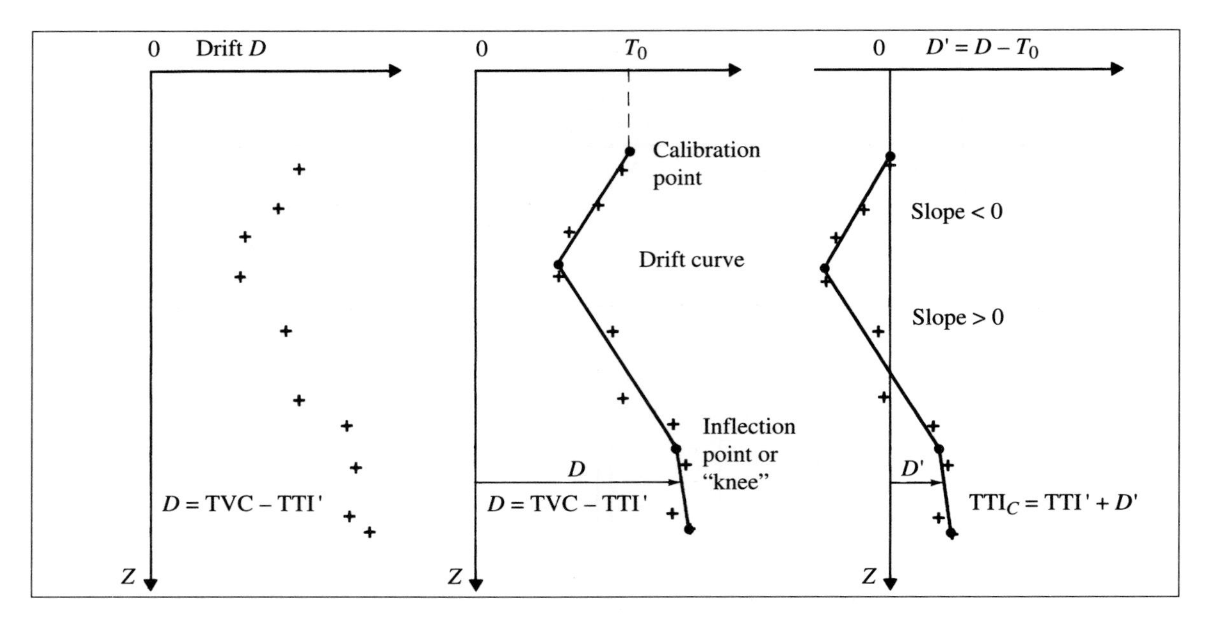

Fig. 3.8 Establishing a drift curve.

At each depth Z, the corrected vertical time $TTI_C(Z)$ calculated from the sonic log therefore corresponds to the TTI' time plus the corresponding D' ($D - T_0$) value derived from the drift curve.

The D' values so obtained can be locally positive or negative. A positive slope (or drift) is said to occur when the differences tend to increase with depth, whereas a negative slope refers to decreasing difference with depth.

3.2.3 Correction of sonic log to the check shot survey

This correction is carried out by progressively redistributing the gains and losses in time within intervals delimited by the "knees" of the drift curve (Fig. 3.9).

Two conventional methods are used:

(1) the block shift method involves the uniform redistribution of the time losses or gains within each interval, by applying a constant correction to the sonic log,

(2) the minimum Δt method, in which the correction is a function of the Δt value itself.

Clearly, more sophisticated methods could be used, such as readjustment of the sonic log using a permanent modification of the Δt but in a non-uniform manner. The correction to add to or subtract from Δt is as follows:

$$\int_{Z_1}^{Z_2} \mathrm{d}\Delta t(z)\, \mathrm{d}z = D_{2-1}$$

where D_{2-1} is the difference in total time (between TTI' and TVC) between two depths Z_1 and Z_2.

a. Block Shift Method

In this method, it is assumed that the correction is distributed uniformly over the whole interval between two inflections points (Fig. 3.10); the Δt is corrected within the interval using a constant value $\mathrm{d}\Delta t$ as follows:

$$\Delta t_{cor} = \Delta t + \mathrm{d}\Delta t \text{ (in } \mu\text{s / ft)}$$

$$\mathrm{d}\Delta t = \frac{1000 \cdot D_{2-1}}{3.28 \cdot Z_2 - Z_1} \text{ (in } \mu\text{s / ft)}$$

where:

D_{2-1} = time difference at the Z_2 depth less the time difference at the Z_1 depth (in ms),

Z_1 and Z_2 = depths of the knees limiting the zone to which the correction is applied (in meters).

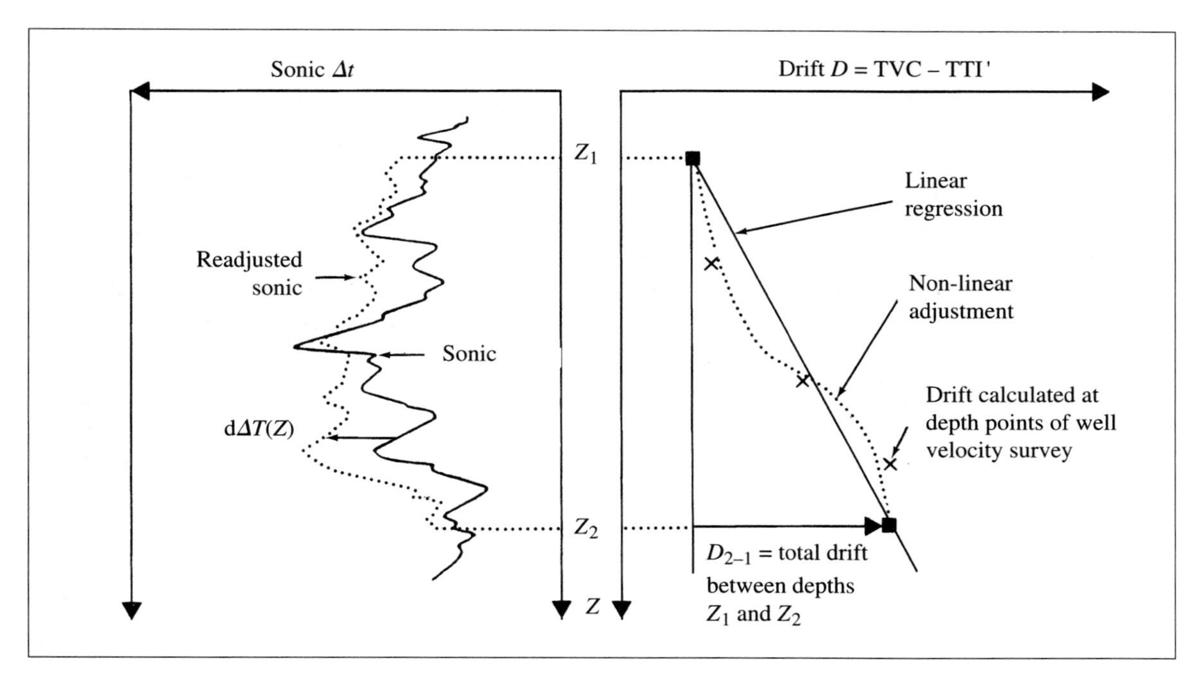

Fig. 3.9 Calibration of the sonic log with a check shot survey.

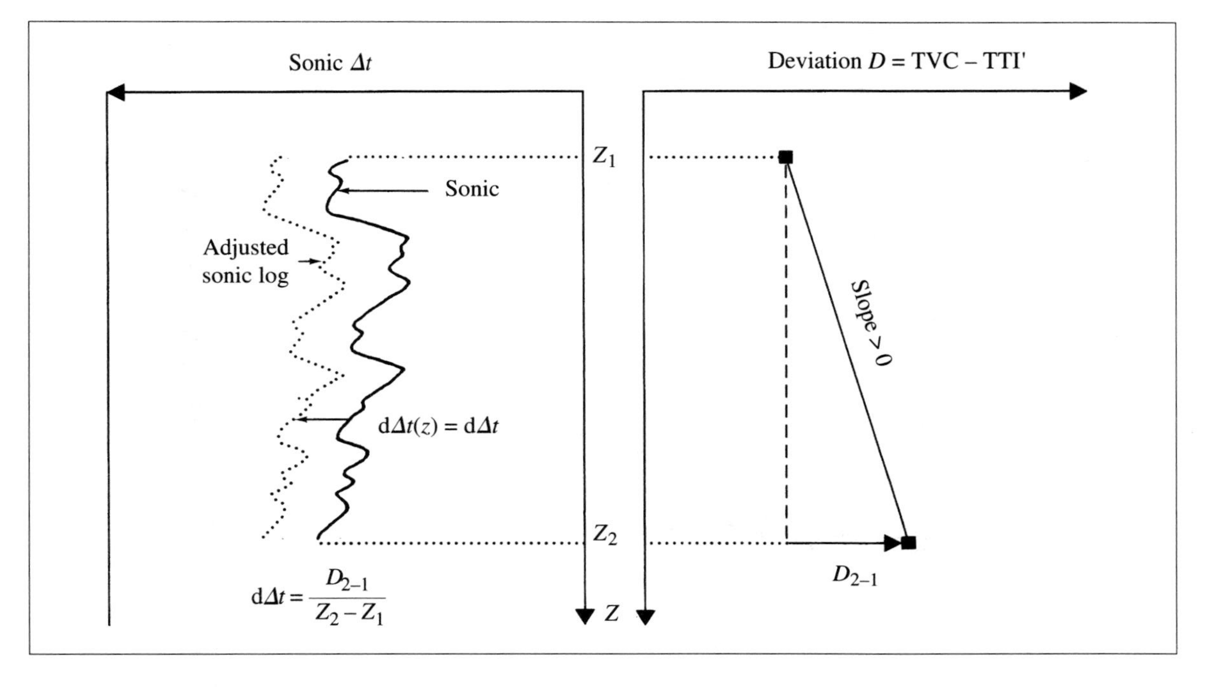

Fig. 3.10 Correction by the block shift method.

This method is generally used when there is a positive drift gradient (drift increasing with depth, $D_{2-1} > 0$) or possibly a negative gradient but with very low Δt values.

b. Minimum Δt method (Fig. 3.11)

In this method, the adjustement is proportional to the value of Δt itself. Below a chosen minimum Δt no correction should be applied. Above this value, the correction is a function of Δt.

Since the overall correction in a zone between two "knees" separated by the distance $Z_2 - Z_1$ is defined as D_{2-1} (in ms), the shift required at each time or depth value may be obtained as follows:

$$\Delta t_{cor} = G(\Delta t - \Delta t_{min}) + \Delta t_{min}$$

where $G = 1 + (D_{2-1} / T)$ and $D_{2-1} < 0$

$$T = \int_{Z_1}^{Z_2} d\Delta t(z)\, dz$$

where $d\Delta t = \Delta t - \Delta t_{min} > 0$.

The choice of Δt_{min} can be made by identifying zones in which there is no discrepancy between TTI' and the seismic TVC, or by using an empirical approach. It is indeed well known that, in shales or poorly-cemented formations of high porosity, the sonic log tends to give high Δt.

This method is utilised when there is a negative drift (negative D_{2-1} value) associated with a strong Δt.

3.2.4 Depth-to-time conversion of corrected sonic log data

Once corrected, the integrated sonic log leads to the calculation of a new Time vs. Depth relation. This allows the conversion of sonic and other types of logs into sections as a function of time (Fig. 3.12).

Figure 3.13 summarises the results obtained from the check shot survey presented in Fig. 2.11 and sonic calibration in Fig. 2.12. Corresponding logs are presented on a time scale in Fig. 3.12. The following abbreviations are used:

(1) one-way times (T),

(2) corresponding depths on the matched sonic log (Z/DP),

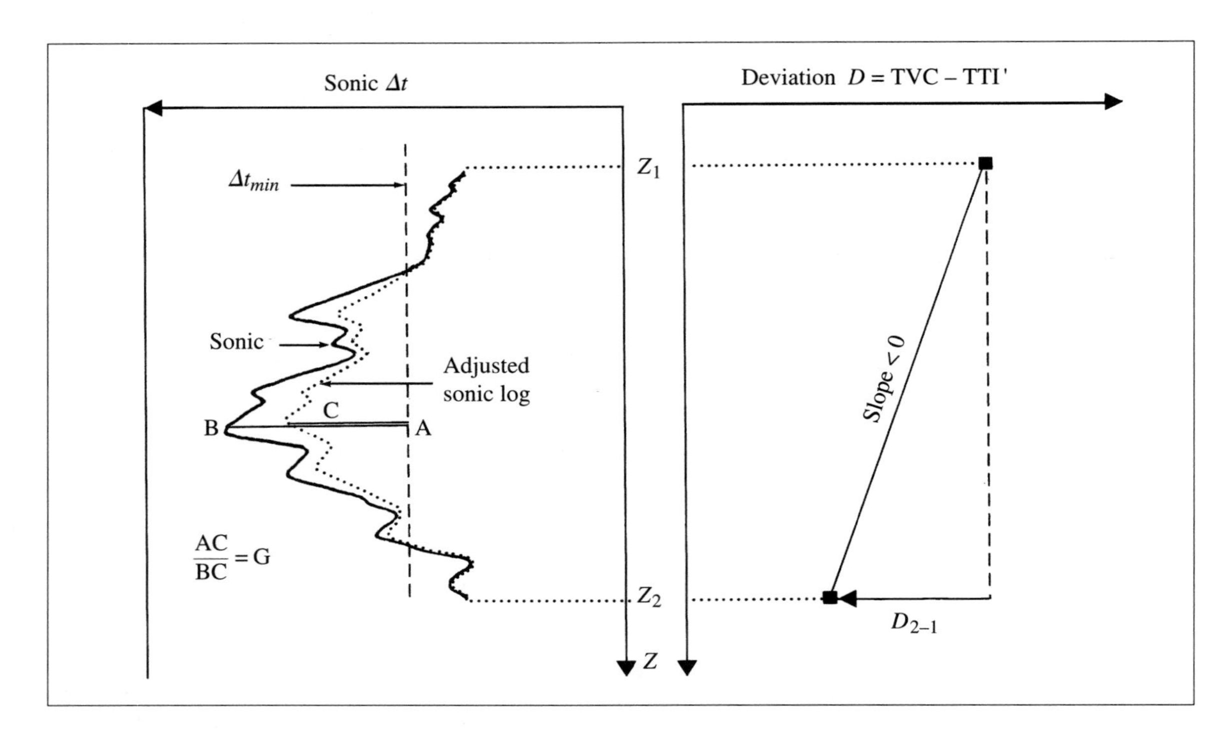

Fig. 3.11 Δt correction by the minimum Δt method.

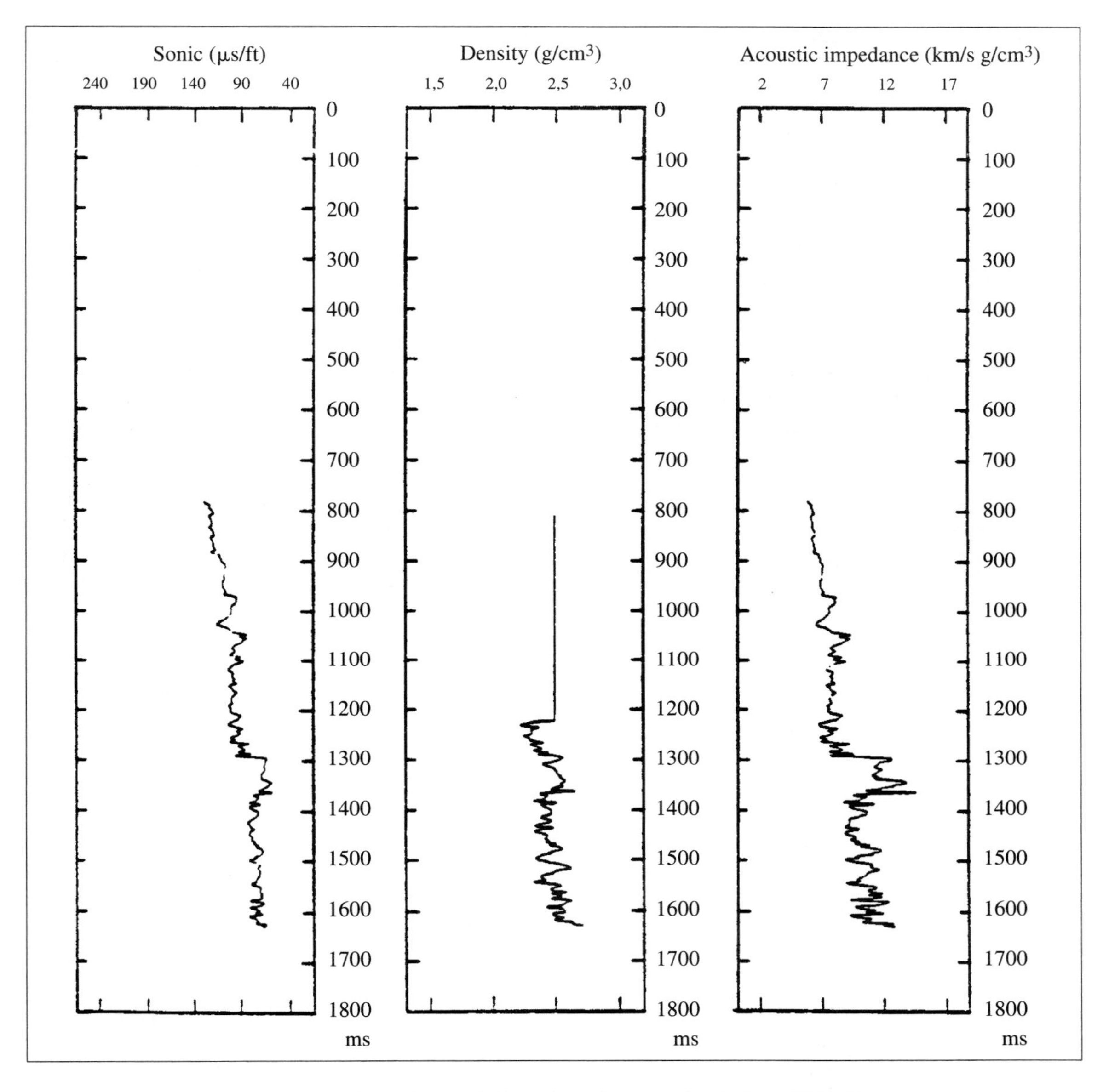

Fig. 3.12 Sonic, density and acoustic impedance logs vs seismic time. *(CGG Document)*

(3) interval velocity (IV),

(4) mean velocity (VM) and root-mean-square velocity (VMQ),

(5) calibrated curve $T = F(Z)$,

(6) calibrated velocity log, a squared function (CVL = continuous velocity log).

Figures 3.14 to 3.16 present an example of logs converted to seismic time. Such an approach allows the geophysicist to achieve a close match between the seismic signatures and the log signatures and to determine which geological units give rise to seismic markers.

The geophysicist calculates an acoustic impedance log, then a series of reflection coefficients in order to produce synthetic seismogram with different wavelets (see Chapter Four: Synthetic Seismic Records). In practice, the wavelets are generally either at zero phase or minimum phase. The synthetic seismograms are plotted with normal and/or inverse polarity.

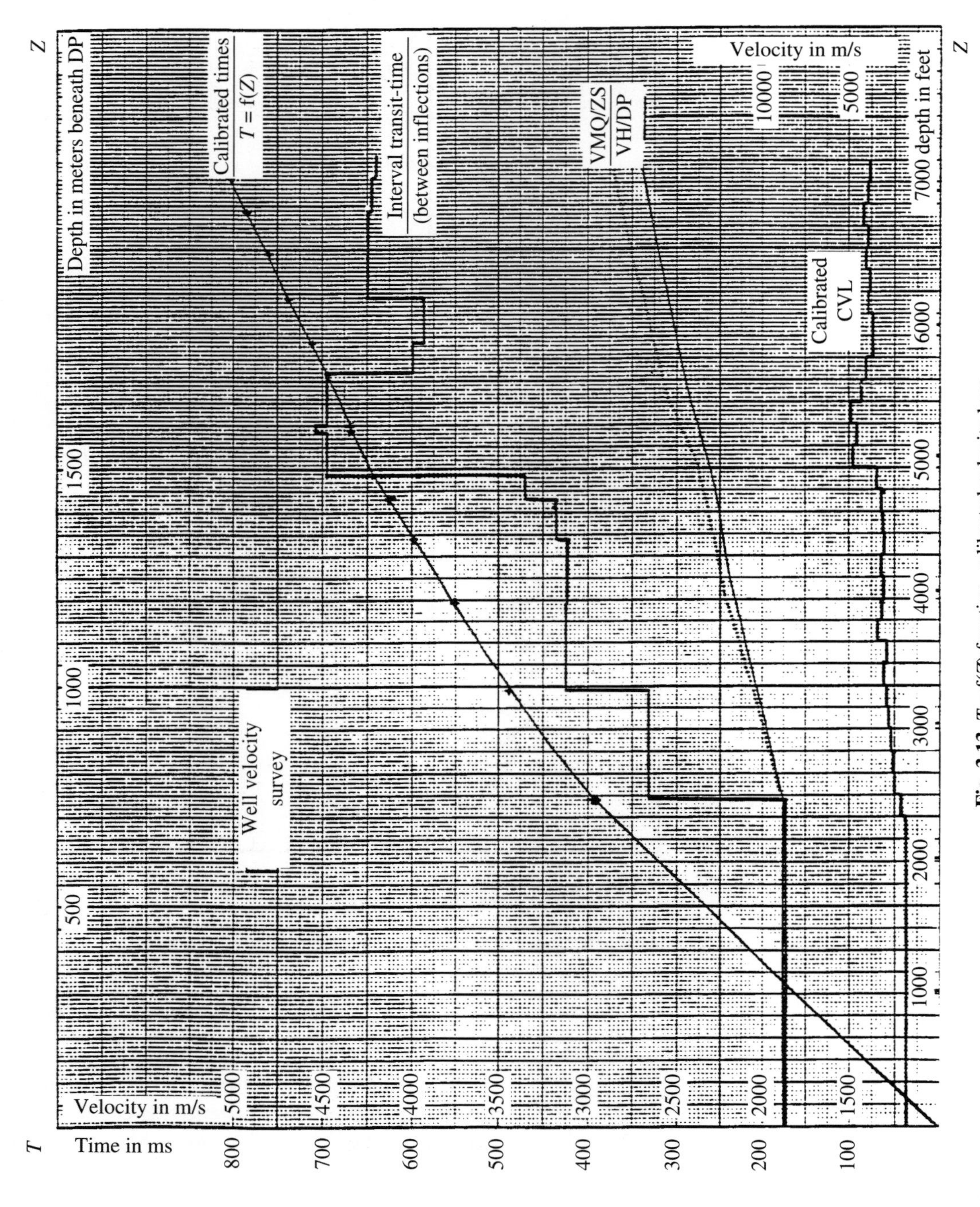

Fig. 3.13 T = f(Z) function, calibrated velocity logs.
(CGG Document)

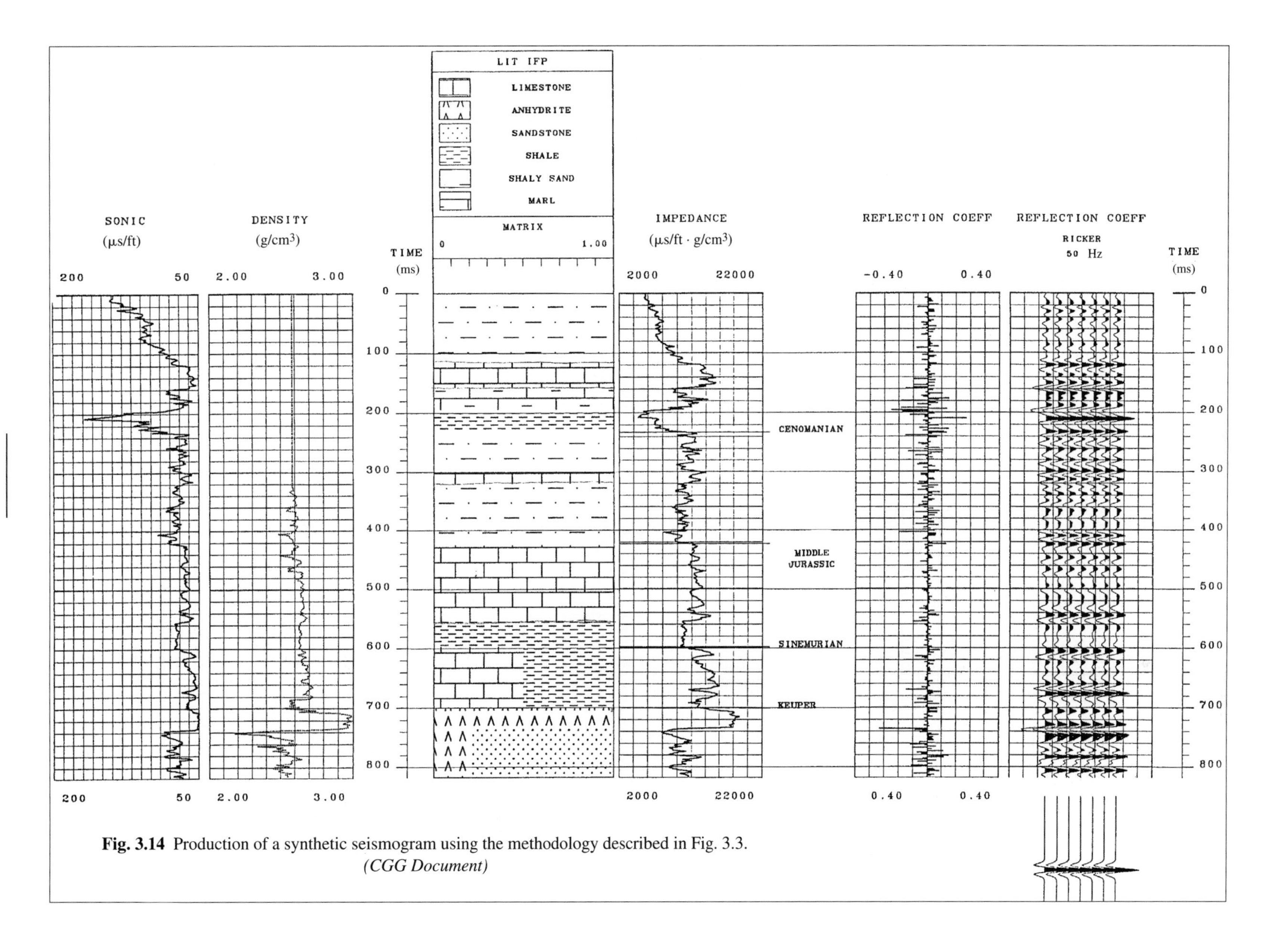

Fig. 3.14 Production of a synthetic seismogram using the methodology described in Fig. 3.3. *(CGG Document)*

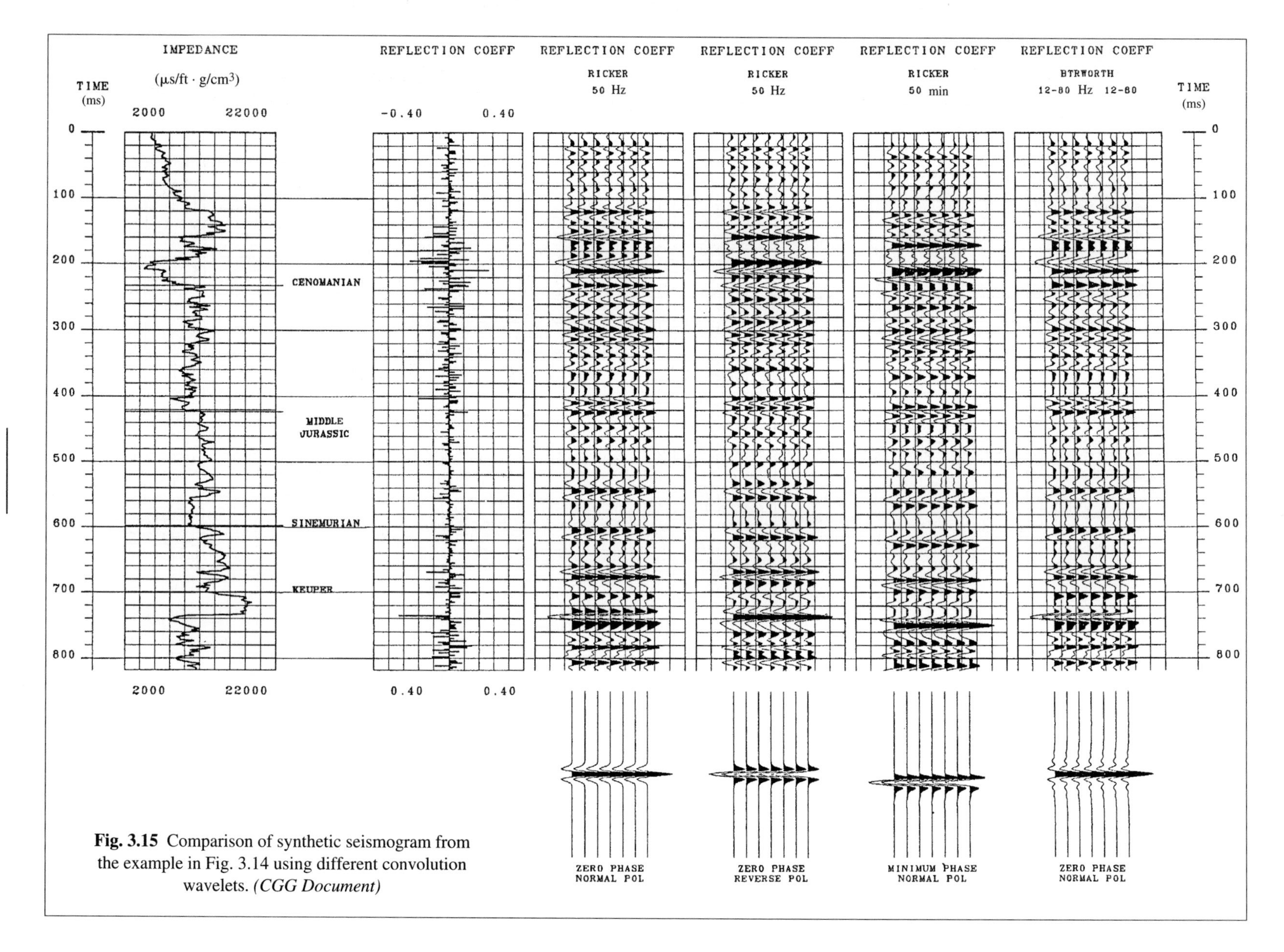

Fig. 3.15 Comparison of synthetic seismogram from the example in Fig. 3.14 using different convolution wavelets. *(CGG Document)*

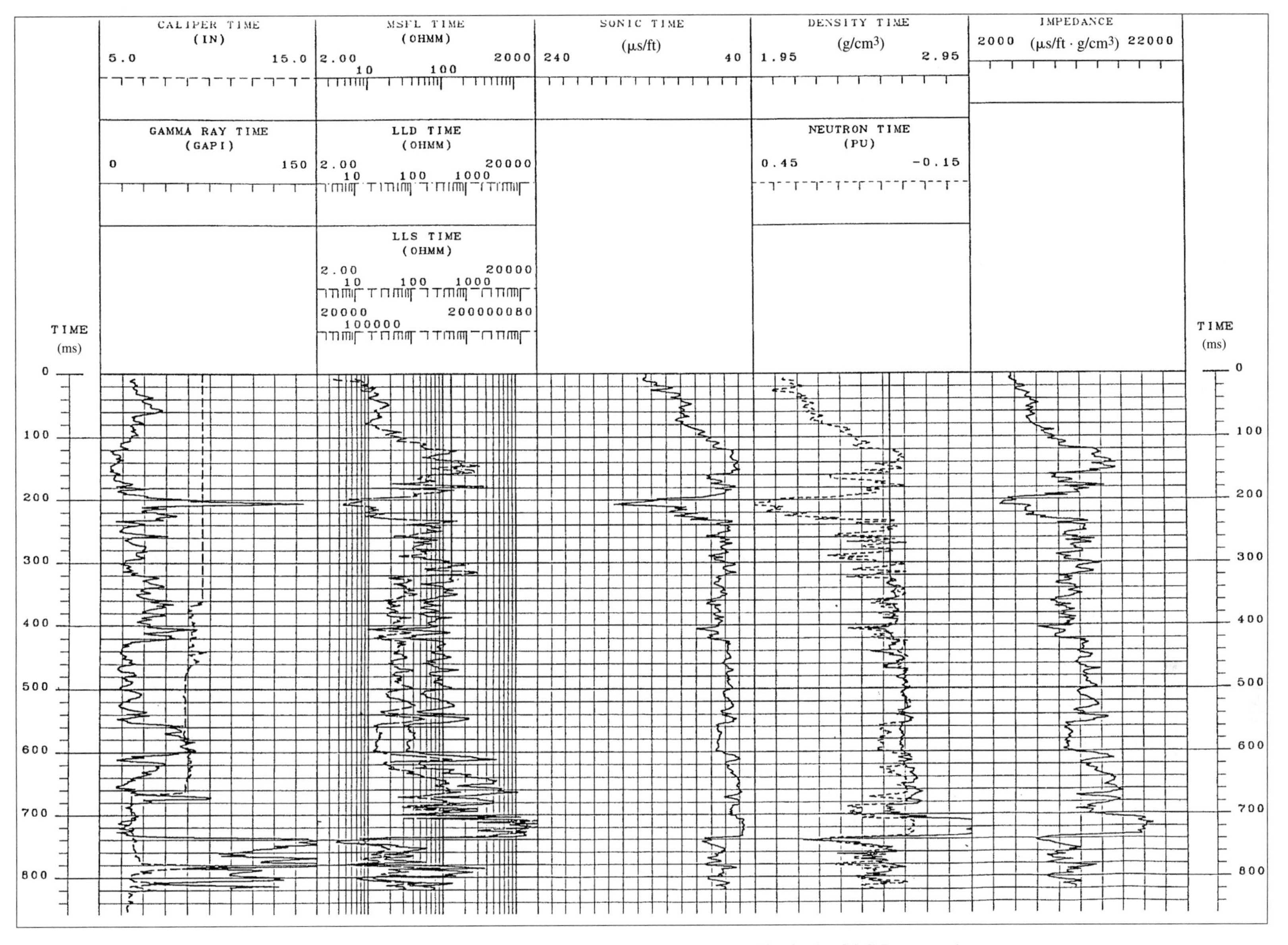

Fig. 3.16 Set of time logs from the well data presented in Fig. 3.14. *(CGG Document)*

Definition: SEG convention

A seismic section or a synthetic seismogram is said to be of "Normal Polarity" (SEG convention) if an increase in the acoustic impedance is represented by a negative phase (or white). Reverse polarity is the opposite case (Fig. 3.17).

The SEG convention is not always respected. It is therefore recommended to verify the polarity definition of each document.

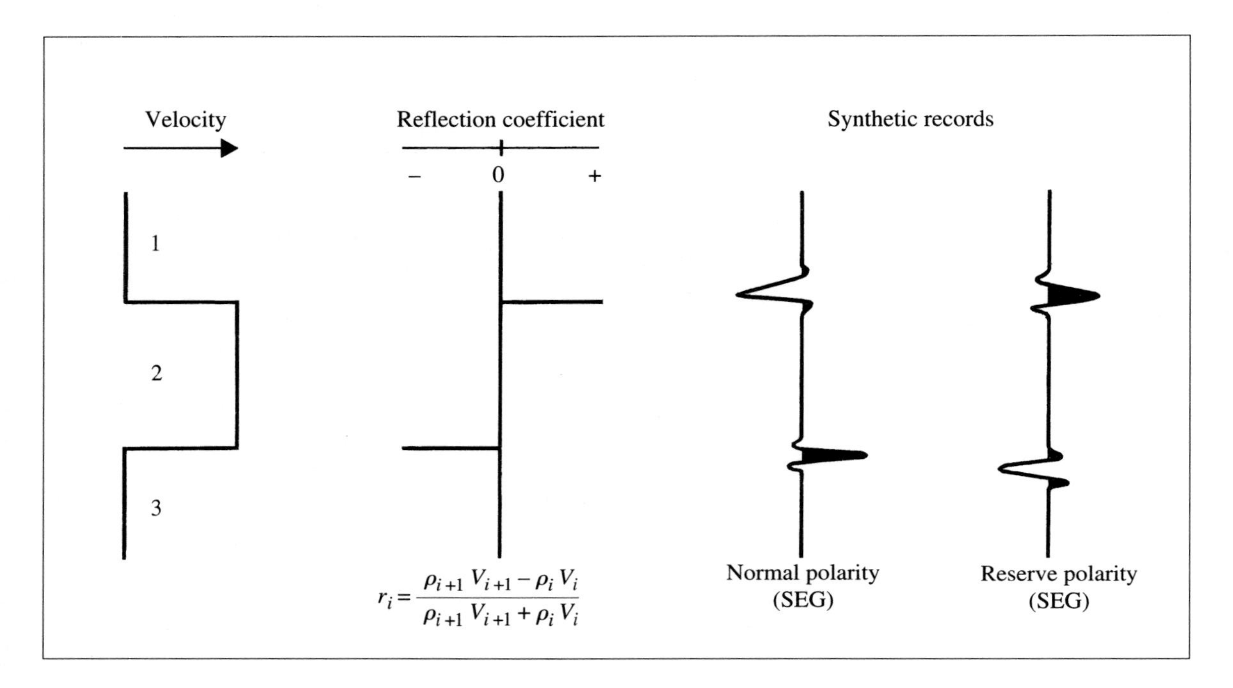

Fig. 3.17 Normal and reverse polarities.

Chapter **4**

SYNTHETIC SEISMOGRAMS

Synthetic seismograms calculated from an acoustic impedance model, or more simply from a velocity model based on a geological model or well logs are simulated traces comparable to the seismic sections obtained after processing of seismic data (i.e. surface seismic reflection after CDP stacking or VSP stacked traces). They are used to identify seismic markers and to calibrate surface seismic events in time and amplitude.

4.1 CONVENTIONAL SYNTHETIC SEISMOGRAMS

Most conventional synthetic seismograms are calculated from well logs.

Once the log (sonic and density) corrections have been made and the sonic data calibrated using the vertical times (TVC) obtained by well velocity surveying, it then becomes possible to convert the depth profiles of corrected velocity $V(Z)$ and density $\rho(Z)$ into $V(t)$ and $\rho(t)$ as a function of two-way travel time (TWT). These logs are sampled with the same interval used in the seismic survey (i.e. 1 or 2 ms). From this, a discrete series of velocities V_i and densities ρ_i can be obtained as a function of time.

Using velocity and density logs (V_i and ρ_i), it is then possible to calculate synthetic seismograms with frequencies and seismic sampling intervals corresponding to different acquisition geometries, with of without multiple events and taking into account transmission losses and/or the intrinsic attenuation of the medium. By acquisition geometry we mean the relative positions of the sources and receivers.

The simplest synthetic seismogram is the series of reflection coefficients r_i:

$$r_i = \frac{\rho_{i+1} V_{i+1} - \rho_i V_i}{\rho_{i+1} V_{i+1} + \rho_i V_i}$$

This series would correspond to a seismic reflection record obtained at normal incidence, with a source and a receiver at the surface, and from a geological structure consisting of planar horizontal layers of homogeneous and isotropic composition. If the source signal actually corresponded to a Dirac impulse, there would be no multiples nor any losses due to intrinsic attenuation and transmission through the traversed layers.

To compute realistic shot-firing, the impulse seismic profile is convolved using a signature representing the source signal (*cf.* Figs. 3.14 and 3.15). The synthetic seismogram so obtained is directly comparable to a seismic reflection trace after CDP stacking.

The synthetic seismogram without multiples enables the identification of reflectors on surface seismic profiles. Since CDP stacking is carried out in order to attenuate the multiples, matching is generally better using a synthetic seismogram without multiples than when multiples are present.

The differences between seismic reflection profiles and synthetic seismograms can have several origins, notably:

- *Shortcomings in the seismic processing*

  These may arise from various phenomena such as noise, certain types of amplitude anomalies or the presence of residual multiples and/or conversion waves on the seismic section. In fact,

P wave seismic sections could well emphasise seismic horizons associated with conversion waves and lead to errors of interpretation. The conversion rate will be that much greater as the offset increases and, as a result, the angles of emergence will be larger. Fig. 4.1 illustrates the phenomenon.

The source used is a detonating cord spread out over 50 m. The spread is an end-on spread, with a 75 m offset and 48 traces spaced 50 m apart. The seismic section is a 24 fold coverage CDP stack. Fig. 4.1 shows successively, from right to left, the complete coverage and partial coverages obtained by selecting the seismic raypaths corresponding to angles of emergence in the ranges 0-10°, 10-20° and 20-30°. Certain seismic horizons appear or disappear according to the angle of emergence. The time horizon A, situated at 550 ms on the complete multiple coverage section, can be attributed to conversion waves since it only appears for angles of emergence greater than 20°. On the 0-10° stack, a time horizon B is seen at 1.060 s which occurs neither on the normal CDP stack nor at larger angles of incidence. The figure shows that correlations between the simplified velocity log and the P wave section for angles in the range 0-10° are much better than correlations with the complete P wave multiple coverage.

- *Imperfections in the log data*

These imperfections have several origins: errors in measurements, uncertainties in the corrections, or quite simply a lack of data. In the calculation of a synthetic seismogram, the lack of data may arise from either the sonic or the density log.

In many cases, a lack of density measurements makes it necessary to calculate a synthetic seismogram using velocities alone (or transit time), or leads to the replacement of a density log by synthetic density values (see Chapter One: density well logging). The absence of velocity information can also lead to the use of other types of logs, for example resistivities (see Chapter One: sonic well logging).

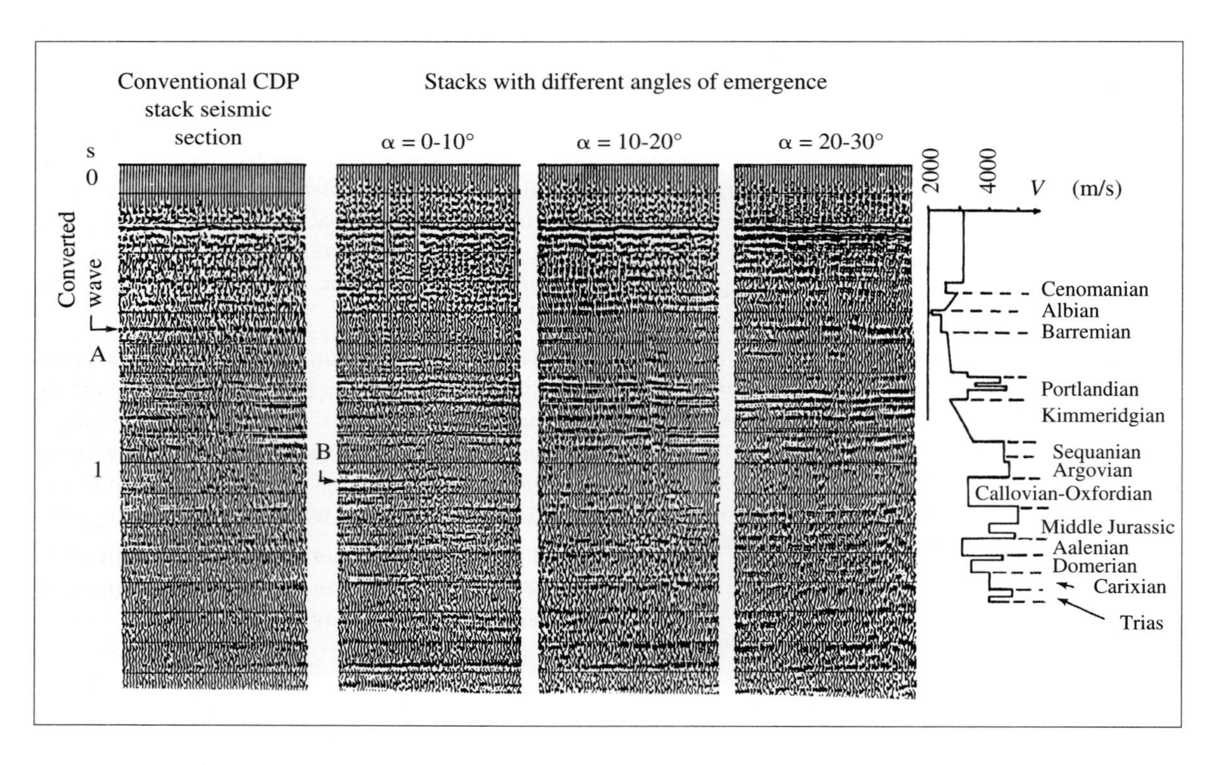

Fig. 4.1 Conversion waves on a P-wave seismic section.
(Coppens et al., 1985)

The procedure of replacing logs can only be regarded as a last effort solution.

- *Calibration errors*

  Calibration errors are basically due to a poor estimation of the drift curve, a poor positioning of the "knees" or a poor choice of reference plane.

- *Poor choice of the seismic signal* used to carry out convolution of the synthetic impulse seismogram.

  In practice, the synthetic pulse is convolved by a given signal chosen in the surface seismics frequency band using a phase hypothesis (zero or minimum phase). Zero phase is most commonly adopted. In order to take account of the variation of the wavelet with depth, a set of synthetic seismograms is produced containing signals associated with different frequency bands (for example: 10-20 Hz, 20-40 Hz, etc.); alternatively, the impulse seismic profile is filtered by a variable frequency filter as a function of time using filtering parameters defined during the processing of the surface seismic profile. There are, however, optimal methods for extracting the wavelet which avoid empirical trial and error, particularly in regard to phase problems. Such a method is described below.

- *Failure to take account of attenuation*

  To compensate for this drawback, either a synthetic seismogram with attenuation can be calculated after estimating an attenuation log, or a compensating attenuation filter (inverse Q filter) can be applied to the surface seismic data.

- *Different lateral investigations* for the surface seismic data and the logs.

  A synthetic seismogram without multiples can be used to calibrate well seismic data and to define the stratigraphic deconvolution operators which, when applied to the surface seismic data, enable a better matching at the well and an increase in the vertical resolution.

  It is equally possible to generate normal incidence seismic profiles with multiples by considering the transmission losses (Baranov and Kunetz, 1960; Wuenschel, 1960), and losses through absorption and dispersion.

4.2 VSP-TYPE SYNTHETIC RECORDS

Synthetic seismograms are generally calculated with the transmitter and receiver situated at the same point on the surface (i.e. using vertical two way integrated transit times). Different acquisition geometries may be envisaged, including the VSP-type configuration (Ganley, 1981).

VSP-type synthetic records enable an evaluation of the acoustic-signal variation with depth, while also indicating the distribution of markers which are the principal generators of multiples. They also give the best definition of the processing parameters to be applied to field VSP data in order to obtain a stacked trace or an optimal impedance log.

A stacked trace is directly comparable to a conventional synthetic seismogram (both transmitter and receiver at the surface) without multiples, and is therefore analogous to a seismic reflection profile at the well after CDP stacking.

The differences between field and synthetic stacked traces may have diverse origins, as already mentioned in the preceding section (*cf.* Conventional synthetic seismograms).

The quality of a stacked trace depends on the performance of the algorithms employed for wave separation and deconvolution, and also on the choice of stacking corridor. The result is strongly dependent on the quality of the data and the signal-to-noise ratio.

Figure 4.2 shows a synthetic seismogram with normal incidence obtained by convolving the series of reflection coefficients derived from a velocity model using a zero-phase signal in the 12-120 Hz frequency band, and a VSP-type synthetic seismic section in normal incidence with and without addition of noise.

In a VSP, the source is at the surface and the well geophone occupies different positions in the well. A synthetic VSP is a series of synthetic sections corresponding to different acquisition geometries (source at surface and well geophone at variable depth positions). The horizontal axis represents the depth of the well geophone, the scale in Z represents the variations in depth between two well geophone positions, while the vertical scale represents the listening time. The impulse utilised

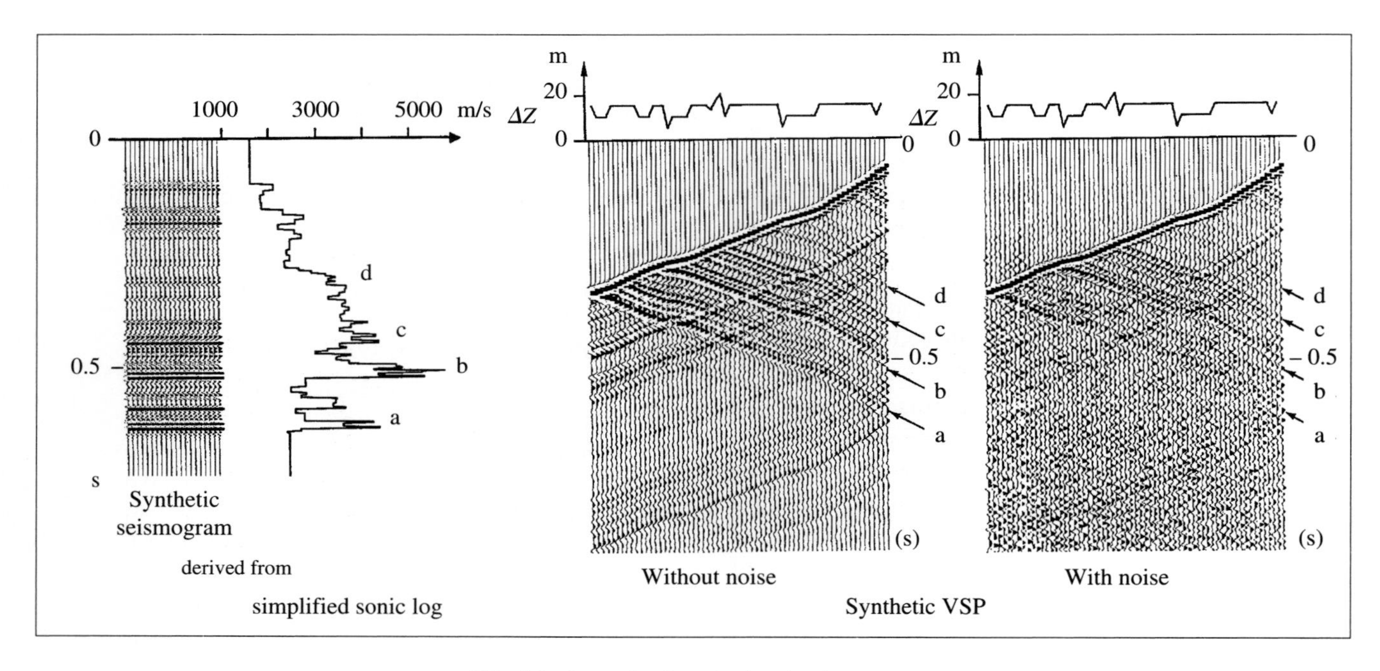

Fig. 4.2 Example of synthetic seismic sections.
(Mari et al., 1990)

for convolution of the different synthetic seismic sections is an impulsive source signature recorded by a reference receiver.

Figure 4.3 shows the observable differences between the synthetic seismogram and the VSP stacked traces of the preceding example; the comparison is illustrated in terms of choice of separation filter and signal-to-noise ratio.

This figure compares the VSP corridor stack obtained using different wave-separation filters with a conventional filtered synthetic seismogram in the seismic frequency band. Trace by algorithms (Mari et al., 1990) are used for processing of the VSP data and wave separation. The trace-by-trace algorithms are either a Wiener filter, or a simple solution filter (averaging or anti-averaging of VSP data after flatterning) or a least-squares method for solving the wave propagation equation in a homogeous medium.

The comparison shows that the results are equivalent as regards the major markers, but slight differences remain in the details.

The VSP corridor stack derived from a field recording is used to calibrate the surface seismic data after matching with the well logging information (generally replotted as a synthetic seismogram).

If major differences are observed, it is advisable to carefully examine the selected processing parameters:

(1) wave separation filters with an excessive number of terms give smoothed solutions and are liable to eliminate markers with very weak amplitudes,

(2) the choice of stacking corridor is equally critical to the extent that markers associated with upgoing markers may appear if the chosen corridor is too wide,

(3) if the deconvolution operator is badly chosen, there is a risk of producing artefacts,

(4) otherwise, differences can be linked to differences in vertical and lateral resolution between the logs and the well seismic data as shown in Fig. 4.4.

Conventionally, the vertical resolution *VR* is taken as equal to one quarter of the wavelength. The lateral resolution *LR* before migration, is estimated as equal to the diameter of the first Fresnel zone.

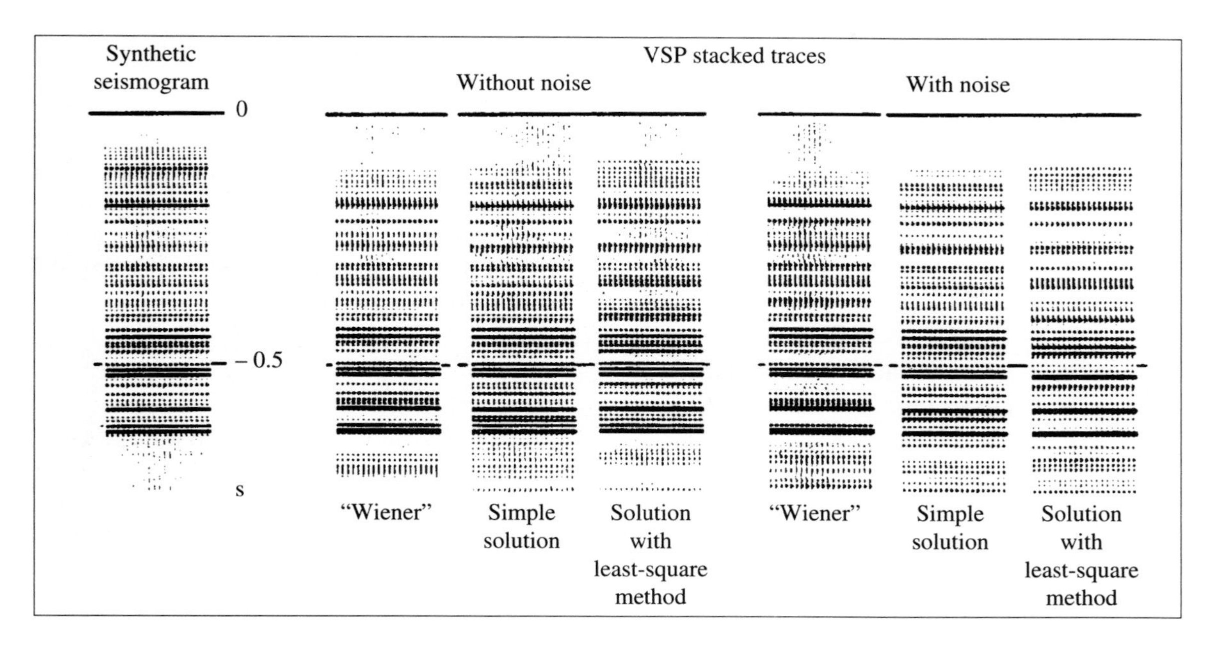

Fig. 4.3 Comparison of different stacked traces, showing influence of wave-separation filtering. *(Mari et al., 1990)*

In VSP, the vertical resolution is of the order of a few meters and the lateral resolution a few hundred meters. In well logging, the lateral resolution (depth of investigation) is a few tens of centimeters and can be taken as equal to the diameter of the first Fresnel zone.

Vertical and lateral resolutions are functions of the frequency used, the source-marker distance *DS*, the receiver-marker distance *DR* and the velocity *V* of the medium. For the sonic tool, it is considered that *DR* = *DS* = half the transmitter-receiver distance. The vertical resolution *VR* is given as follows:

$$VR = \frac{\lambda}{4} = \frac{V}{4f}$$

V = Interval velocity

f = Dominant frequency

λ = Wavelength

Lateral resolution *LR*:

$$LR = 2\sqrt{\frac{V}{f}\frac{DS \cdot DR}{DS + DR}}$$

where *DS* is the source-marker distance,
DR is the receiver-marker distance.

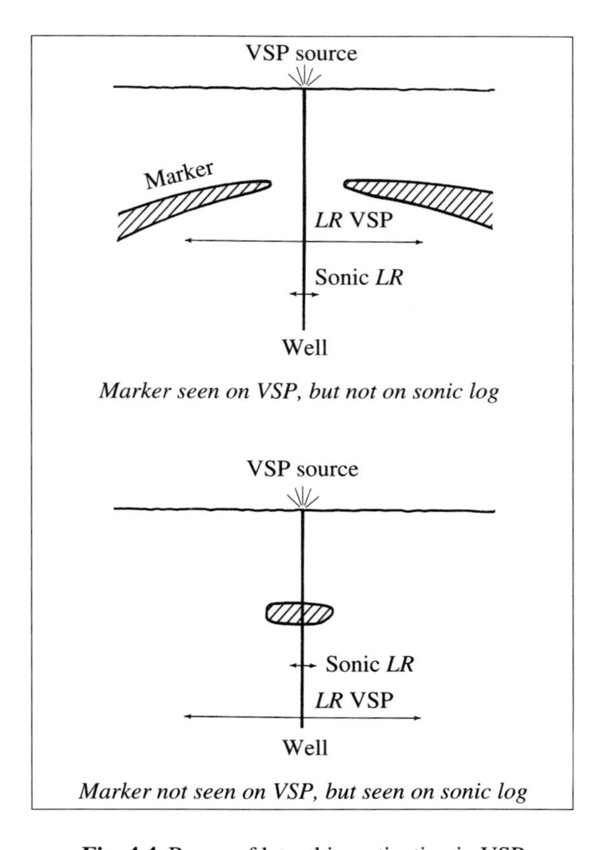

Fig. 4.4 Range of lateral investigation in VSP and sonic logs.

Examples:

VSP:

$V = 2500$ m/s

$f = 75$ Hz $\quad VR \simeq 8$ m

$DS = 1000$ m $\quad LR \simeq 50$ m

$DR = 200$ m

Sonic log:

$V = 2500$ m/s

$f = 15$ kHz

$DS = DR = 0.5$ m $\quad LR \simeq 0.2$ m

$DS = DR = 1$ m $\quad LR \simeq 0.6$ m

$DS = DR = 2$ m $\quad LR \simeq 0.8$ m

$DS = DR = 5$ m $\quad LR \simeq 1.3$ m

4.2.1 Seismic traces and synthetic seismograms

The seismic trace $\underline{Y}$ obtained after multiple coverage type processing and migration where necessary, can be taken as a set of coefficients $\underline{R}$ convolved by the wavelet $\underline{W}$. The $\underline{Y}$ vector represents the stacked trace vector composed of $(N+1)$ samples denoted y_j to y_{j+N}, y_j being the first sample considered. The $\underline{R}$ vector represents the vector series of reflection coefficients composed of $(N+1)$ samples denoted r_j to r_{j+N}. The wavelet is represented by a $\underline{W}$ vector composed of $(m+1)$ samples denoted W_0 to W_m. Using these notations, the relationship linking the stacked trace $\underline{Y}$, the wavelet $\underline{W}$ and the series of $\underline{R}$ coefficients can be written in the form of a matrix:

$$\underline{\underline{R}}\,\underline{W} = \underline{Y}$$

$$\begin{bmatrix} r_j & r_{j-1} & \cdots & r_{j-m} \\ r_{j+1} & & & \vdots \\ \vdots & & & r_j \\ \vdots & & & \vdots \\ r_{j+N} & \cdots & \cdots & r_{j-m+N} \end{bmatrix} \begin{bmatrix} W_0 \\ \vdots \\ W_m \end{bmatrix} = \begin{bmatrix} y_j \\ \vdots \\ y_{j+N} \end{bmatrix}$$

To take account of the wavelet with depth, this equation can be applied locally for a given time interval.

Estimating the $\underline{W}$ wavelet to calculate a synthetic seismogram

The $\underline{W}$ wavelet is extracted from the seismic section by deconvolving the seismic trace $\underline{Y}$ using the set of reflection coefficients $\underline{R}$. The set of reflection coefficients (source and receivers in the same position) is obtained from density ρ_b and velocity V logs recorded at the well using the following relation $\underline{R}\,(r_j)$ with:

$$r_j = \frac{\rho_{j+1} V_{j+1} - \rho_j V_j}{\rho_{j+1} V_{j+1} + \rho_j V_j}$$

Over a given interval, the optimal wavelet which minimises the least-square deviations between the observed seismic trace and the synthetic records (i.e. the convolution of the set of reflection coefficients using the wavelet) is obtained from the following formula:

$$\underline{W} = (\underline{\underline{R}}^T \underline{\underline{R}})^{-1} \underline{\underline{R}}^T \underline{Y}$$

where $\underline{\underline{R}}^T$ is the transposed matrix derived from matrix $\underline{\underline{R}}$.

The quality of the estimated wavelet can easily be evaluated by calculating a simulated coefficient between the recorded seismic trace $\underline{Y}(y_j)$ and the synthetic seismogram $\underline{X}(x_j)$. A coherence coefficient C can be provided by the relation:

$$C = \frac{\sum_{j=1}^{N} x_j y_j}{\sqrt{\sum_{j=1}^{N} x_j^2 \sum_{j=1}^{N} y_j^2}}$$

Example of a synthetic seismogram produced using an optimal wavelength

Figure 4.5 illustrates an example of a synthetic seismogram obtained from an acoustic impedance log produced from velocity and density logs.

The following steps are presented in succession:

A. Acoustic impedance log as a function of time (calculated from density and velocity logs).
B. Well-vicinity seismic profile (WVP).
C. Synthetic seismic section calculated with a time varying wavelet.
D. Series of reflection coefficients derived from acoustic impedance log.
E. Seismic trace at the well.

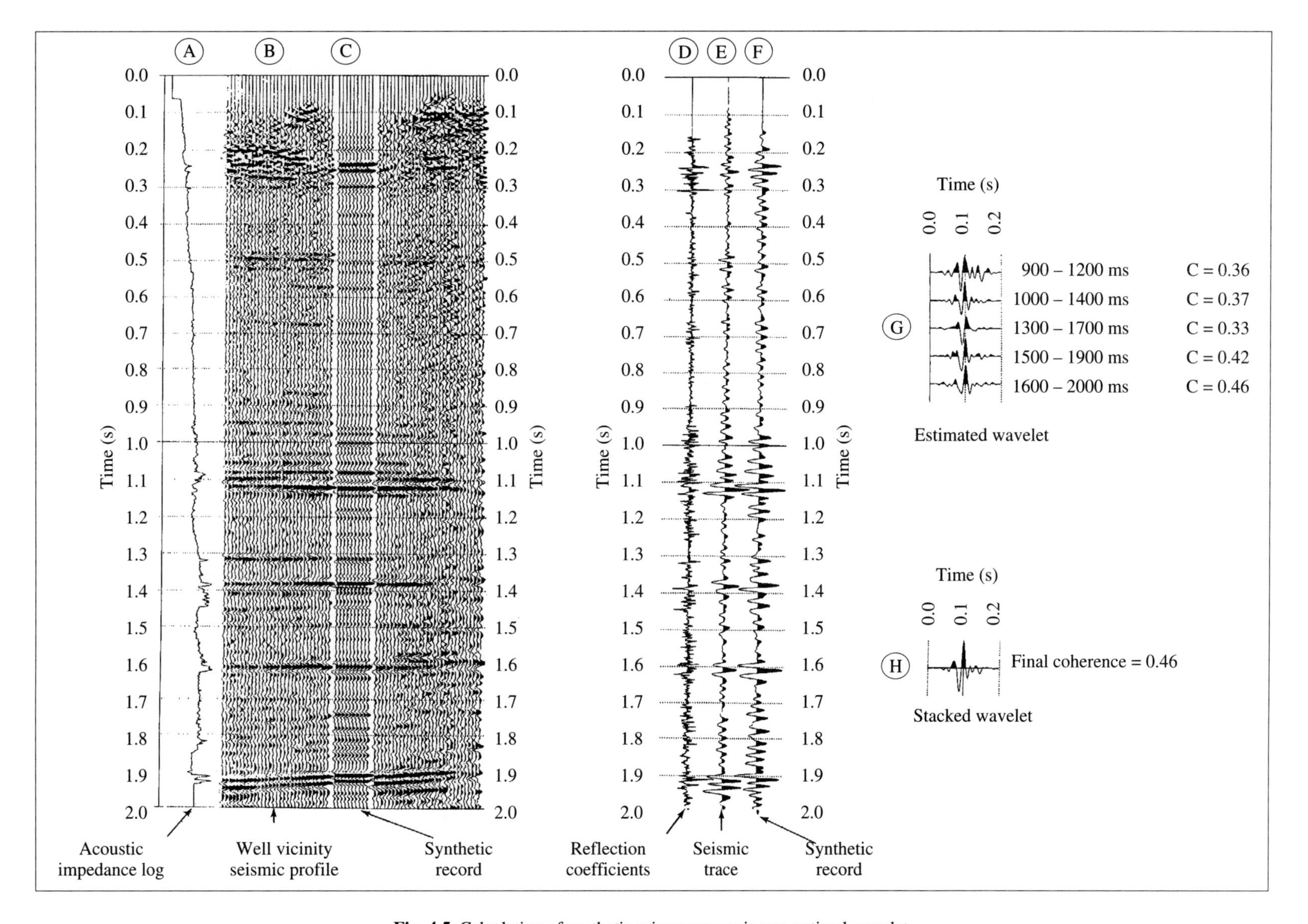

Fig. 4.5 Calculation of synthetic seismogram using an optimal wavelet. *(After Redanz et al., 1986)*

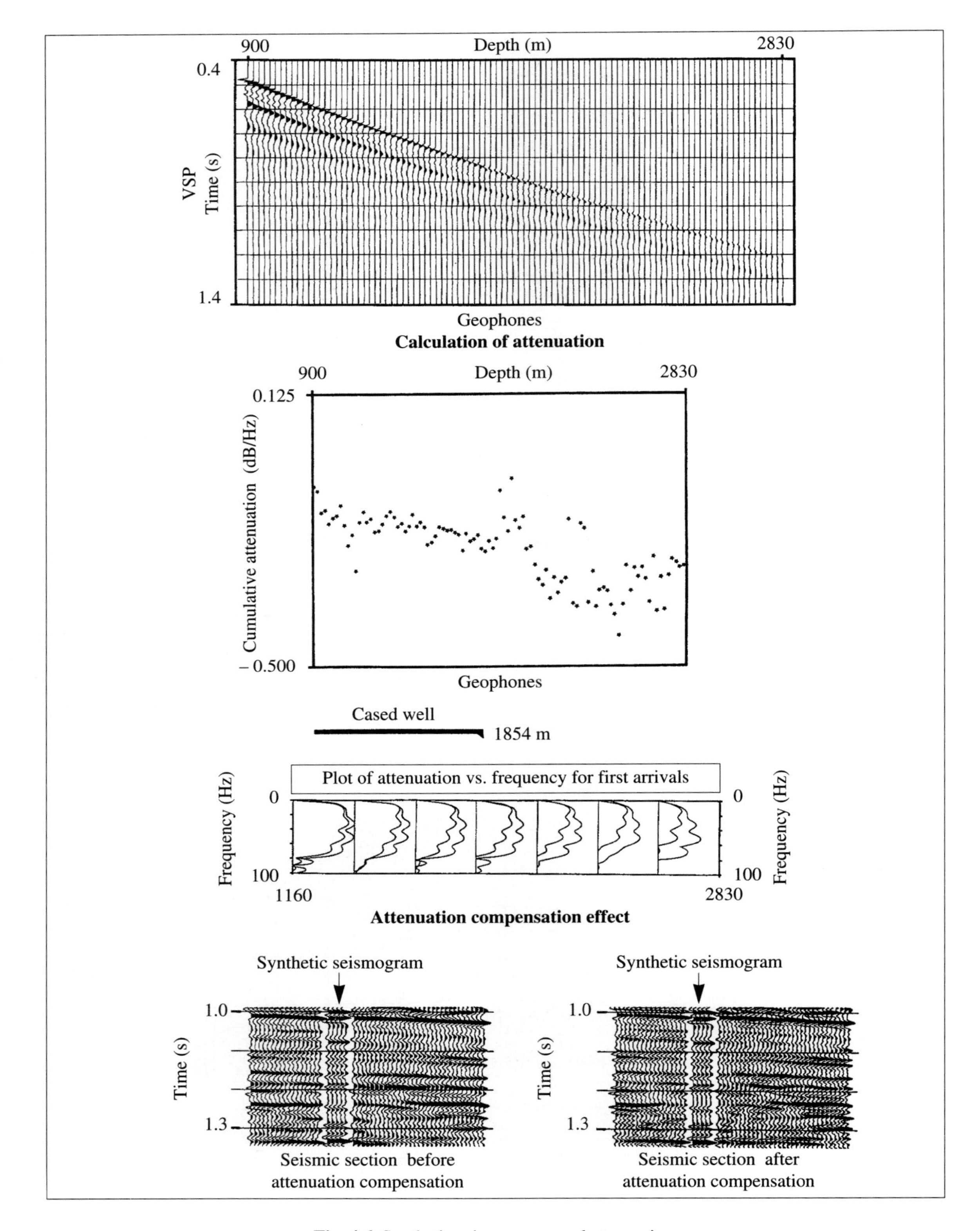

Fig. 4.6 Synthetic seismograms and attenuation.
(After Aleotti et al., 1989)

F. Synthetic seismogram.
G. Different estimates of the wavelet as a function of time.
H. Stacked wavelet.

The synthetic seismogram thus calculated with an optimal wavelet enables a good time fit for the seismic section at the well, even in zones having a poorer signal-to-noise ratio (particularly in the interval 1.7-1.9 s).

The wavelet calculated in the interval 0.9-1.2 s enables a good fit of the seismic horizons expressing a strong impedance anomaly situated around 1.15 s. It can also be seen that the amplitude of the seismic horizons situated at 0.25, 1.45 and 1.6 s are weaker in the real data than in the synthetic data. This could be linked to the fact that attenuation has not been taken into account in calculating the synthetic data.

4.2.2 Synthetic seismograms and attenuation effects

Aleotti and Angeleri (1989) have shown that taking account of the effect of attenuation improves the correlation between surface seismic data and downhole data (VSP and logs).

The example presented in Fig. 4.6 is taken from a seismic study carried out in the Adriatic Sea. Attenuation is measured on the basis of VSP data using the spectral ratio method.

The figure shows, from the top downwards, a VSP composed of hundreds of geophone positions situated between 900 and 2830 m, the cumulated attenuation curve expressed in dB/Hz as a function of depth, amplitude spectra of some VSP traces before and after recovery of the attenuation effect and also a comparison between the synthetic seismogram and a portion of the seismic section.

The cumulated attenuation curve shows an evident increase in the attenuation measurements when the well geophone is clamped in an open hole. This is linked to poor coupling of the well geophone. After compensation for attenuation effects, an overall improvement can be noted in the matching of phase and amplitude between the seismic section and the synthetic records. The effect is particularly marked at the top and base of the geological interval studied for the reflectors situated at 1.0 and 1.3 s. The compensation for attenuation is obtained by using attenuation compensatory filters based on the spectral equalization technique (Coppens and Mari, 1985; Gelius, 1987).

4.3 STRATIGRAPHIC DECONVOLUTION AND INVERSION PROCEDURES

4.3.1 Stratigraphic deconvolution

The aim of carrying out stratigraphic deconvolution on seismic sections is to arrive at the best possible series of reflection coefficients by eliminating the wavelet effect. If the stratigraphic deconvolution operator is denoted as $\underline{W}^{-1}$, the deconvolved seismic trace $\underline{\tilde{R}}$ can be expressed as:

$$\underline{\tilde{R}} = \underline{\underline{Y}}\,\underline{W}^{-1}$$

where $\underline{Y}$ represents the initial seismic trace.

At the well, the optimal operator minimising the least-square deviations between the series of reflection coefficients R and the deconvolved seismic trace $\underline{\tilde{R}}$ is written as follows:

$$\underline{W}^{-1} = (\underline{\underline{Y}}^T \underline{\underline{Y}})^{-1} \underline{\underline{Y}}^T \underline{R}$$

$\underline{W}^{-1}$ minimises the deviations between $\underline{R}$ and $\underline{\tilde{R}}$. The series of reflection coefficients $\underline{R}$ derived from the logs is filtered in the seismic frequency band.

This operation is equivalent to the calculation of a series of reflection coefficients $\underline{\tilde{R}}$ such that the seismic objective function Js is minimal:

$$Js\,(\underline{\tilde{R}}) = \left\| \underline{Y} - \underline{\underline{\tilde{R}}}\,\underline{W} \right\|^2 = Js\,(\underline{Y}, f(\underline{\rho V}, \underline{W}))$$

the deviation having been defined by a least square method and $\underline{W}$ representing the wavelet.

The $\underline{\tilde{R}}$ solution so obtained has a limited frequency content and is influenced by the signal-to-noise ratio. Before carrying out any stratigraphic deconvolution, it is necessary to improve the signal-to-noise ratio by extracting the noise from the seismic section. To this purpose, a technique of spectral matrix filtering can be used (Mari and Glangeaud, 1988).

The $\underline{\tilde{R}}$ solution can be integrated as a function of time using the formula:

$$(\rho V)_{j+1} = (\rho V)_j \, \frac{1 + \tilde{r}_j}{1 - \tilde{r}_j}$$

This leads to an estimation of the acoustic impedance log filtered in the seismic frequency band. The low-frequency component of the impedance log is lost during this processing.

The $\underline{W}^{-1}$ operator can either be applied as a unique value to the whole seismic section or can be varied from trace to trace. This stratigraphic deconvolution may be performed, by replacing the reflectivity function or the impedance values from well logs by equivalent logs derived from VSP recordings using the VSP stacked trace.

Example of stratigraphic deconvolution using well logs

Figure 4.7 presents:

A. Acoustic impedance log obtained from sonic and density logs.

B. Impedance log after application of a low-cut frequency filter designed to eliminate the very low frequencies (<5 Hz) which cannot be recorded by the surface geophone.

C. Surface seismic trace obtained at the reflection point nearest the well (around 200 m).

D. Acoustic impedance log obtained by stratigraphic deconvolution.

A comparison of the impedance logs presented in B and D shows that stratigraphic deconvolution enables the matching in time and amplitude of the seismic and logging data in the same frequency band as the surface seismic survey. Outside the seismic band, the amplitudes are no longer respected, notably in the 550-650 ms interval where the very-low-frequency component of the log could not be reconstituted.

Figure 4.8 shows the surface seismic data calibrated by well log data.

Figure 4.9 shows, from the top downwards, the unprocessed seismic section, the impedance section obtained after stratigraphic deconvolution with the operator calculated in the 300-600 ms interval and the reflectivity section convolved with a zero-phase signal. On the impedance section, it can be seen that horizons at 500 and 600 ms have an acoustic impedance that is practically constant laterally. In the 600-650 ms interval, acoustic impedance shows a major lateral variation marked by a change in the frequency content. Such trends may be linked not only to intrinsic acoustic impedance variations but also to the effects of varying bed thickness at the limits of seismic resolution.

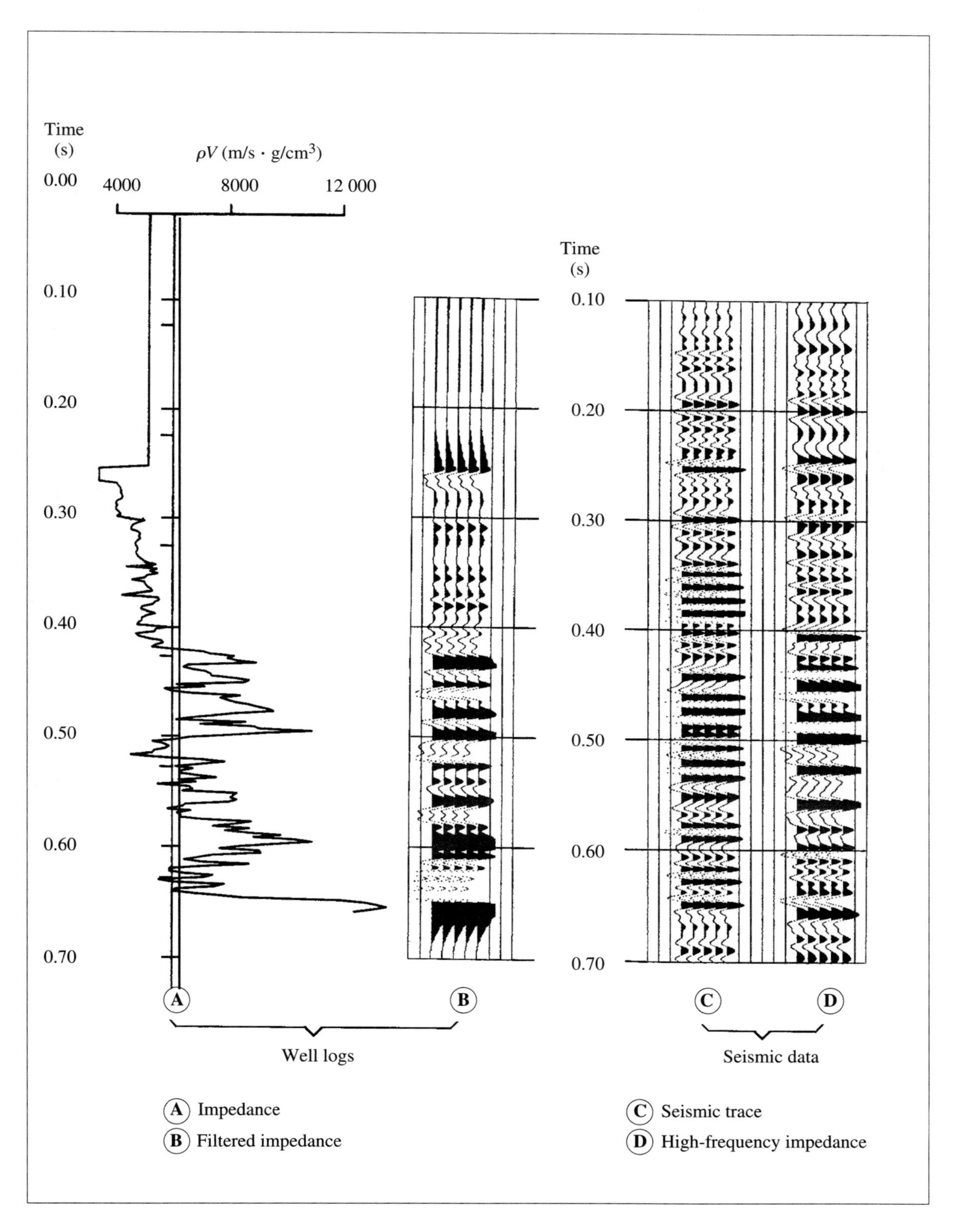

Fig. 4.7 Example of stratigraphic deconvolution of well data.
(Gaz de France-IFP Document)

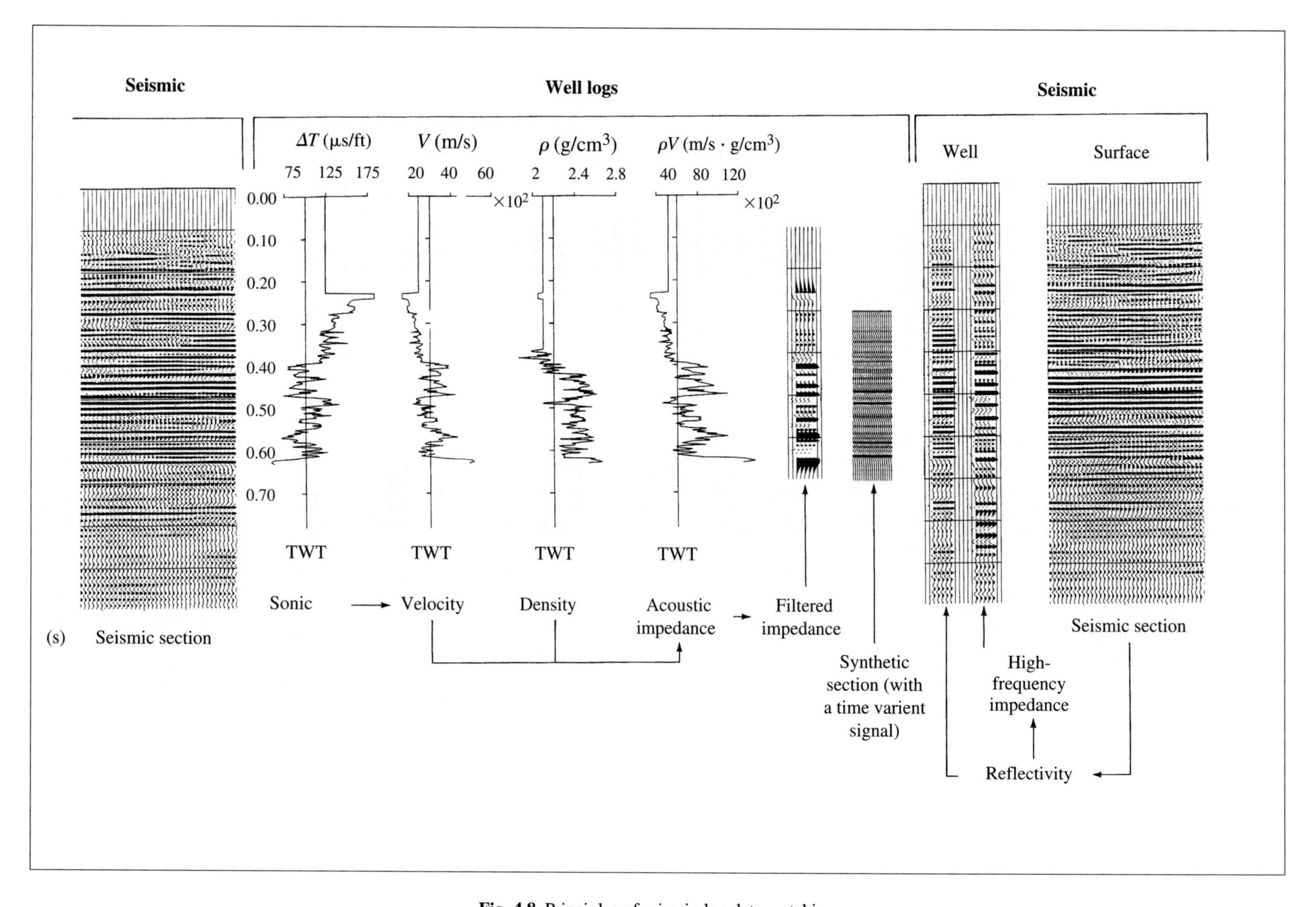

Fig. 4.8 Principles of seismic-log data matching. *(Gaz de France-IFP Document)*

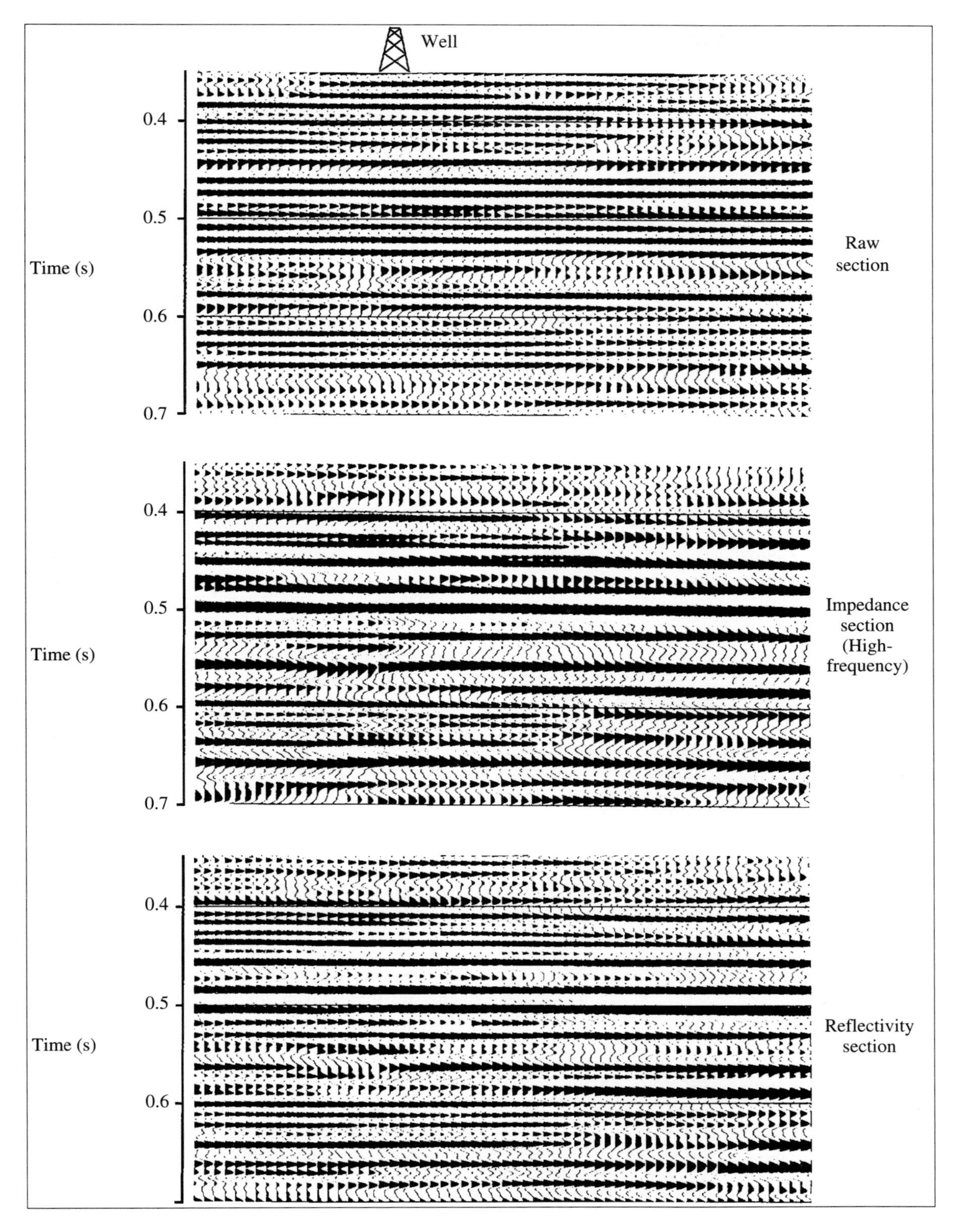

Fig. 4.9 Stratigraphic deconvolution applied to a seismic profile passing close to a well (Fig. 4.7).
(Gaz de France-IFP Document)

Example of stratigraphic deconvolution with VSP stacked trace

It is always useful to compare logs and seismic information in order to ensure a good interpretation. Figure 4.10 presents a complete correlation between the downhole data and the surface seismic data. The downhole data are made up of VSP measurements (see Chapter Two: Section 2.2.2: processing of VSP data, p. 89) and sonic logs. The surface seismics are made up of a conventional multiple coverage survey and a Well Vicinity Profiling (WVP) before and after deconvolution. The WVP is a seismic reflection profile with a fixed receiver array (Mari et al., 1987).

The stratigraphic matching indicated on the depth axis of the VSP is based on log data.

The following geological units are plotted

Kc	Kimmeridgian (limestone),
Km	Kimmeridgian (marl),
Oxf	Oxfordian (limestone),
Cal	Callovian (marl and limestone),
M. Jur.	Middle Jurassic (compact limestone),
Toa	Toarcian (shale),
Dom.-Het.	Domerian to Hettangian (interbedded clay/limestone),
Trias q.	Reservoir zone with interbedded sands and shales,
Trias d.	Dolomitic zone,
Trias g.	Shaley/dolomitic sandy zone.

It can be seen on the VSP that the seismic horizon corresponding to the top of the dolomitic Trias is masked by multiples stemming from the marker corresponding to the base of the Middle Jurassic, which is itself attenuated and deformed by multiples created by the horizon situated at 0.4 s in the Oxfordian. Both of these phenomena are plainly evident on the unprocessed Well Vicinity Profiling (WVP). By taking account of multiples, stratigraphic deconvolution enables the identification of events which are themselves affected by the multiples. In this way, correlation between surface seismics and well data is facilitated by the use of a deconvolved WVP section. Correlation between WVP and conventional surface seismic data proceeds via the deconvolved VSP profile then the untreated WVP. It is noteworthy that vertical seismic and well-vicinity profiles have equivalent vertical resolutions. This can only be achieved in both types of acquisition by using seismic sources with the same signal spectrum.

On both the CDP stack and the raw VSP, the H4 and H7 horizons are identified. They correspond to the stratigraphic picking carried out by the geologist.

4.3.2 Inversion or stratigraphic deconvolution using constraints based on non-seismic data

As shown above, stratigraphic deconvolution is equivalent to finding a series of reflection coefficients $\underline{R}$ (or the series of impedances $\underline{\rho V}$) and the wavelet $\underline{W}$ such that the objective function J ($\underline{R}$) is minimal.

The objective function J ($\underline{R}$) = J ($\underline{\rho V}$) represents the deviation between the seismic trace $\underline{Y}$ obtained after processing and the synthetic seismogram calculated from $\underline{\underline{R}}\ \underline{W}$ = f ($\underline{\rho V}$, $\underline{W}$). The function f ($\underline{\rho V}$, $\underline{W}$) represents the synthetic seismogram without multiples calculated with the impedance distribution $\underline{\rho V}$ and the wavelet $\underline{W}$ using a convolution model or the 1-D propagation equation.

Since the seismic surface data has a limited frequency content without any information about very low frequencies, a solution in terms of impedance does not yield an estimation of the low-frequency component of the impedance log. Such a solution can only be found by introducing prior assumptions with the help of an initial model ($\underline{\rho V}$o). Thus, a solution is sought that minimises the objective function:

$$J(\underline{\rho V}) = Js\,(\underline{Y}, f(\underline{\rho V}, \underline{W})) + \left\|\underline{\rho V} - (\underline{\rho V})o\right\|^2$$

The first term of the objective function ensures the optimal search for an acoustic impedance distribution that minimises the discrepancies between the observed seismic profile and the calculated synthetic section.

The second term places constraints on the solution, stabilises the response of the deconvolution or inversion to various phenomena (e.g. sensitivity to noise) and enables a recovery of the low-frequency signal.

The initial impedance distribution model ($\underline{\rho V}$)o is built using well log information (velocity and density)

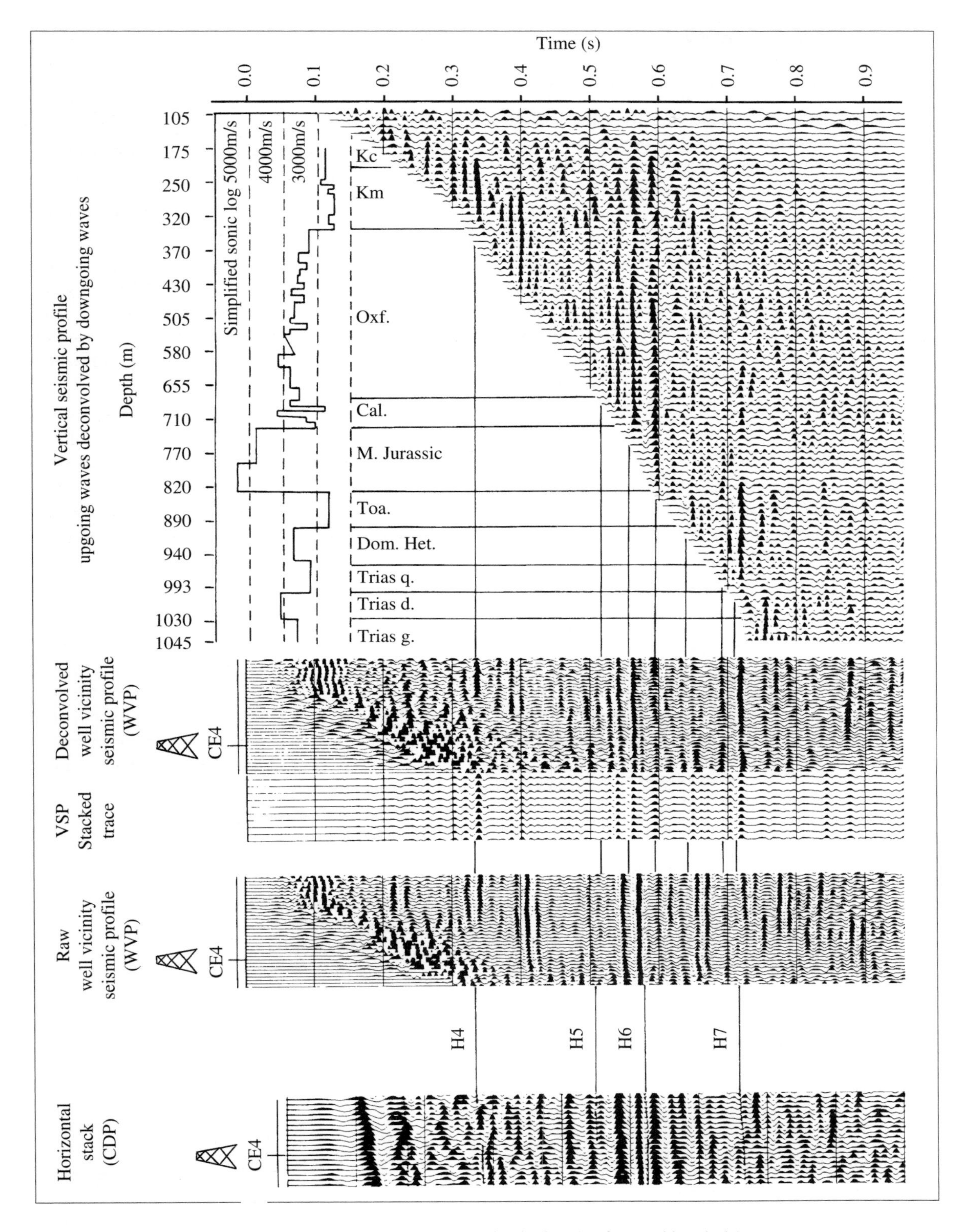

Fig. 4.10 Comparaison between seismic data (surface and borehole).
(After Mari et al., 1987)

and by picking seismic horizons that characterise the boundaries of the different layers in the geological model. This serves to produce a stratigraphic model.

The stratigraphic model is built from the seismic section after estimating the optimal wavelet and carrying out stratigraphic deconvolution. Velocity and density values are first attributed to each bed using the log data for a given well, and then the values are interpolated between different wells. Bed thicknesses are calculated using well log velocities and the seismically picked times corresponding to the bottoms and tops of beds. Once the initial model is defined and the wavelet estimated, computation of the impedance section can be carried out by perturbing the model and running successive iterations to minimise the objective function J ($\underline{\rho V}$), as shown in Fig. 4.11.

Example of inversion

Figure 4.12 illustrates the application of the *Interwell* (*IFP* registered trade mark) inversion software to a seismic profile operated on a carbonate reservoir (Brac et al., 1988).

Section A shows the seismic section and the position of the wells numbered 1 to 4; the arrow indicates the carbonate reservoir. The initial impedance distribution model assumes constant impedance layer-by-layer and is constructed using only the impedance values from well 1, the limits of the layers being obtained by picking after stratigraphic deconvolution.

Section B presents the distribution of optimal acoustic impedance after inversion.

Figure C illustrates a comparison between the acoustic impedance section derived from inversion (dotted line (a)) and the acoustic impedance from downhole logging (solid line (b)) in well 4. The excellent match between the two logs indicates the high quality of the processing utilised.

Section D shows the seismic reflectivity section associated with the optimal acoustic impedance distribution presented in section B. The wavelet used has a wide band and a zero phase. The section calculated in this way shows a distinct improvement in vertical resolution associated with a good signal-to-noise ratio.

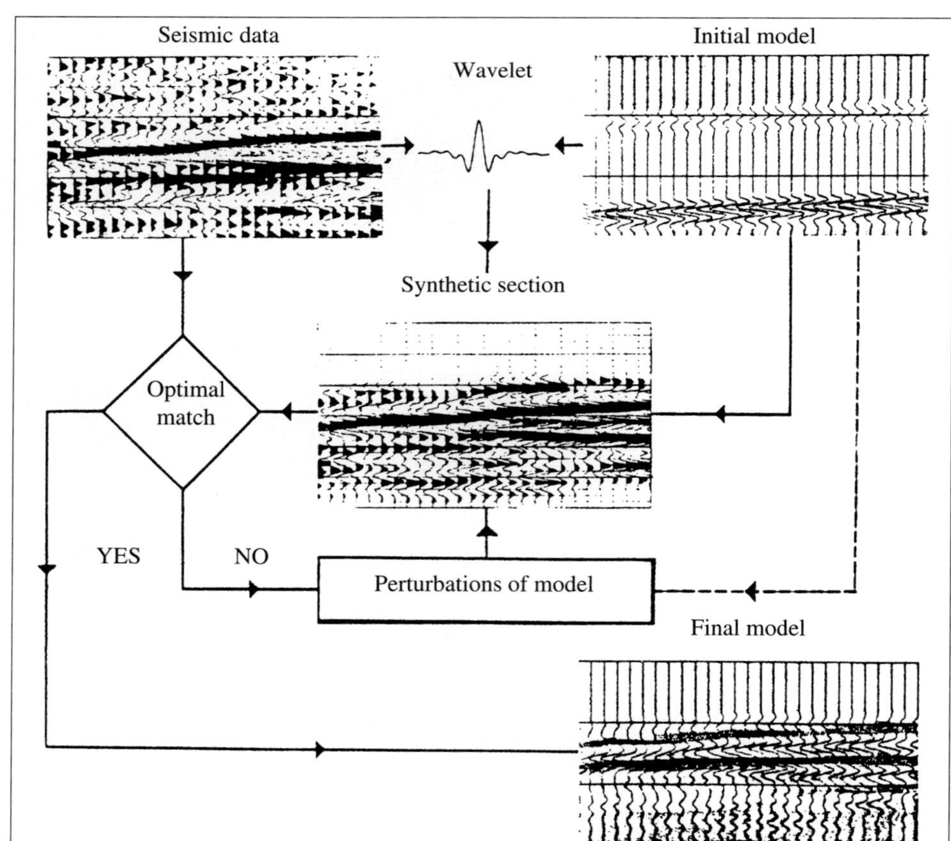

Fig. 4.11 Principle of the inversion method. *(Gelfand et al., 1984)*

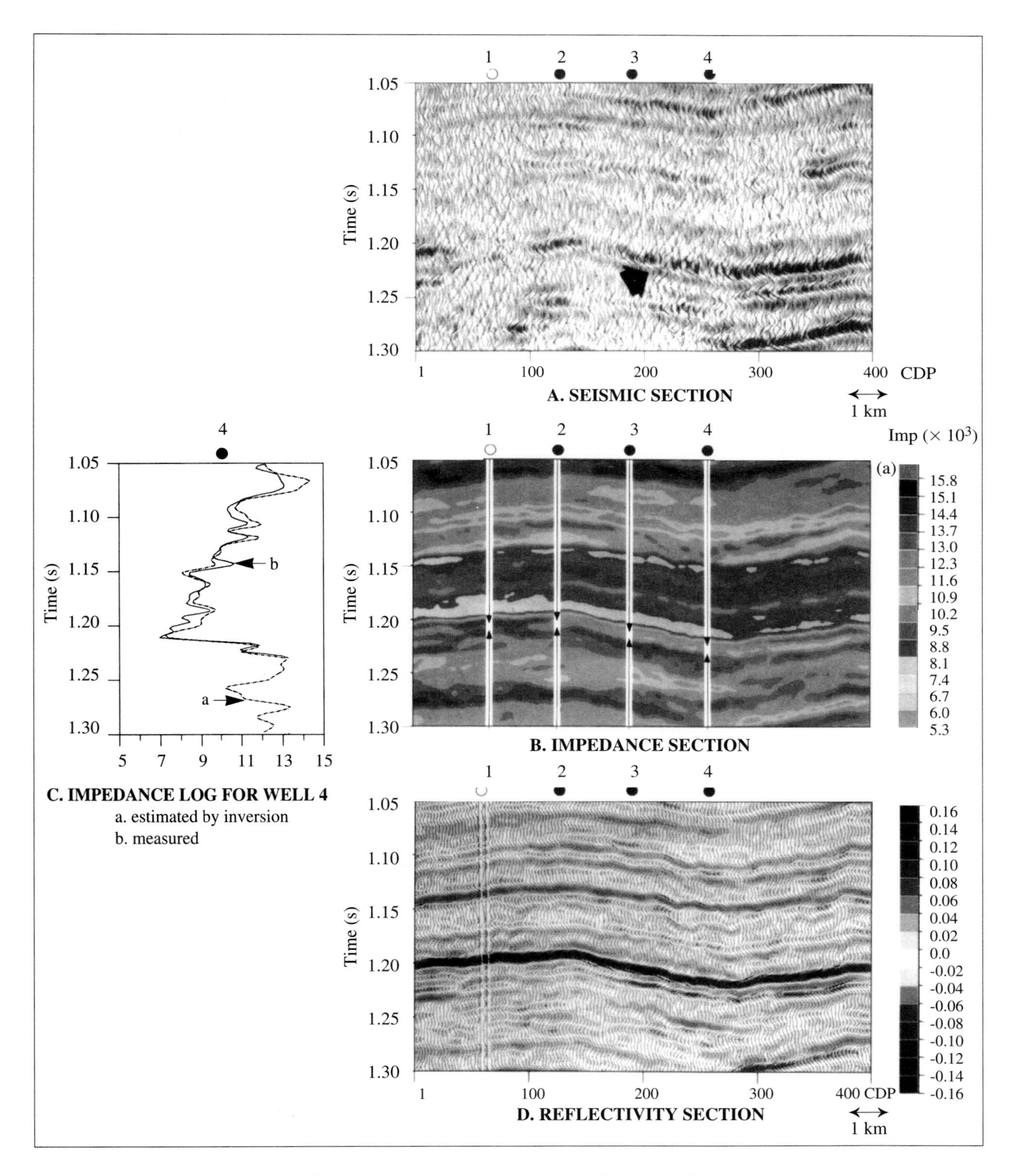

Fig. 4.12 Example of stratigraphic inversion using *Interwell* software.
(Brac et al., 1988)

Chapter **5**

SEISMIC AMPLITUDE OBSERVATIONS

The maximum resolution obtainable by seismic methods on a target at a given depth depends on the dominant frequency and bandwidth of the seismic signal involved, as well as the signal-to-noise ratio. For layers less than or equal to the half-wavelength of the seismic wave, there is an interference between reflections created at the top and base of the reservoir which leads to variations in the phase and amplitude of the seismic signal. In order to achieve a simultaneous determination of the variations in acoustic impedance and thickness of the layers, it is necessary to use seismic processing techniques that take into account wavelet interferences.

Widess (1973) discussed the effects of this type of interference in the simple case of a wavelet at normal incidence onto a thick and seismically fast layer in a homogeneous medium (*cf.* Fig. 5.1).

When the layer thickness is equal to one eighth of the wavelength, the composite wavelet is the time derivative of the incident wavelet. Widess proposed a simple formula for estimating the thickness of a thin bed, according to the frequency and amplitude of the composite wavelet.

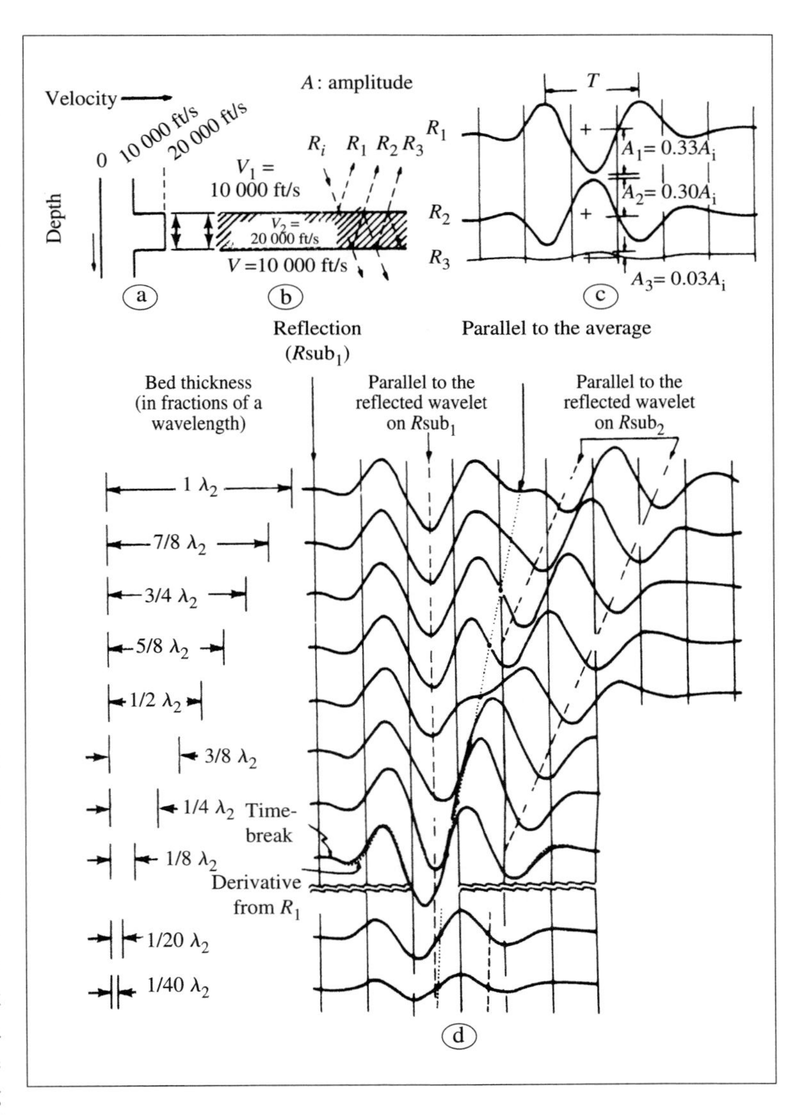

Fig. 5.1 Influence of layer thickness on seismic response.
(Widess, 1973)

To a first approximation, the amplitude of a reflection coming off a thin bed is equal to $4\pi\, Ab/\lambda_b$, where b is the thickness of the bed and λ_b is the wavelength calculated from the interval velocity and the dominant frequency of the signal. The amplitude of a reflection from a thick bed is represented by A.

5.1 MAIN CAUSES OF AMPLITUDE ANOMALIES

Apart from wavelet interference, various other phenomena may be at the origin of the amplitude anomalies observed on seismic sections. Some of the better known causes include:

(1) spherical divergence,
(2) "tuning effect",
(3) attenuation due to the receiver array spread,
(4) seismic noise,
(5) variation of reflection coefficient as a function of offset,
(6) curvature of reflectors,
(7) interface scattering,
(8) transmission losses,
(9) attenuation,
(10) instrument problems,
(11) shortcomings of seismic data processing.

The decrease in amplitude due to spherical divergence can be recovered by applying a compensation factor, function of time, offset and the velocities of media, which is based on the equations proposed by Newman (1973).

The tuning effect is analogous to the thin bed effect. The curvature of the seismic reflection time-distance curves decreases with depth, so the time difference between two closely-spaced reflectors decreases as the offset increases. As a result, interference occurs between the wavelets associated with each reflection. This effect may manifest itself either by an increase or a decrease in the amplitude as a function of the offset and the thickness of the beds.

Attenuation due to the geometry of the source/receiver configuration is linked to increased spacing of the arrays. This is carried out to achieve spatial filtering, which is essential to ensure an adequate distance sampling (x) of the seismic function. The effect is greater for shallow reflectors and large offsets, while it decreases for deep reflectors and small offsets. Attenuation of this type can be taken into account by a compensation factor depending on time, velocity, the dominant frequency of the signal and the length of the array (Ostrander, 1984).

The influence of noise can be reduced by stacking or by the application of special filters designed to separate signal and noise from seismic data (e.g. matrix filtering techniques: Mari and Glangeaud, 1988).

5.2 AVO EFFECT AMPLITUDE VARIATION AS A FUNCTION OF OFFSET

Changes in the amplitude of a reflected wave are directly related to the variation of the reflection coefficient as a function of offset, in accordance with the Zoeppritz equations. The variation of reflection coefficient with angle of incidence depends on the petrophysical parameters (compressional wave velocity V_P, density ρ_b, Poisson's ratio σ, etc.) of the media situated on either side of the discontinuity giving rise to the reflection. The variation of the reflection coefficient as a function of angle of incidence is particularly sensitive to the ratio between the Poisson coefficients in the two media (σ_1/σ_2).

Figure 5.2 shows the variation of reflection coefficients as a function of angle of incidence for different values of σ_1/σ_2.

A study of the amplitude versus offset (AVO) relationship enables an indirect estimate to be made of Poisson's ratio. It also provides quantitative information on the physical parameters of the media traversed, as well as some indication concerning the presence of hydrocarbons in reservoir rocks.

In practice, different simplified forms of the Zoeppritz equations (Shuey, 1985; Pan and Gardner, 1987; Silva et al., 1987) are used to obtain these parameters.

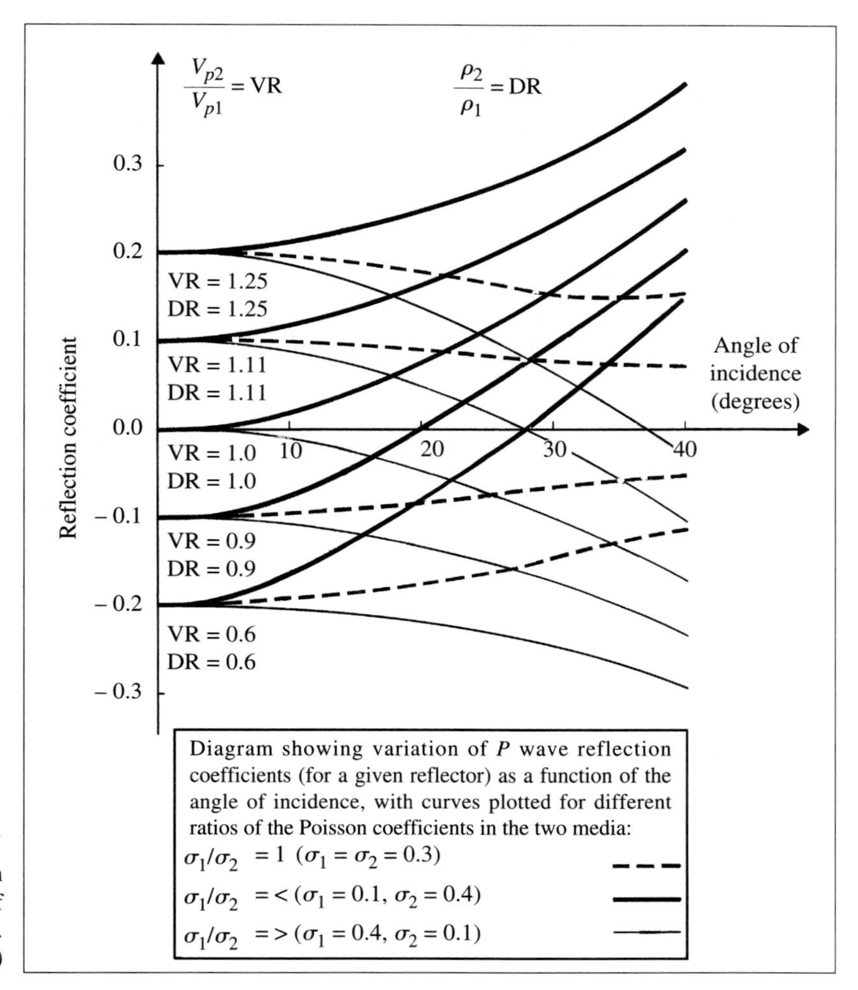

Fig. 5.2 Plot of reflection coefficient vs. angle of incidence, and σ_1/σ_2 ratio. *(Ostrander, 1984)*

Shuey (1985) proposed the following linear approximation:

$$R(\theta) = A + B \sin^2(\theta)$$

where:

$$A = \frac{1}{2}\left(\frac{\Delta V_p}{V_p} + \frac{\Delta\rho}{\rho}\right)$$

and:

$$B = A A_0 + \frac{\sigma_2 - \sigma_1}{\left(1 - \frac{\sigma_2 + \sigma_1}{2}\right)^2}$$

$$A_0 = \beta - \frac{4V_s^2}{V_p^2}(1 + \beta)$$

and:

$$\beta = \frac{\frac{\Delta V_p}{V_p}}{\frac{\Delta V_p}{V_p} + \frac{\Delta\rho}{\rho}}$$

Coefficient B, which corresponds to the slope, can be approximated by the following expression:

$$B = \frac{1}{2}\left(\frac{\Delta V_p}{V_p} - \frac{\Delta\rho}{\rho} - 2\frac{\Delta V_s}{V_s}\right)$$

For seismic traces sorted according to common depth-point collections, the adjustment of amplitude variation using a linear regression as a function of $\sin^2\theta$ is able to yield the A and B parameters for any given distinct reflection. These parameters can then be reworked into a seismic section.

The A-parameter section corresponds to the compressional wave seismic section, while the (A-B)-parameter section corresponds to the shear wave seismic section.

Silva et al. (1989) have described a parabolic approximation linking amplitude with the mean angle between the incident and refracted rays. These authors propose the use of the following equation:

$$R(i)\cos^2(i) = \frac{1}{4}\left(\frac{\Delta\rho}{\rho} + \frac{\Delta M}{M}\right)$$
$$-\frac{1}{4}\left(\frac{\Delta\rho}{\rho} + 4\frac{\Delta\mu}{M}\right)\sin^2(i) + 2\frac{\Delta\mu}{M}\sin^4(i)$$

where:

i = average angles of incidence and refraction at the interface

$R(i)$ = reflection coefficient

$\Delta\rho/\rho$ = density coefficient

$\Delta M/M$ = compressibility coefficient

$\Delta\mu/M$ = ratio of shear modulus increment to bulk modulus

Using a common depth point collection for a distinct reflection, the parabolic adjustment of the amplitudes can be carried out according to a relation of the type:

$$y = a + bx + cx^2$$

with $x = \sin^2(i)$ and $y = R(i)\cos^2(i)$.

This enables an evaluation of the parameters a, b and c which are themselves related to mechanical properties by the following equations:

$$a = \frac{1}{2}\left(\frac{\Delta\rho}{\rho} + \frac{\Delta V_p}{V_p}\right)$$
$$b = -\frac{1}{2}\left\{\frac{\Delta\rho}{\rho} + 4\frac{V_s^2}{V_p^2}\left(\frac{2\Delta V_s}{V_s} + \frac{\Delta\rho}{\rho}\right)\right\}$$
$$c = 2\frac{V_s^2}{V_p^2}\left(\frac{2\Delta V_s}{V_s} + \frac{\Delta\rho}{\rho}\right)$$

where:

$$\frac{\Delta V_s}{V_s} = \left(\frac{V_{s_2} - V_{s_1}}{V_{s_2} + V_{s_1}}\right)$$

is the shear wave velocity coefficient;

$$\frac{\Delta\rho}{\rho} = \left(\frac{\rho_2 - \rho_1}{\rho_2 + \rho_1}\right)$$

is the density coefficient;

$$V_p = \left(\frac{V_{P_1} + V_{P_2}}{2}\right)$$

is the average P wave velocity at the interface;

$$V_s = \left(\frac{V_{s_1} + V_{s_2}}{2}\right)$$

is the average S wave velocity at the interface.

Parameters a, b and c can then be reworked in the form of seismic sections. The direct combination of these parameters makes it possible to obtain seismic sections in terms of density coefficient or compressibility coefficient, i.e.:

$$\Delta\rho/\rho = -2(b + c)$$
$$\Delta M/M = 4a + 2(b + c)$$

Silva et al. (1989) have carried out a very thorough AVO study of a gas-bearing sandy reservoir prospect in North Sea.

Ostrander (1984) has reported a very clear example of AVO anomalies observable on seismic sections in relation to the presence of hydrocarbons. The seismic section in this case was recorded in a gas-field located in the Sacramento valley.

The sandy reservoir, situated at a depth of 6700 feet is characterised by amplitude anomalies on the seismic section at around 1.75 s. The reservoir is a Cretaceous deep-sea fan type deposit, while the trap is both structural and stratigraphic (i.e. faults and wedge-outs). The velocities and densities of the gas-bearing sands are significantly less than those measured in the surrounding clays, thus giving rise to strong reflections at the top and at the base of the reservoir.

Figure 5.3 shows a seismic section, as well as common depth-point gathers and partial stacks used to improve the signal-to-noise ratio. For a given depth-point gather, close offsets correspond to angles of incidence around 5°, whereas the farther spaced offsets correspond to angles of 35°. A strong increase in amplitude as a function of offset can be seen for common depth-points belonging to gather A and B; this indicates the presence of gas. However, the gather from point C shows very little AVO variation, which is characteristic of the absence of gas. A water-gas contact is found in a well that can be projected structurally near common depth-point 120, thus confirming the absence of gas in the C zone.

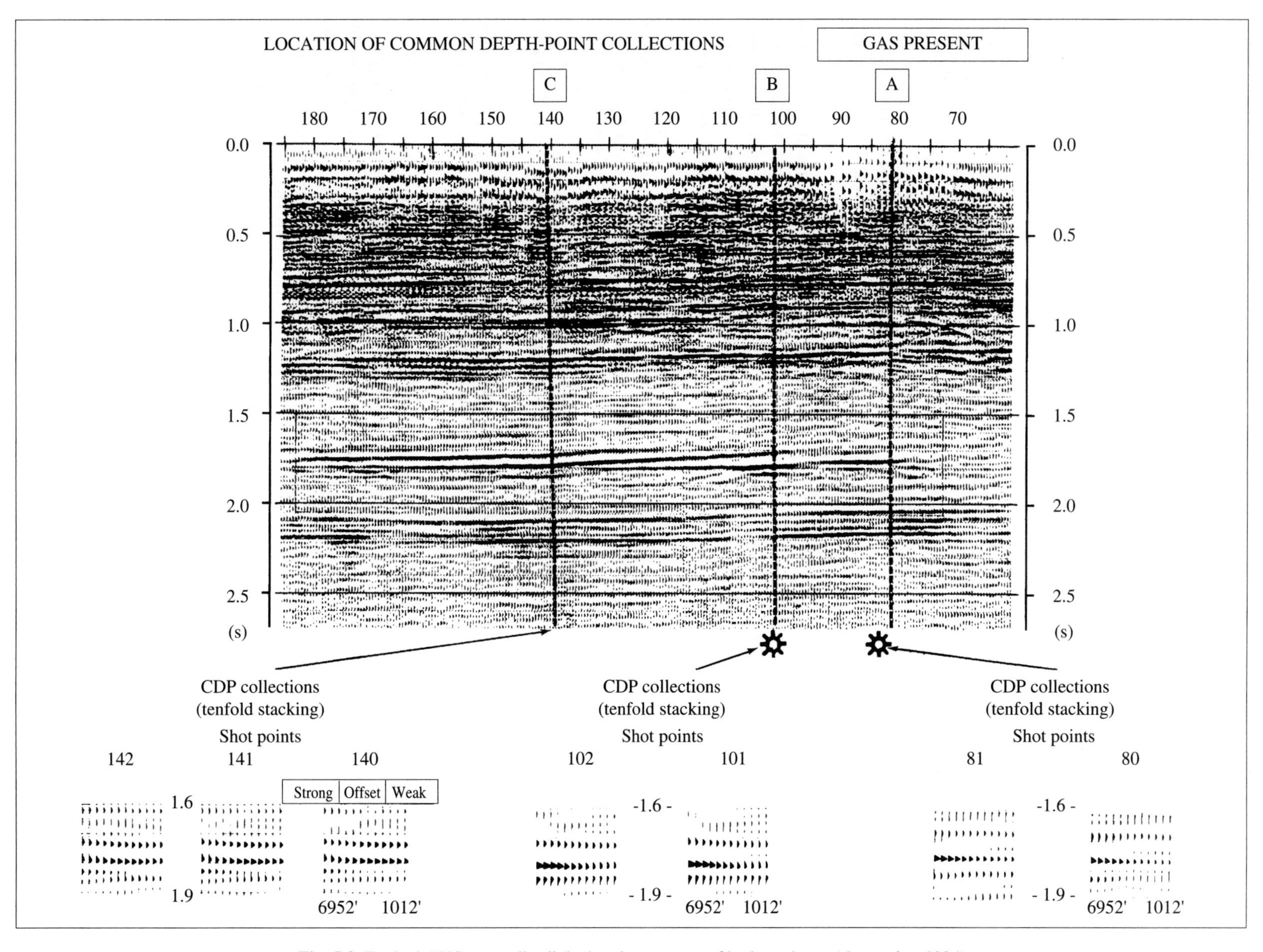

Fig. 5.3 Typical AVO anomalies linked to the presence of hydrocarbons. *(Ostrander, 1984)*

5.3 DIRECT HYDROCARBON INDICATORS (DHI'S)

Numerous authors have studied how the presence of gas in rocks can influence the behaviour of seismic waves. Dilay (1982) has presented a detailed review of the various phenomena generally referred to as Direct Hydrocarbon Indicators (DHI's). These indicators are linked to changes in the accoustic properties of the reservoir, notably the decrease in *P* wave velocity with respect to the same reservoir satured in water.

The presence of hydrocarbons may be indicated by:

(1) Amplitude anomalies: an abnormally strong reflection (or bright spot) may be observed above the gas reservoir. This phenomenon is due to the weak acoustic impedance of gas-bearing rocks. In some other cases, a fading of the amplitude (or dim spot) may be observed (Blondin and Mari, 1984). An example of the dim spot phenomenon is presented in Chapter Six.

(2) Flat spots: isochronal reflections can be associated with the gas-oil or gas-water interface.

(3) Polarity changes, which may be observed in reflections at the boundaries of the gas reservoir, are due to changes in the reflection coefficient at the top or base of the reservoir.

(4) Diffraction effects are produced by abrupt changes in the reflection coefficient at the top or base of the reservoir (at its lateral extremities).

(5) Time delays may be recorded in reflections coming from beneath the gas reservoir due to a relative increase in the transit times of acoustic wave traversing the reservoir between the gas and water zones. An example of this phenomenon is presented in Chapter Six.

(6) Zones of weak amplitude beneath the gas reservoir are due to the loss of energy of reflections as they traverse the gas reservoir.

(7) Lowering of frequency: a zone of low-frequency reflections may be observed within the reservoir unit. This phenomenon is due to the absorption and low velocity of acoustic waves in the gas phase.

Unfortunately, DHI's do not always reflect the presence of hydrocarbons. They can also arise from stratification of the medium or the presence of seismically slow interbeds within a much faster formation. Since fluids are unaffected by shear stress, *S* waves can offer the possibility of discriminating real DHI's from artefacts; real DHI's are only observed on *P* wave seismic sections, whereas false DHI's will appear both on *P* and *S* wave sections.

Ensley (1984) has reported some very clear examples of real and false DHI's obtained from seismic surveys and synthetic seismograms. Fig. 5.4 shows a real example of a bright spot related to the presence of gas in a reservoir. In this case, a bright spot is observed on the *P* wave section but not on the *S* wave section. In Fig. 5.5, the bright spot is of lithological origin and is observed on both the *P* and *S* wave sections.

After matching with the downhole measurements, a comparison of *P* and *S* wave seismic sections can provide — at least theorically — some indications on the porosity of rocks impregnated with hydrocarbons (e.g. by means of V_P/V_S ratios; see Chapter Two).

Moreover, a combined investigation of *P* and *S* waves can provide an effective means of evaluating the relative anisotropy of a medium.

A medium is said to be anisotropic if the velocities of the *P* and/or *S* waves are different in the vertical and horizontal directions (transverse anisotropy). The ratio between these vertical and horizontal velocities yields the anisotropy coefficient.

The velocities derived from seismic recordings provide velocities along the layer in a horizontal direction. Given two geological horizons identified from *P* and *S* wave sections — based on downhole logging and vertical seismic profiling — it is possible to determine the horizontal layer velocity and the vertical transit time using seismic data. From this, the anisotropy coefficients can be calculated.

In summary, the comparison of *P* and *S* wave sections is able to provide information on lithology, porosity and anisotropy. In addition, this approach enables a distinction to be made between real DHI's and false DHI's. Such results can only be reliably obtained on the basis of good quality seismic data with a satisfactory signal-to-noise ratio and high resolution.

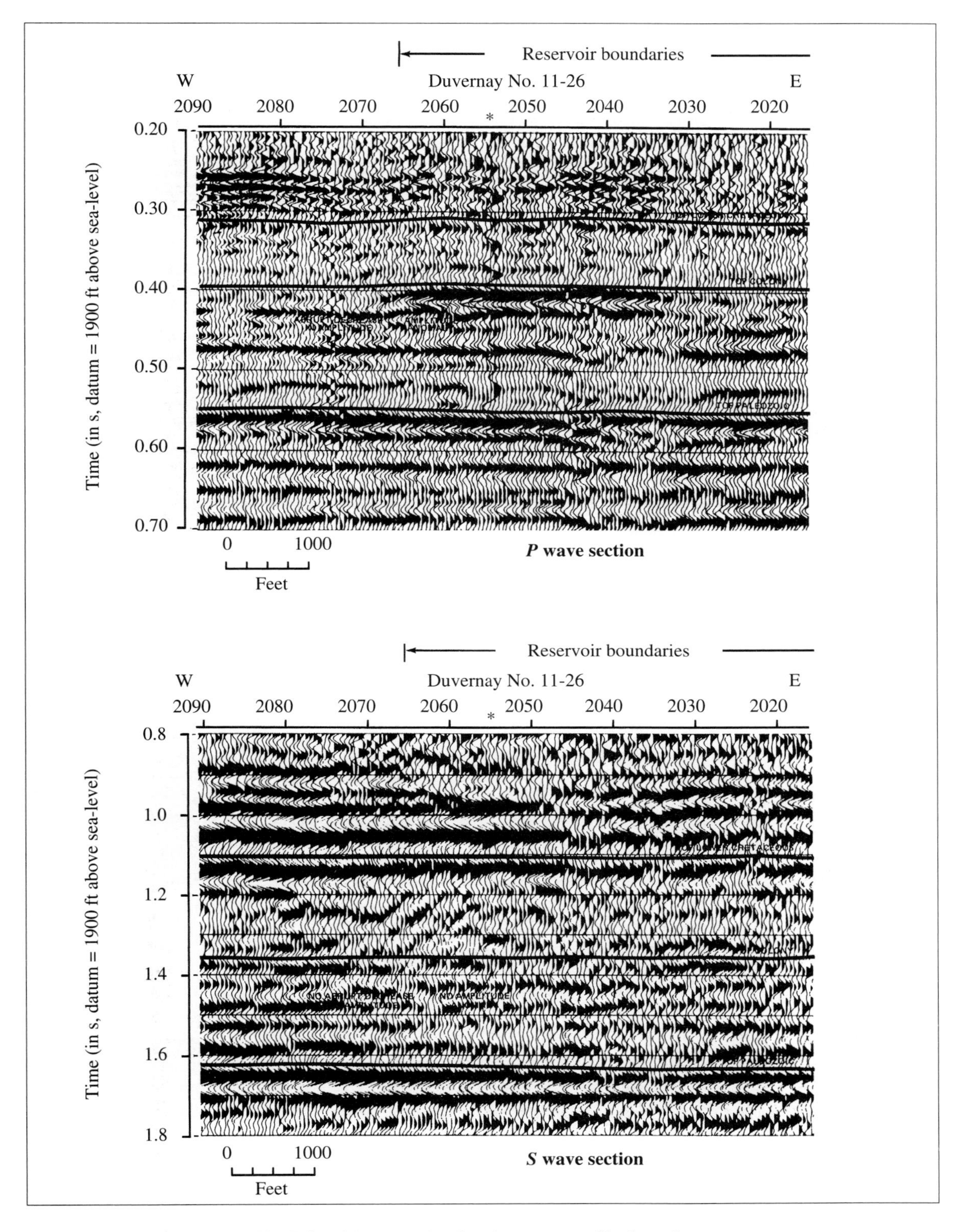

Fig. 5.4 Bright spot related to the presence of hydrocarbons.
(Ensley, 1984)

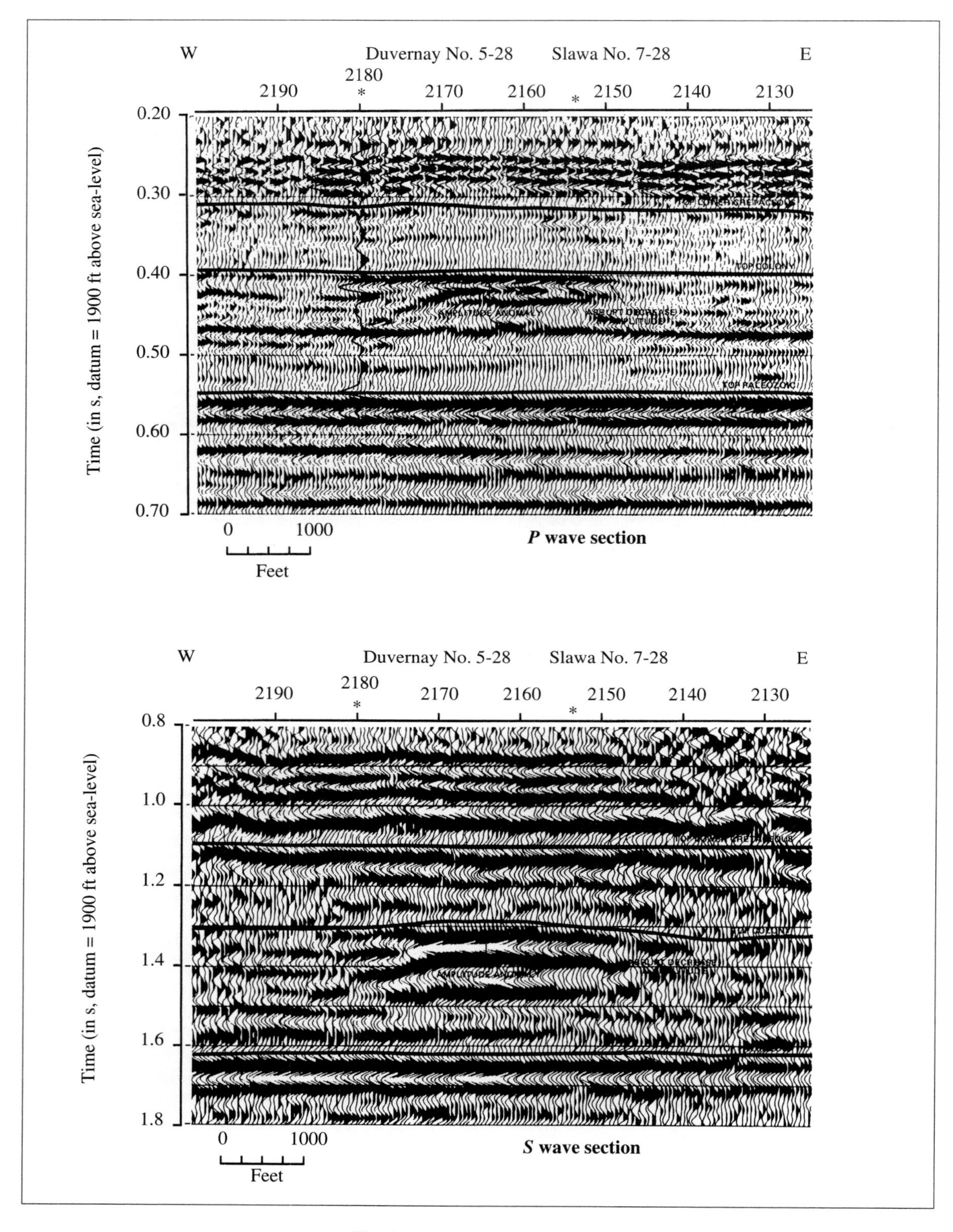

Fig. 5.5 Lithological bright spot.
(Ensley, 1984)

Chapter **6**

CONCLUSION AND CASE STUDIES

In order to optimise the use of conventional well log measurements (cuttings, core descriptions, mud logging, wireline logging, seismic well surveying) and geophysical surveys (e.g. surface seismics), with a view to improving the characterisation of reservoirs (petrophysical parameters, structure, geometry and facies analysis), it is prudent to carry out matching between all the different datasets. Thus, it is preferable to have a thorough knowledge of the methods of correlation while keeping in mind the reliability that can be attributed to the measurements used. This requires a good understanding of the tools and data processing techniques, as well as a certain number of checking procedures and corrections carried out on the raw data, before proceeding with the various processings that have to be applied to the matched data.

Although supplementary information can be found in the preceding chapters of this text, the main points are briefly summarised here under four headings:

(1) logging and borehole seismic operations,
(2) processing of data and application of results,
(3) relationship between well logging and borehole seismic surveys,
(4) relationship between downhole data (logging and seismic well surveys) and surface seismic data.

Implementation	
Well logging ***and***	***seismic borehole surveying***
Main characteristics	
• *Sampling rate:*	
1/2 ' (15 cm)	Well velocity survey: 100 m (or more if necessary) VSP: 5-20 m
• *Vertical resolution:*	
0.5 m for conventional tools	meters to 10's of meters
• *Lateral investigation:*	
10's of cm to meters	10's of meters to a few 100's of meters
Precautions	
Choice of tool type Tool calibration Measurement conditions (borehole condition and fluids in well) Depth matching of all logs Identification of anomalous zones	Tool clamping and cable slackening Choice of data acquisition parameters • distance between geophone positions as a function of formation velocity and seismic frequency, • offset as a function of target. Optimal pre-processing

Well logging	***and***	***seismic borehole surveying***
	Processing	
Editing • depth corrections, • elimination of anomalous zones. Environmental corrections Verticalization of log Validation of integrated travel time Establishment of a continuous T vs. Z relation		VSP (in different cases) • picking of first breaks and computation of vertical travel times (T vs. Z relation), • separation of upgoing and downgoing waves, • extraction of wavelets, • stacked trace in the case of zero-offset VSP, • migrated trace in the case of offset VSP, • inversion with or without constraints.
	Main applications	
Imaging (microseismic sections) Measurement of P and S wave velocities Structural data (dipmeter) Petrophysical and mechanical		Detailed well vicinity seismic profiling Identification of primary and multiple arrivals Below well depth prediction Structural data Petrophysical and mechanical properties

Relationships between well logs and well seismic surveys

- Matching of T vs. Z relation from acoustic logging to well seismic survey data (e.g. Block Shift, Δt minimum methods, etc.).
- Adjustment of sonic log using borehole seismic data and depth-time conversion.
- Calculation of acoustic impedance log and reflectivity profile.
- Construction of synthetic seismogram using an appropriate wavelet.
- Comparative analysis of seismic traces (seismogram calculated from log data and seismic traces) and/or acoustic impedance logs (derived from well logs and borehole seismic surveys) leading to the identification of markers.
- Combined use of logging and borehole seismic data at the well, in the vicinity of the well and beneath the well (information on structural geology, under-compaction, etc.).

Relationships between surface seismic and downhole data (seismic well surveying and well logs)

- Deconvolution or inversion of surface seismic data using constraints provided by downhole data.
- Obtaining reflectivity or acoustic impedance logs.
- Analysis of seismic amplitudes and acquisition of geological information.
- Conversion of seismic sections into profiles of rock mechanical/petrophysical parameters with depth (e.g. porosity sections).

In the following, a number of examples are given of applied studies which combine seismic and logging approaches:

- Determination of open fractures and assessment of vertical resolution of surface seismic using sonic log data. (P. Gaudiani, 1989).
- Calibration of surface seismic by means of well logs. (J.C. Lecomte, 1990).
- Acoustic impedance in relation to porosity. (M. de Buyl, T. Guidish and F. Bell, 1988).
- High-resolution seismic survey for monitoring an underground gas storage. (E. Blondin and J.L. Mari, 1984).
- Subwell prediction and detection of overpressure zone. (S. Brun, P. Grivelet and A. Paul, 1985).
- Use of wireline logs in the interpretation of seismic lines across salt-bearing formations (P. Renoux, 1989).
- Applications of VSP and dipmeter surveys in the development of a structural model. (J.L. Mari, P. Gavin and F. Verdier, 1990).

Determination of open fractures and assessment of the vertical resolution of surface seismic using sonic log data

(P. Gaudiani, 1989)

The sonic log presented in this example was run in a well traversing a sandstone formation in northern Canada. The objective of the study was to locate mechanical-type anomalies; in particular, the presence of open fractures that could bring the safety of the rig into question during the re-drilling of the borehole with a larger diameter.

The large-diameter borehole has now been drilled in close proximity to the hole used for sonic measurements. The predicted fractures were encountered and the drillers were able to inject the zones of highly open fracturing to avoid flooding of the bottom of the well.

Figure 6.1 shows a full waveform sonic log corresponding to a common-offset section of the 300-380 m depth interval, with the total calculated energy log plotted alongside.

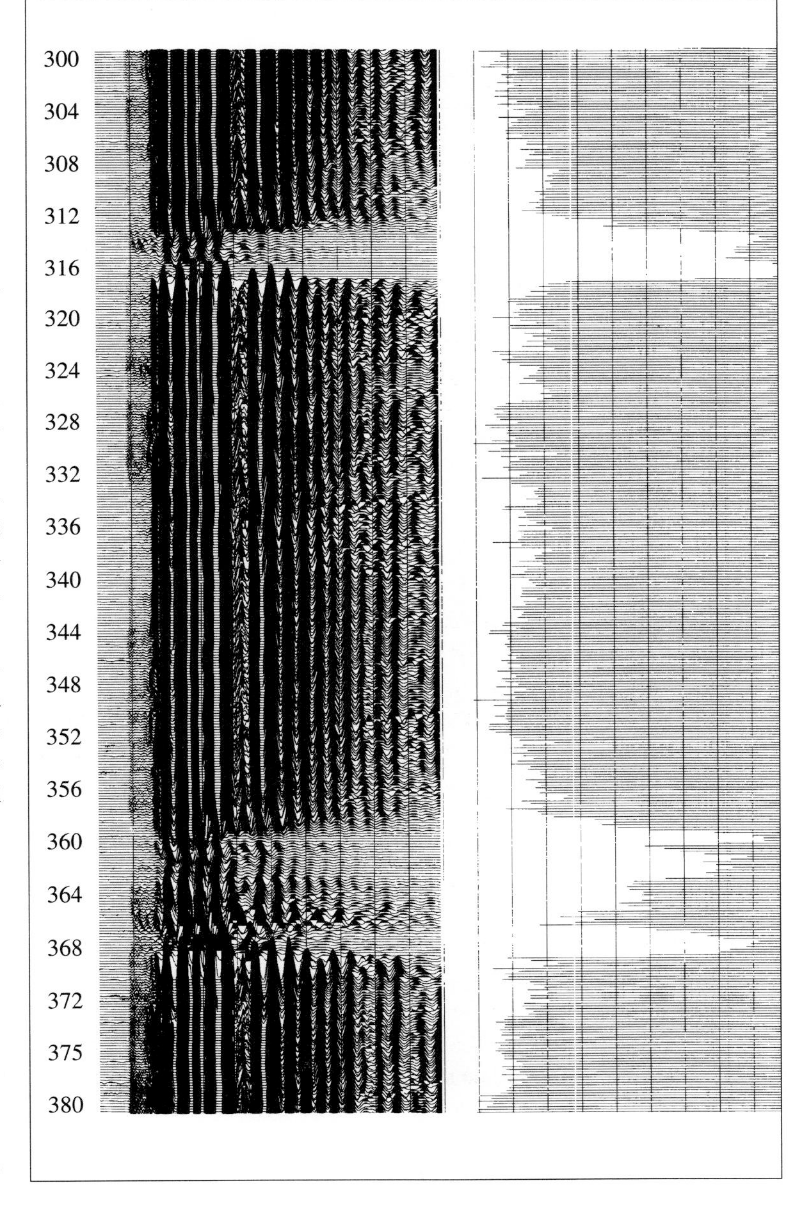

Fig. 6.1 ▶
Sonic log: full waveform recording.
(Document by courtesy of SEMM)

Presentation of results

Figure 6.2 shows the following logs:

(1) Compressional wave slowness Δt expressed in μs per 0.25 m (the spacing between the two receivers of the *SEMM* tool used) is relatively constant over most of the well.

(2) The shape factor I_c (see Chapter One) is relatively well anticorrelated with the energy ratio calculated from the full wave on two receivers (R1 and R2).

(3) The Stoneley wave attenuation is obtained by full wave measurements, but is calculated within a time window centred on these waves. There is clear evidence for two zones of major energy loss in the 300-375 m interval, characterised on the full waveform log (Fig. 6.1) by the disappearance of Stoneley waves. These two zones

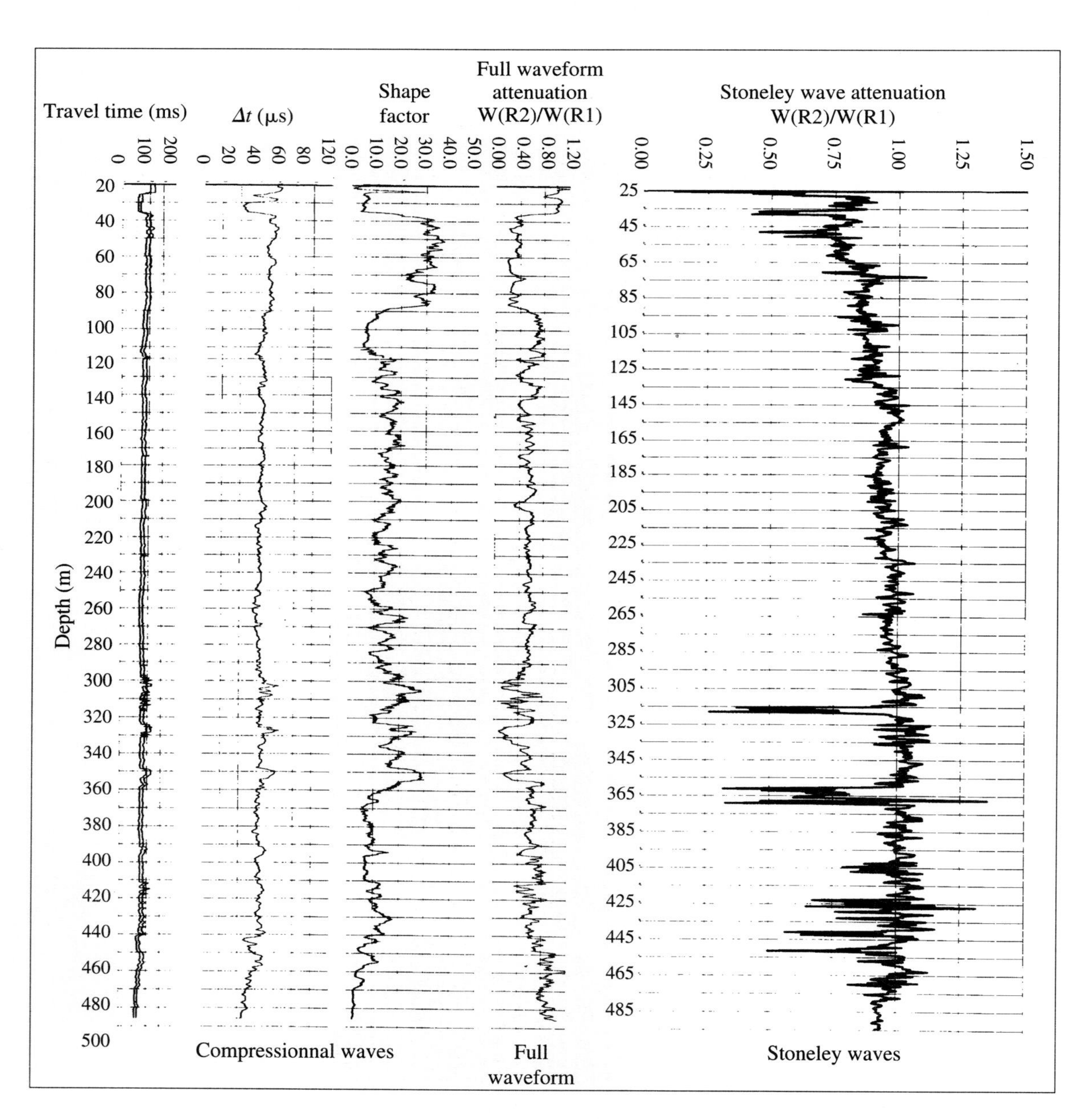

Fig. 6.2 Logs obtained from the recording presented in Fig. 6.1. *(Document by courtesy of SEMM)*

correspond to the intervals with open fractures encountered during drilling.

In addition to enabling the identification of fractured zones, the present example also illustrates the possibility of extracting geophysical information from sonic logs. The standard measurements (Fig. 6.3) usually include a time-depth curve derived from integrated transit-times, a mean velocity log, an interval velocity log (each interval normalised to a transit-time of 0.5 ms) and a set of synthetic seismic records calculated for a range of dominant wavelet frequencies. This information makes it possible to evaluate vertical resolution as a function of the distribution of emitted frequency and acoustic impedance contrast. The use of such borehole data before the acquisition of seismic data can enable a geophysicist to predict the quality of the surface survey in terms of the geological objective; it may also help to decide the actual survey operation parameters.

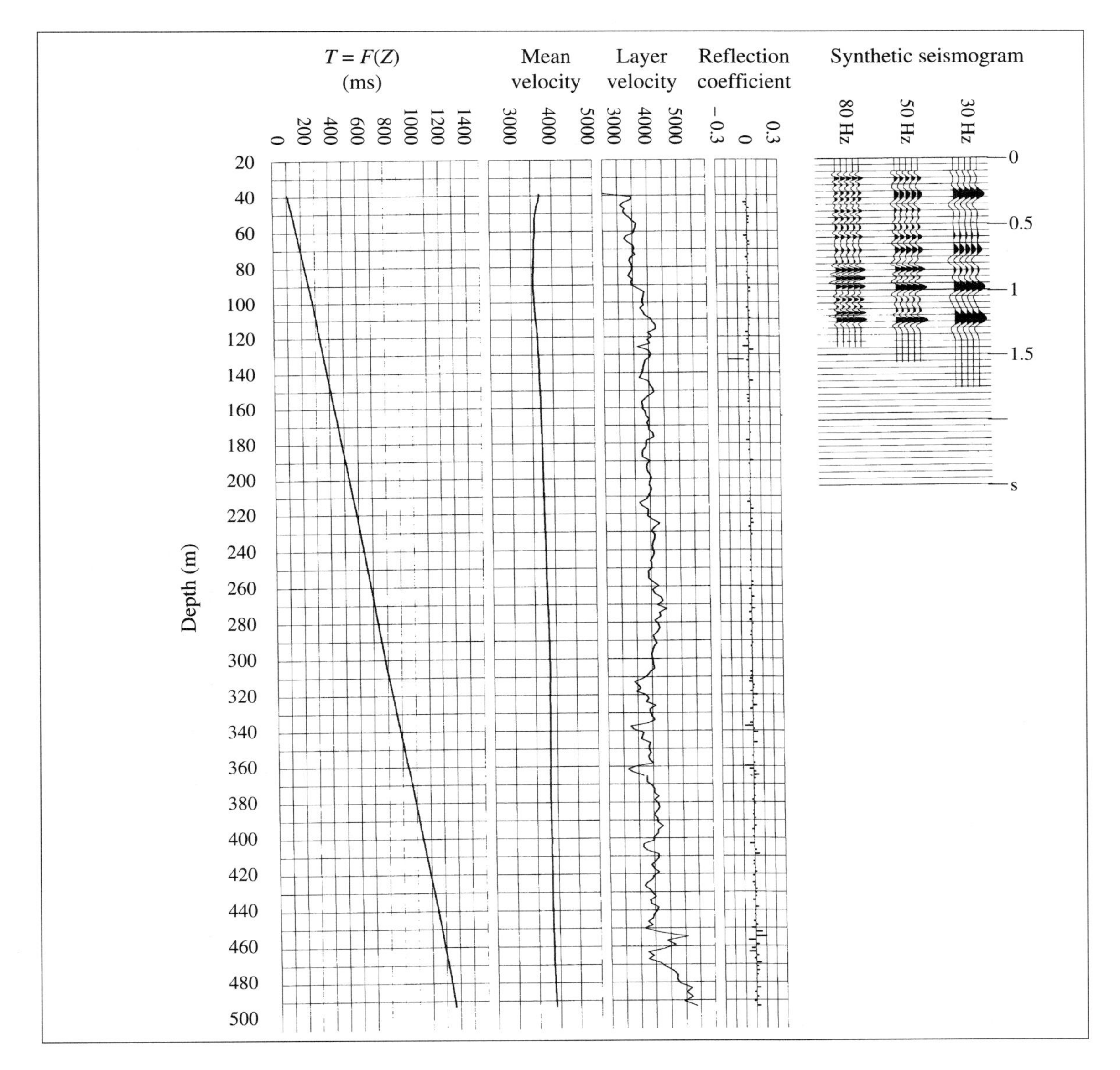

Fig. 6.3 Time vs. depth curve, calculated velocity logs (in units of m/s) and synthetic seismogram section derived from sonic log data on Fig. 6.1. *(Document by courtesy of SEMM)*

Calibration of surface seismic data by means of well logs

(J.C. Lecomte, 1990)

In most cases, exploration wells are located according to seismic data in order to check geological hypotheses and confirm interpretations (i.e. position of a fault, culmination of a structural high or saddle, etc.).

The information derived from downhole data (well logs, etc.) is used to calibrate surface seismic data. However, wells are not necessarily situated on seismic lines. As a result, an offset of only 50 m from the line will be sufficient to cause problems if there are major lateral variations in the geology of the reservoir on this scale. In certain cases, it may be necessary to link the well to the nearest profile by carrying out a supplementary surface seismic survey (Mari et al., 1987). In other cases, the accuracy of the calibration has to be evaluated.

The example presented here shows how data (sonic and density) can be used to calibrate a profile that passes close to a well.

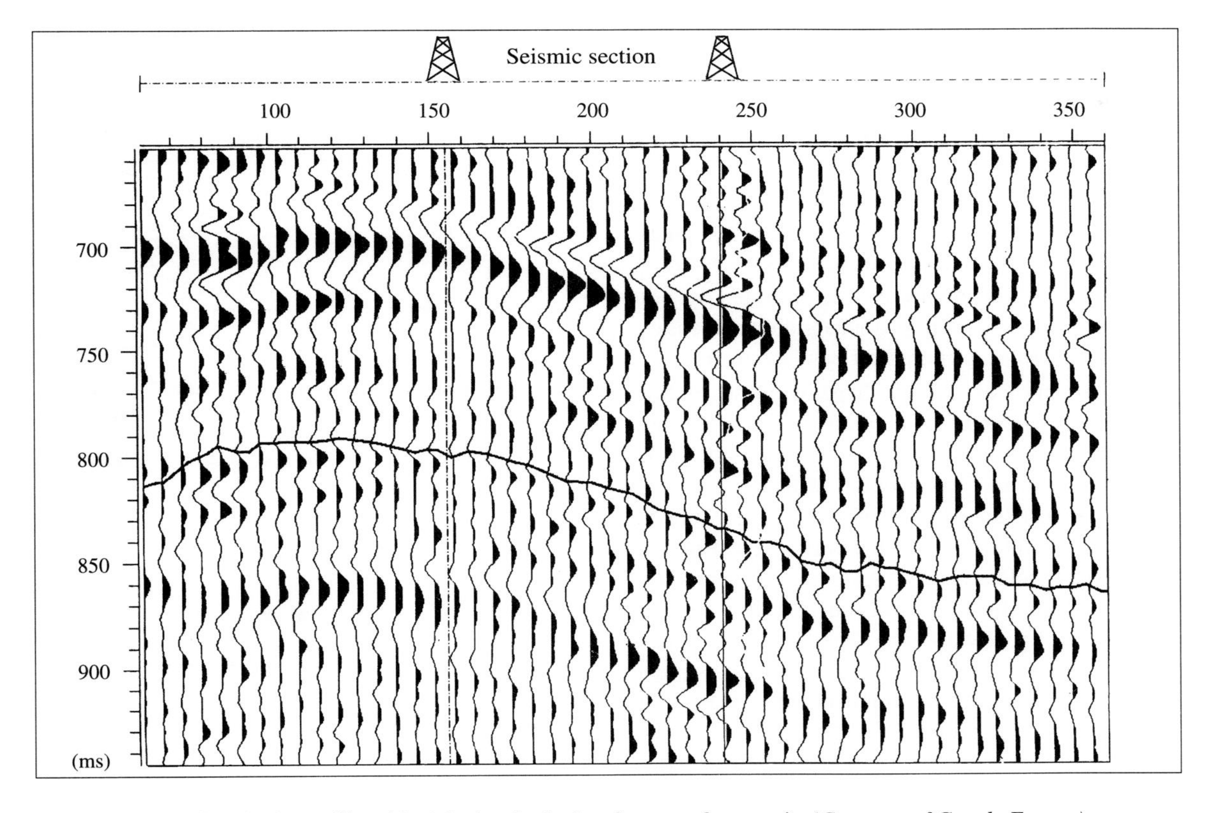

Fig. 6.4 Seismic profile with picked arrivals showing top of reservoir. *(Courtesy of Gaz de France)*

The calibration procedure consists of estimating the acoustic impedance distribution and the optimal wavelet in a time-window situated around the geological target on the seismic profile and making use of the acoustic impedance data provided by the well logs. Due to geological variations, the acoustic impedance variation on the profile is not the same as that observed in the well.

The example presented in Fig. 6.4 is taken from a seismic survey of a reservoir carried out on a structure drilled by *Gaz de France* (Lecomte, 1990).

The reservoir is a shale-sandstone sequence characteristic of a fluvial sedimentary environment where lateral facies variations are extremely rapid. The thickness of the reservoir is of the order of 100 m, corresponding to 80 ms in two-way time (TWT).

The seismic line is composed of traces spaced at 10 m intervals on either side of the reflection points. The time sampling rate is 2 ms. On the seismic line, the picked horizon is associated with the reservoir top.

The calibration is performed using the wavelet estimation module (Richard and Brac, 1988) in the *IFP* software package "Interwell". This program enables transformation of seismic sections into acoustic impedance logs. The wavelet is estimated on the basis of all the reflection points occuring on the seismic section, according to a three-step procedure (see Fig. 6.5):

(1) Statistical estimate of the signal amplitude spectrum and the random noise. The zero-phase wavelet is obtained only from surface seismic.
(2) Filtering of impedance profile provided by well logs, using a band-pass appropriate to seismics. Calculation of a filtered reflectivity profile.
(3) Modification of wavelet shape by phase rotation in order to obtain an optimal synthetic seismogram, thus ensuring the best possible correlation with the seismic traces of the surface survey.

At this final step of the calibration, the synthetic section calculated on the basis of the estimated wavelet can yield a highly satisfactory fit (Fig. 6.6). If this is not the case, the calibration can be continued by modifying the well acoustic impedance log values (Fig. 6.7). This readjustment of the log data should only be undertaken if justified by arguments of a geological nature (structural or stratigraphic evidence) or for reasons connected with logging operations (sonic log drift, poor log depth matching, "yo-yo" effect, etc.).

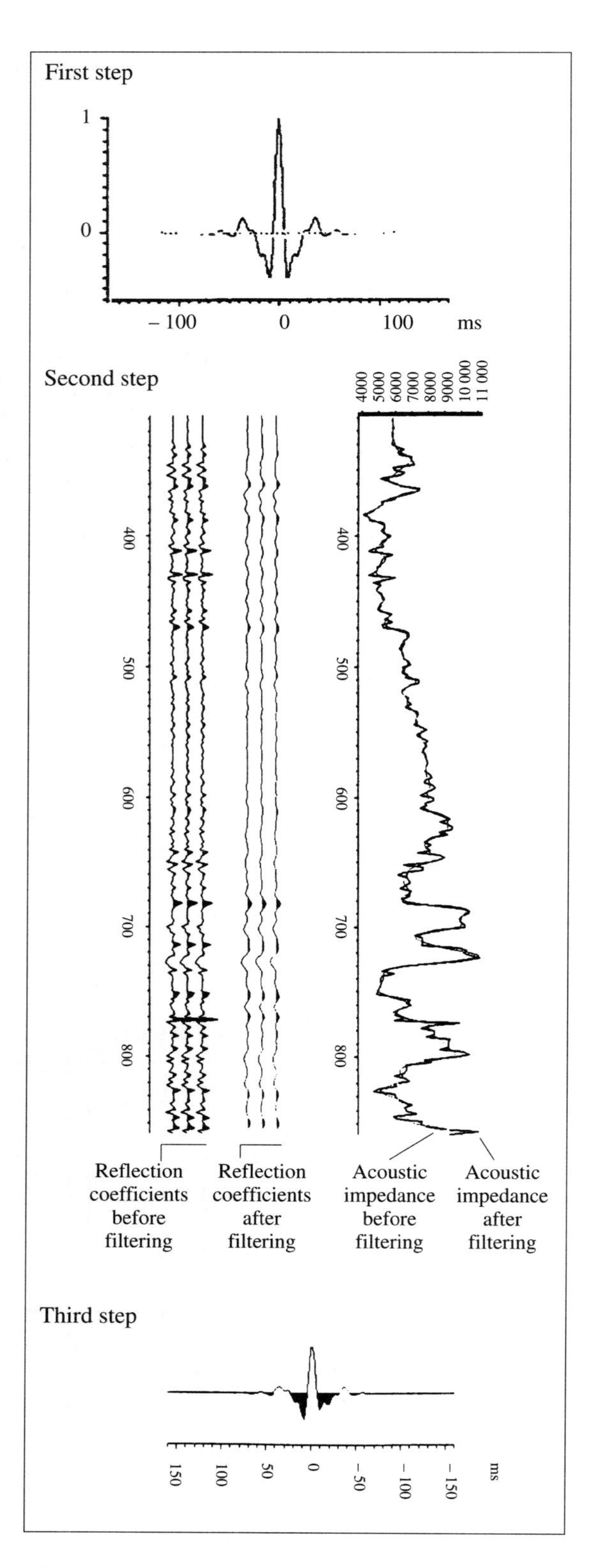

Fig. 6.5 Steps in wavelet estimation performed by the "Interwell" software package. *(Lecomte, 1990)*

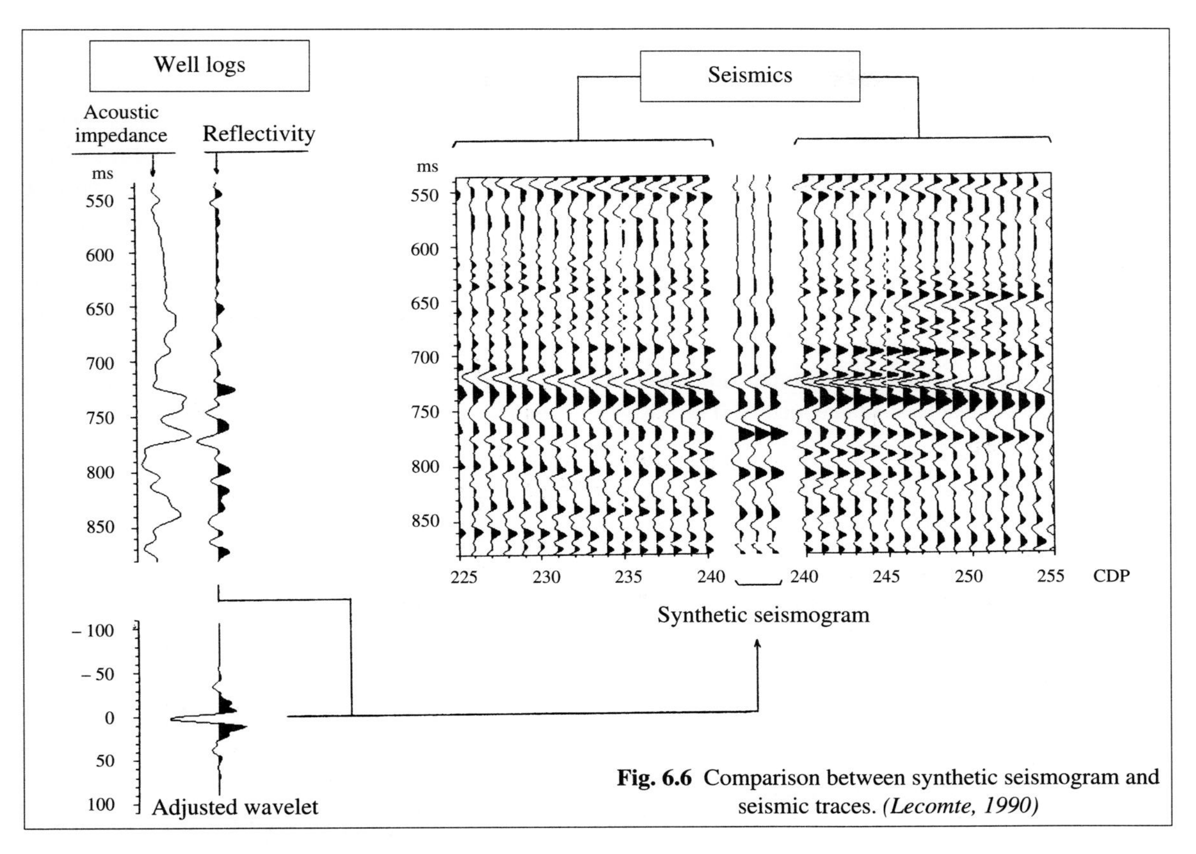

Fig. 6.6 Comparison between synthetic seismogram and seismic traces. *(Lecomte, 1990)*

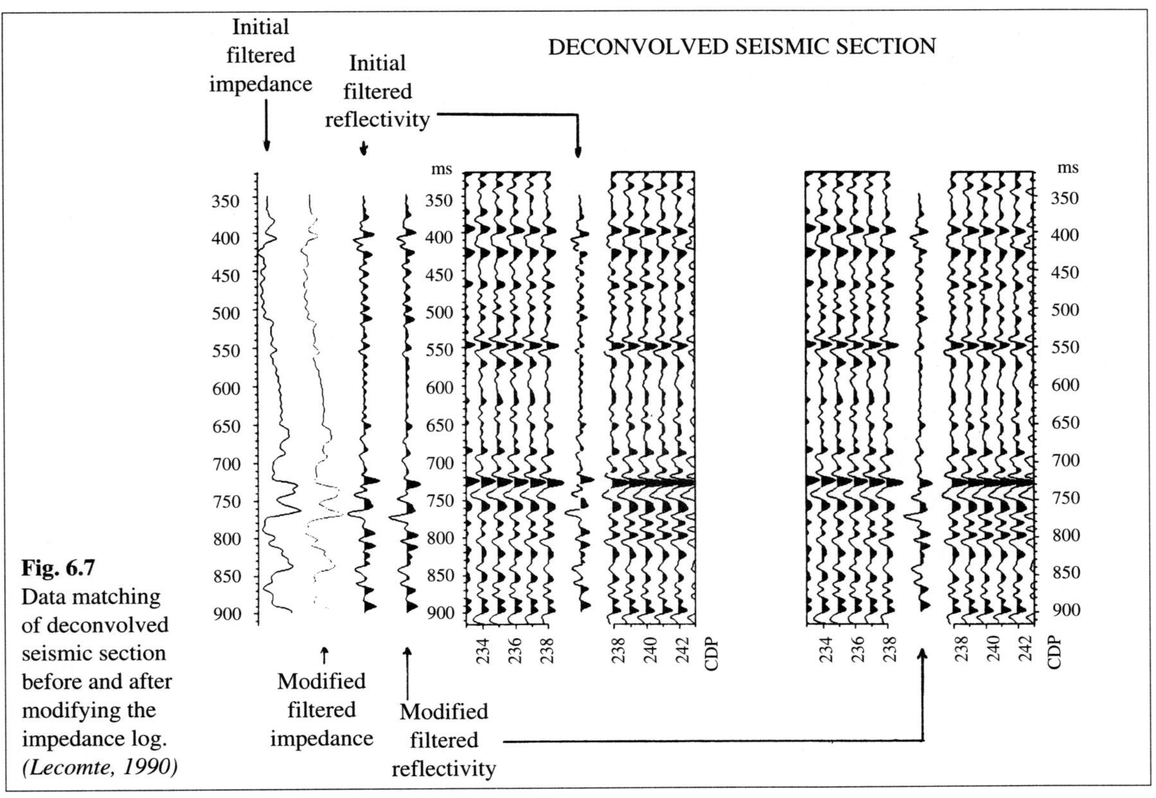

Fig. 6.7
Data matching of deconvolved seismic section before and after modifying the impedance log.
(Lecomte, 1990)

The wavelet is estimated by reiterating steps 2 and 3 of the procedure until a satisfactory result is considered to have been obtained.

If the fit remains unsatisfactory, the well may be judged inadequate for calibration of the seismic data. This may be caused by excessive geological variations between the well and the seismic line that cannot be assessed due to lack of information.

In final processing, a satisfactory fit can be improved even further by applying constraints and adjusting the wavelet with a least-square method (Fig. 6.8).

Conclusion

The matching of surface seismic data against well logs is a very delicate procedure. Data mat ching should not be carried out by applying operators that are designed to force a purely mathematical resemblance at any price. Rather, the matching of datasets should take reliable geological information into account. Certain wells cannot be used because of a lack of information on geological variations between the well and the seismic line.

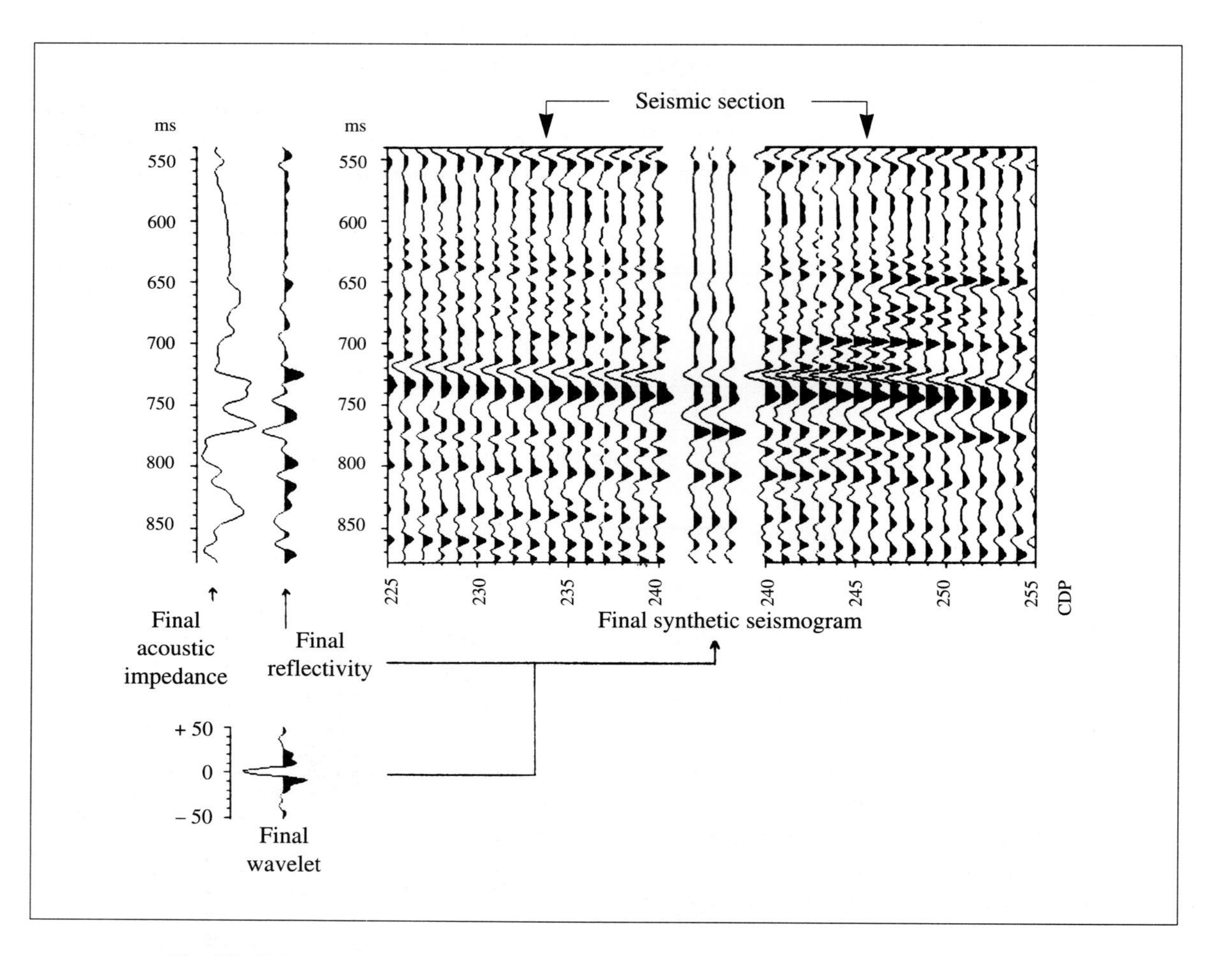

Fig. 6.8 Seismic calibration achieved after wavelet adjustment with a least-square method. *(Lecomte, 1990)*

Acoustic impedance in relation to porosity

(M. de Buyl, T. Guidish and F. Bell, 1988)

The Wyllie equations used to obtain Δt and ρ from sonic and density logs, respectively, can be combined to describe the relation of acoustic impedance to the petrophysical parameters of the medium, in particular the porosity Φ. In fact, we can write the following:

$$\Delta t = \Phi \, \Delta t_f + (1 - \Phi - V_{sh}) \, \Delta t_{ma} + V_{sh} \, \Delta t_{sh}$$

$$\rho = \Phi \, \rho_f + (1 - \Phi - V_{sh}) \, \rho_{ma} + V_{sh} \, \rho_{sh}$$

The impedance Z_a and the porosity are obtained as follows:

$$Z_a = \rho V = \rho / \Delta t$$

$$\Phi = \frac{(1 - V_{sh})(\rho_{ma} - Z_a \Delta t_{ma}) + V_{sh}(\rho_{sh} - Z_a \Delta t_{sh})}{Z_a(\Delta t_f - \Delta t_{ma}) - (\rho_f - \rho_{ma})}$$

where:

Φ	= porosity
V_{sh}	= percentage of clay
ρ_{ma} and Δt_{ma}	= density and interval transit-time of matrix
ρ_{sh} and Δt_{sh}	= density and interval transit-time of shale
ρ_f and Δt_f	= density and interval transit-time of formation fluid

M. de Buyl et al. (1988) have presented an example of the processing of seismic sections for obtaining high-resolution acoustic impedance profiles in an oil-producing sandy reservoir (Taber/ Turin zone, Canada). This study, which takes into account all the downhole data acquired during a 3D seismic reflection survey, enables an improved description of the reservoir and leads to the establishment of a set of maps showing the associated petrophysical parameters, notably the distribution of thickness and porosity.

The results of this study are reported in Fig. 6.10. The SLIM inversion program (Gelfand and Larner, 1984) was used to model the seismic sections in terms of acoustic impedance.

In the present example, a comparison of the thicknesses obtained from acoustic logging with those calculated from seismic data inversion demonstrates that the interval velocities in the reservoir zone (approx. 20 m thick) are precise to about 5%, whereas the thickness is precise to 5 m. These uncertainties are due to the differences in vertical resolution between seismic and wireline logging methods (Fig. 6.9).

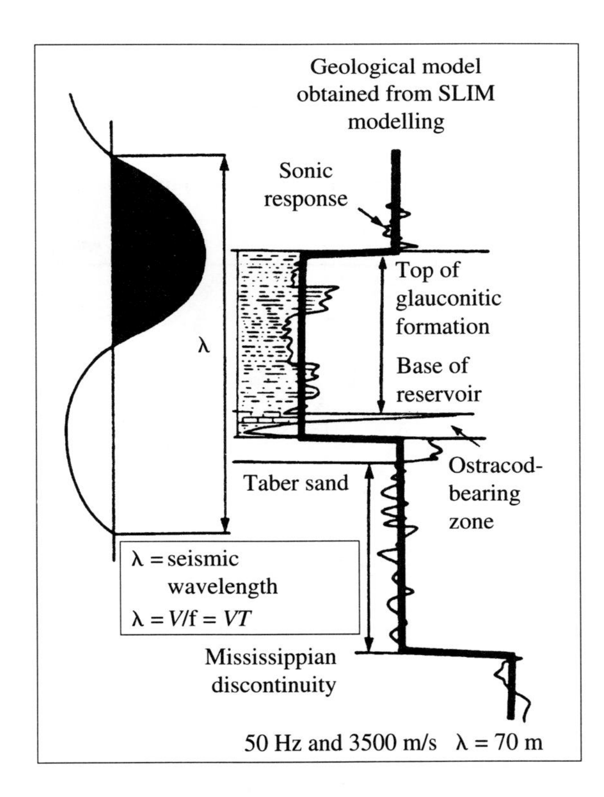

Fig. 6.9 Comparison of vertical resolution between well logs and seismic survey.

The reservoir zone is a slow interval limited at its base by an ostracod-bearing unit. In places, the thickness of the reservoir zone may be overestimated if the inversion method incorporates the ostracod bed into the total thickness of the reservoir.

A comparison of the thicknesses and porosities estimated in the well from log and seismic data shows very good correlation between the methods over intervals in the order of one quarter of the seismic wavelength (i.e. 20 m) (Fig. 6.10a).

The map of reservoir porosities derived from log data alone can thus be modified with the help of seismic data to yield a more precise reservoir model (Fig. 6.10b).

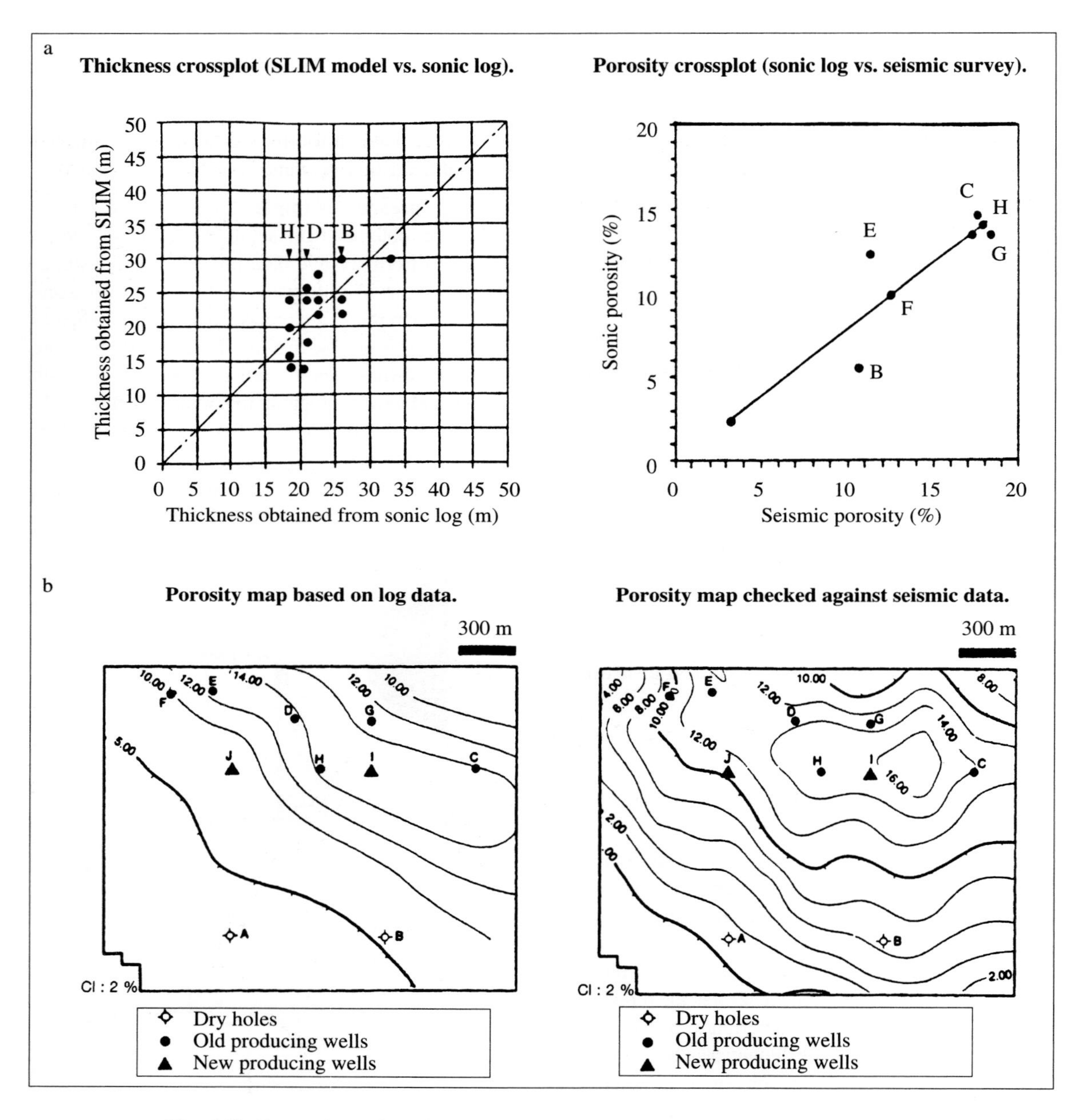

Fig. 6.10 Comparison of results obtained by well logs and by seismic survey inversion.

High-resolution seismic well survey of a reservoir used for underground gas storage

(E. Blondin and J.L. Mari, 1984)

An underground gas storage facility belonging to *Gaz de France* and situated at Gournay-sur-Aronde in the Paris Basin has served as a field experiment to test out the capacity of seismic methods to trace the water-gas interface in a reservoir bed.

Geological structure

The geological structure is a NW-SE-trending anticline with two culminations. Closure of the 65 m contour towards the Northwest delineates an area of 25 km^2 (12 km long and 2 km wide).

The main structural high of the reservoir is situated at a depth of 730 m, while its average thickness is 45 m. The reservoir is made up of Sequanian sand beds. The sandstone has a average total porosity of 20% and an average permeability of 0.6 Darcy. The total storage capacity is estimated at 2500 million cubic meters (N). Calcareous clays of Kimmeridgian age form an impermeable cap rock (180 m thick) to the reservoir.

Figure 6.11 shows the isobaths for the top of the upper part of the reservoir, in addition to the position

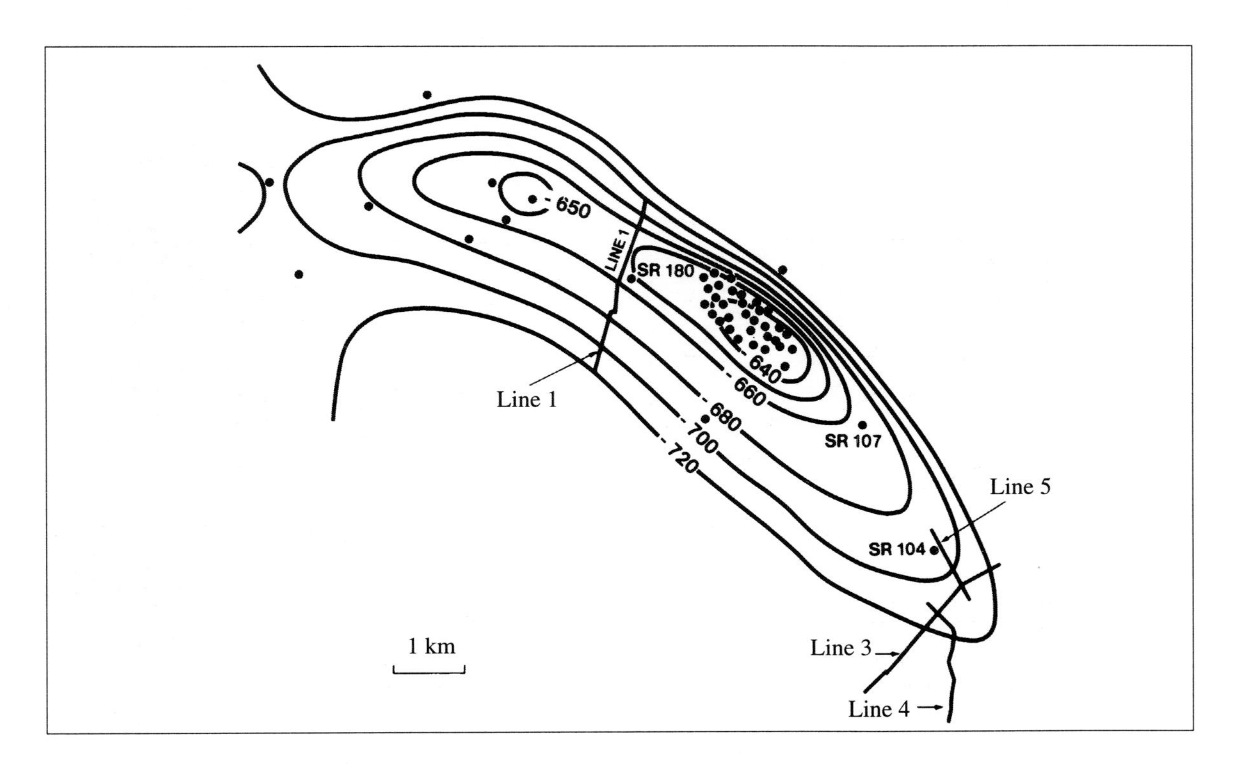

Fig. 6.11 Map of isobaths at the top of the upper reservoir zone, with location of seismic lines and some wells. *(Courtesy of Gaz de France)*

of wells and certain seismic lines. From 1976 onwards, natural gas has been injected into the reservoir from wells situated within the main structural culmination.

Checks carried out during filling of the reservoir show that the gas migrates preferentially towards the southwestern end of the anticline, in the vicinity of seismic lines 3, 4 and 5. Seismic line No. 1 is located on the secondary high, in the northwestern sector, where gas migrates less readily.

The main geological horizons making up the seismic markers were identified in well SR 180 by comparing the lithology log with the downhole seismic data (sonic and acoustic impedance logs, vertical seismic profiles) (Fig. 6.12). These are, from top to bottom:

(1) top and base of the Portlandian limestones (M1 and M2),
(2) top and base of the Sequanian limestones (M3 and M4),
(3) interval (M5) comprising oolitic limestone and unconsolidated sandstones interbeds (M5),
(4) upper reservoir unit (M6),
(5) top of the lower reservoir unit (intermediate cap, M7),
(6) top of the Rauracian limestones (base of reservoir, M8).

The lower reservoir unit is bounded by horizons M7 and M8.

Seismic data acquisition

Seismic lines 3, 4 and 5 were run in the southeastern part of the storage structure, near well SR 104. These three profiles were each repeated twice for *P* wave acquisition —with the conditions maintained as similar as possible— in order to monitor any potential migration of gas between the recording times. The first survey was carried out in April-May 1981 and the second in January 1982.

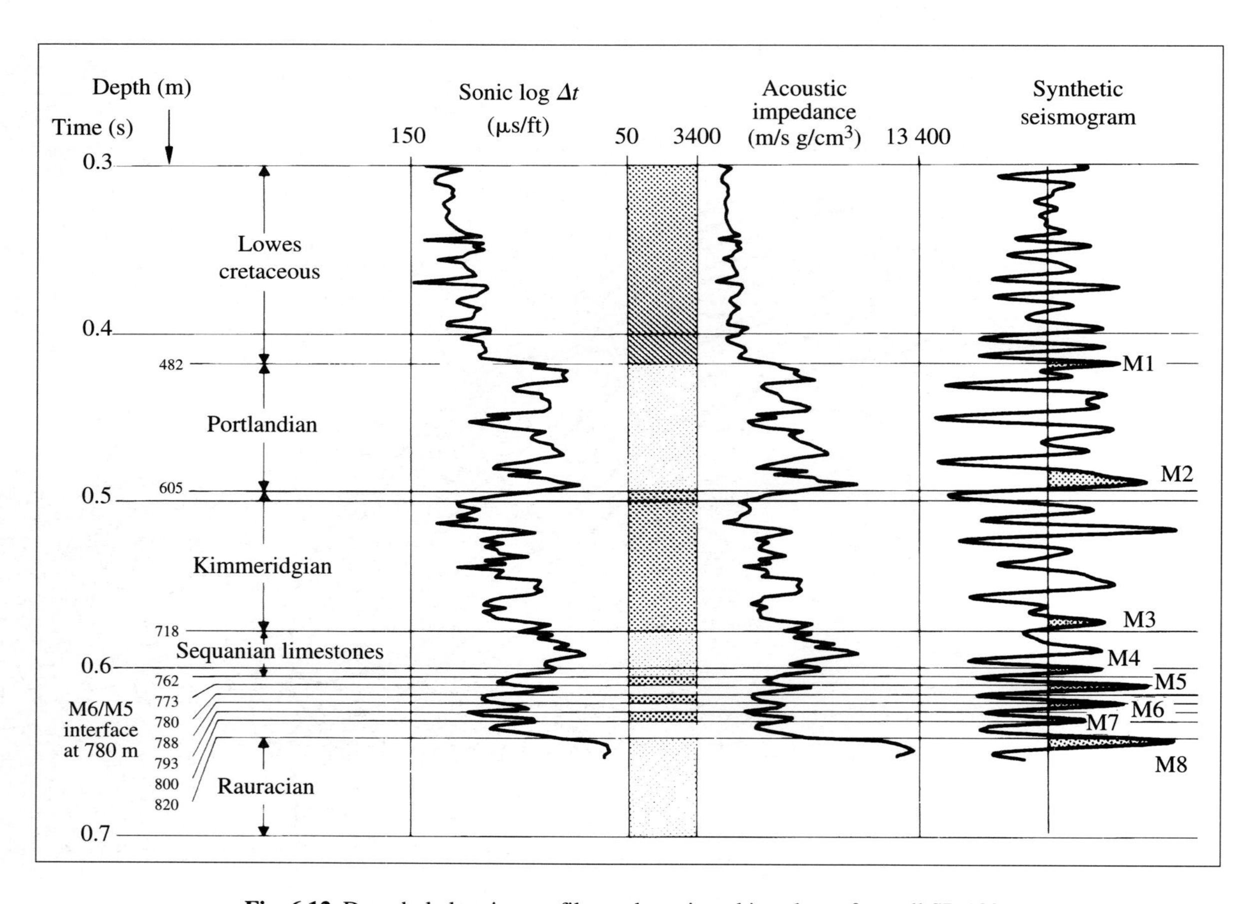

Fig. 6.12 Downhole logging profiles and stratigraphic column for well SR 180.

Line No. 1 was recorded in December 1982 on the northwestern limb of the main structural high. A falling weight source (Sourcile, *IFP*) was used to generate *P* waves, with data from three drops stacked for each emission point. The sampling rate was 1 ms, while the spacing of traces was 10 m using 9 geophones per trace. A 24-fold coverage was employed for line No. 1 and 48-fold for lines 3, 4 and 5.

Data processing

The basic processing of the seismic data consisted of editing, amplitude recovery, deconvolution, static and move out corrections, residual static corrections, common depth-point stacking, filtering and display.

For seismic line No. 1, particular attention was paid to the basic static corrections, which were determined using an algorithm based on a delay method with an automatic picking procedure (Coppens, 1983). The precision on the static corrections thus obtained is very good. In fact, the amplitude of the static correction residuals calculated after application of a dynamic time correction is no more than 2 ms. The seismic section recorded from line No. 1 is presented in Fig. 6.13.

To facilitate calibration of the seismic section with the well data, an operator was applied to the seismic section with the aim of converting the emitted signal to a zero-phase signal. The zero-phase signal section was then migrated to improve the lateral range of investigation and recover the lateral variations in amplitude of seismic markers.

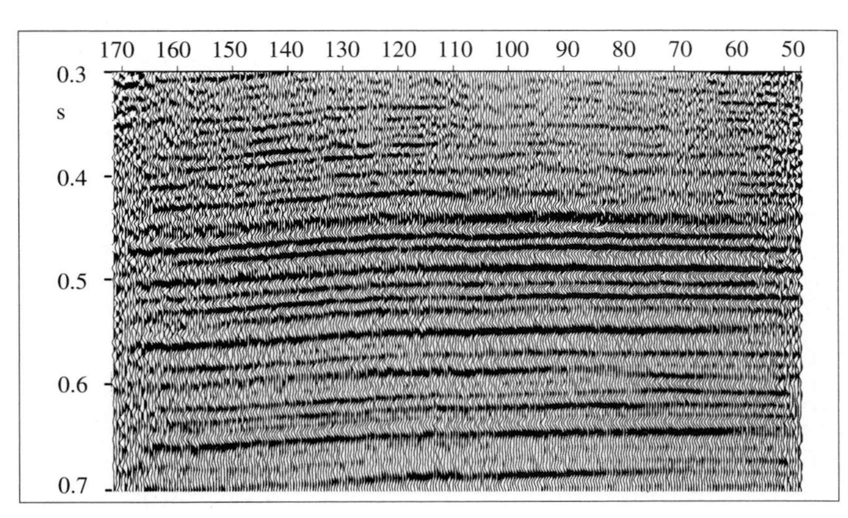

Fig. 6.13
Seismic section (48-fold common depth-point coverage).

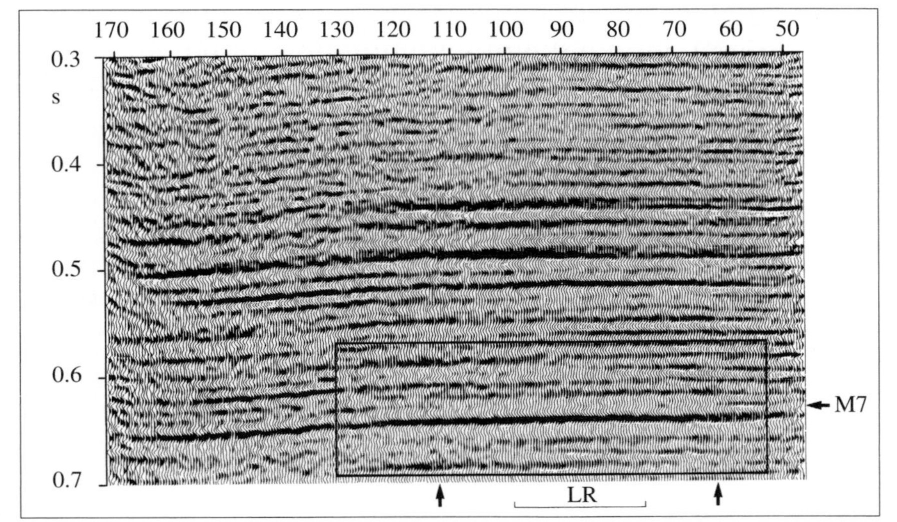

Fig. 6.14
Migrated seismic section.

A decrease in amplitude of the M7 horizon can be observed on the migrated section (Fig. 6.14); this is associated with the top of the lower reservoir unit situated at 0.63 s between shot points 65 and 115.

Results

The decrease in amplitude observed on the migrated section between shot points 65 and 115 at 0.63 s is associated with a loss of reflectivity of the M7 marker corresponding to the top of the lower reservoir unit. Generally speaking, when gas replaces water in an unconsolidated sandy reservoir surrounded by shale, the main effect is an increase in amplitude (bright spot) of the reflection associated with the shale gas-sand interface (Domenico, 1975) and a decrease of seismic velocity in the reservoir.

In the present example, when gas replaces water in the sandy reservoir layers, the acoustic impedance in both lower and upper units falls to values similar to those measured in the intermediate cap rock. As a result, the acoustic impedances of the upper reservoir, the intermediate shale cap rock and the lower reservoir become almost equal, while the top of the lower reservoir shows a decrease in reflectivity. This hypothesis is supported by the synthetic seismogram shown in Fig. 6.15.

Inspection of Fig. 6.15 also enables a judgement to be made on the quality of correlation between the real and synthetic seismogram. The amplitude anomaly observed at the top of the lower reservoir places limits on the position of the gas bubble, which is itself associated with a high degree of gas saturation.

The development of the gas bubble with time can be followed by comparing seismic sections recorded at different times. However, an analysis of seismic sections obtained from conventional data processing shows that the quality of seismic data was better in

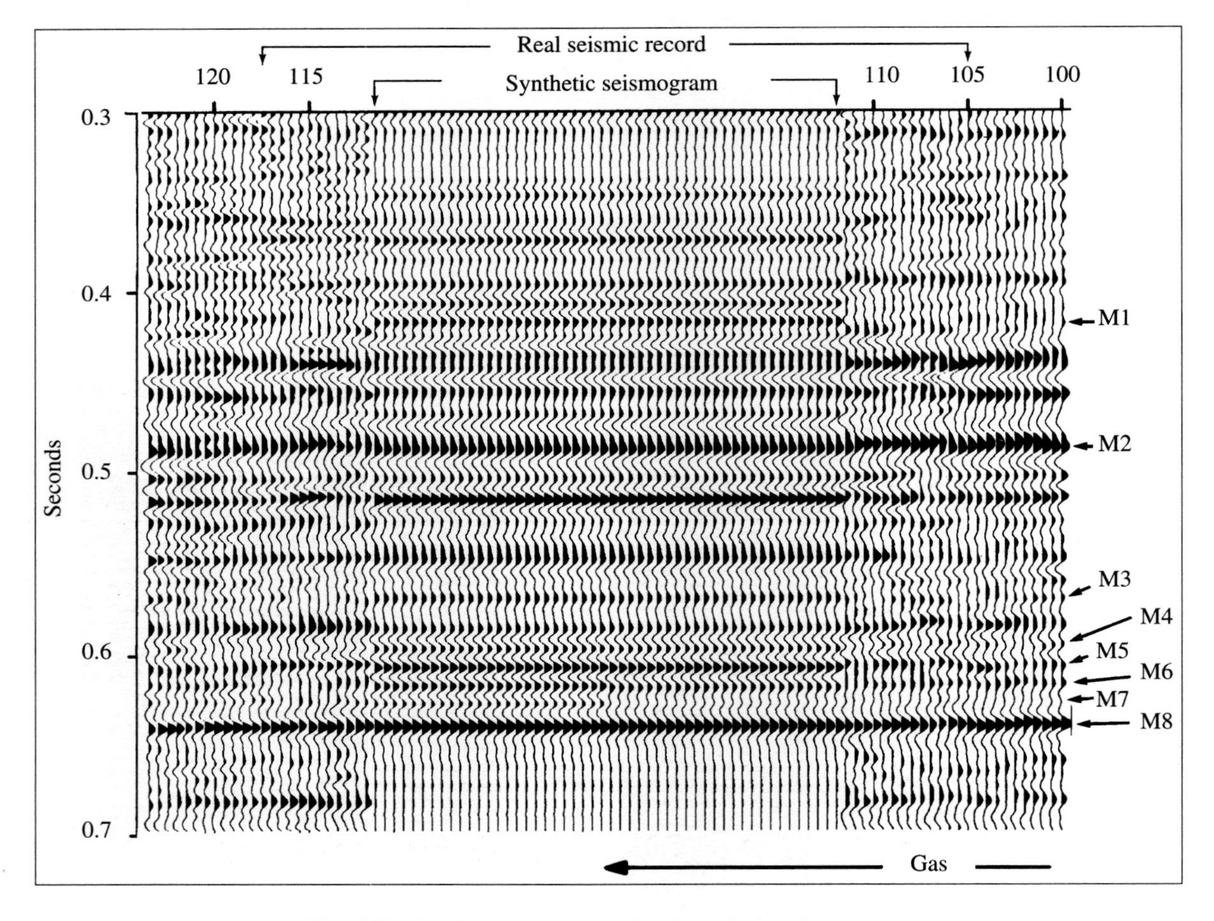

Fig. 6.15 Correlation between real and synthetic seismograms.

April 1981 than in January 1982. This difference is mainly due to climatic variations and changes in the surface conditions.

In a sandy reservoir, the compressional wave velocity decreases as gas replaces the water. This brings about an increase in the travel times of reflectors situated beneath the reservoir, which can then be estimated by measuring the residual time shift based on two CDP traces measured at the same site but at different periods. In, fact the difference between the time shifts ΔT_2 and ΔT_1 —measured respectively beneath and above the reservoir— takes into account the temporal variations due to surface conditions and geophone coupling. ΔT_2 incorporates the same temporal variations as ΔT_1, but shows an additional effect due to the velocity decrease caused by the replacement of water by gas.

Residual time shift curves ($\Delta T = \Delta T_2 - \Delta T_1$) were calculated for each of the profiles recorded in April 1981 and January 1982 in the southeastern part of the anticline. In order to eliminate erroneous time shift residuals, the curves were filtered (median values taken on five points) and then smoothed. Fig. 6.16 shows the seismic sections obtained from line No. 5 at bot periods.

The residual time shift curve presented on Fig. 6.17 shows and anomaly —with a mean amplitude of 0.6 ms— situated between shot points (SP) 30 and 65. According to Toksöz (1976), as an indicative example, a sandstone of porosity 16% at a depth of 1000 m shows a decrease in P wave velocity from 4100 m/s to 3900 m/s as gas replaces the brine in the reservoir formation. For a reservoir of thickness 20 m, this decrease in velocity leads to an increase of 0.5 ms in the arrival times of waves coming from reflectors situated beneath the reservoir zone. The anomaly in time shift residuals apparent on Fig. 6.17 is thus probably related to a change in gas volume in the reservoir between April 1981 and January 1982.

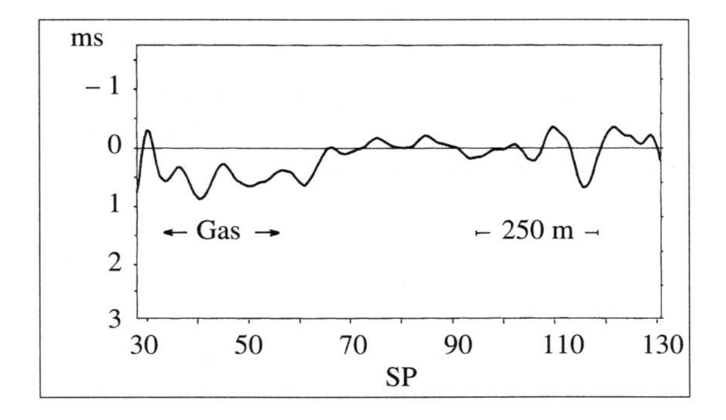

Fig. 6.17 Variation of residual time shifts along seismic line No. 5.

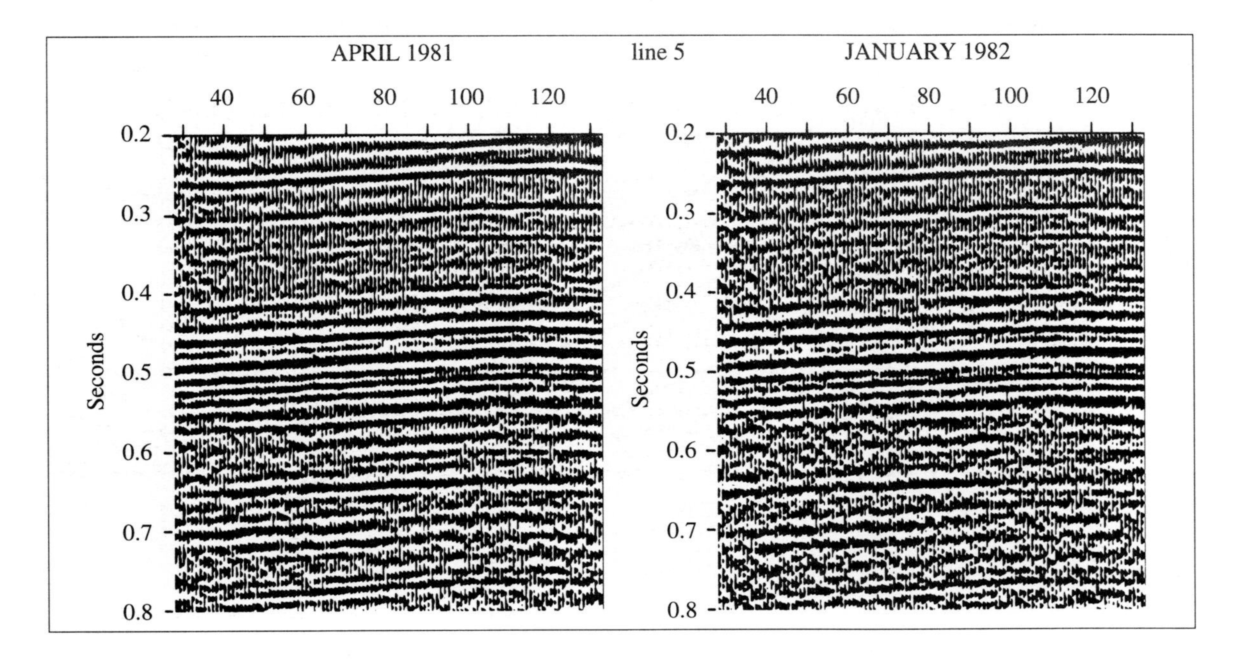

Fig. 6.16 Comparison of seismic sections obtained at different periods.

The structure contour sketch map presented on Fig. 6.18 shows the location of anomalies in time shift residuals observed for lines 3 and 5, as well as the gas bubble front positions for April 1981 and January 1982 derived from water table depth readings. In April 1981, the water table at well SR 104 was at a depth of – 695 m (datum plane = mean sea-level) and the total volume of gas stored was 900 million cubic meters (N). No significant residual time shift was detected on line No. 4. Thus, it may be noted that anomalies in the residual time shifts are a good indication of the migration of the gas bubble front between April 1981 and January 1982.

Conclusion

P wave vertical seismic profiling has enabled a detailed survey of the top of the Sequanian reservoir zone between 700 and 800 m depth. In certain cases, it is even possible to trace the intermediate cap —several meters thick— that is situated between the upper and lower units of this reservoir.

In the case of the Gournay-sur-Aronde reservoir, the acoustic impedance in both the upper and lower units decreases as gas replaces the water in the sand deposits, falling to values typical of those recorded in the intermediate cap rock. As a result, the top of the lower reservoir unit shows a loss of reflectivity. Thus, the presence of gas leads to a decrease in the amplitude of the reflected signal.

Gas bubble movement from one period to another can be pointed and by comparing carefully seismic section recorded at different time period. Correlations based on an appropriate time-window indicate that gas impregnation of the reservoir zone appears to be characterised by a statiscally significant delay. Since the corresponding time delays are of the order of a fraction of a millisecond, the detection of Δt anomalies is only possible due to the high quality of the seismic data (high resolution and good signal to noise ratio) and very careful application of the data processing procedure. Moreover, since the *P* wave residual time shifts is observed over a fairly wide area, it becomes possible —with a certain degree of confidence— to explain the phenomenon in terms of gas saturation of the reservoir.

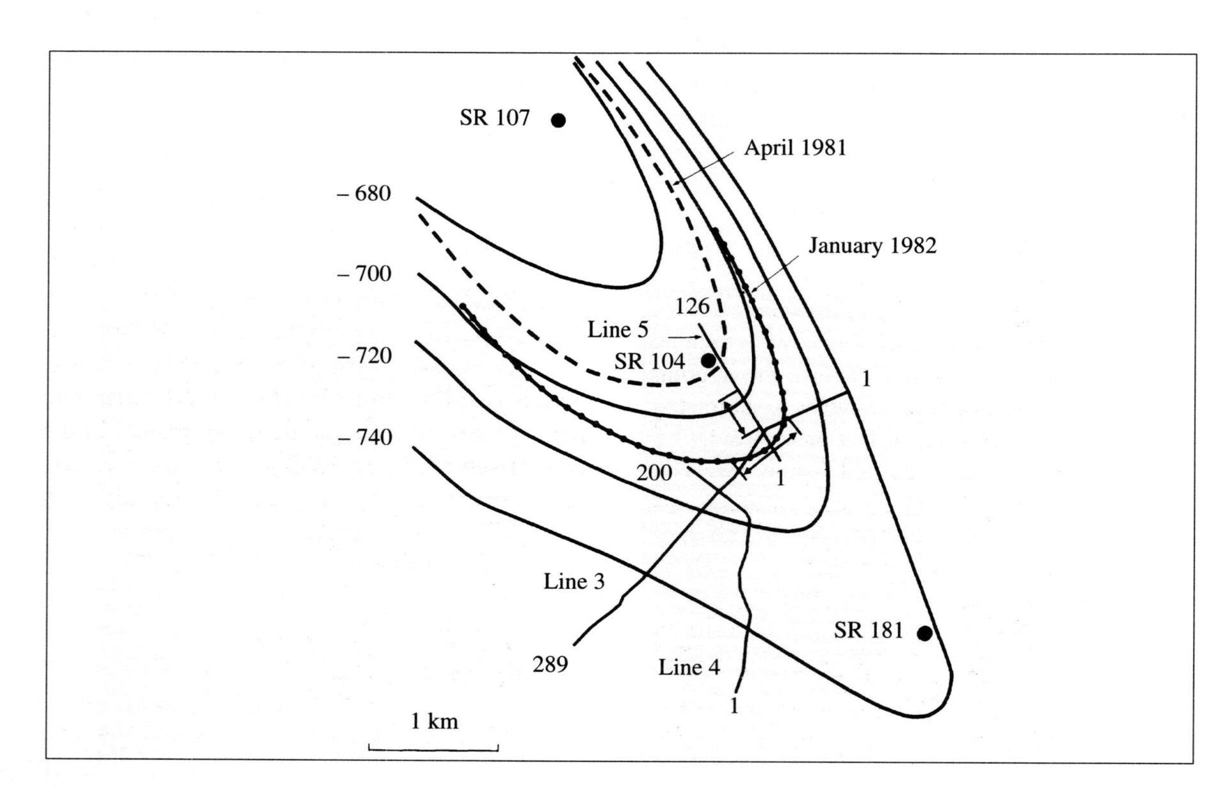

Fig. 6.18 Gournay-sur-Aronde gas storage reservoir: position of gas bubble fronts and values of ΔT anomalies. *(Courtesy of Gaz de France)*

Subwell prediction and detection of overpressure

(S. Brun, P. Grivelet and A. Paul, 1985)

In the context of this case study, Brun et al. (1985) have used Vertical Seismic Profiles in the detection of undercompacted zones corresponding to an acoustic impedance anomaly. The subwell prediction of these acoustic impedance profiles was based on an inversion technique (Grivelet, 1985).

The general principles of VSP inversion are presented in previous chapters of this text. The method employed in this case study consists of firstly determining the layer boundaries and then attributing an acoustic impedance to each layer.

A horizontally layered model is established for depths beneath the VSP trace, in such a way that the downgoing incident trace produces an upgoing synthetic trace —including multiples— that is as close as possible to the field trace obtained at the depth of the selected recording. The first iteration is performed using an initial model which generally contains less than 50 layers. By using a downgoing wavelet as a reference, the strongest reflection coefficient is sought from the upgoing trace and the corresponding wavelet is subtracted. The strongest remaining reflection coefficient is then picked out and discarded in its turn. This process is continued until the chosen number of boundaries is attained, thus limiting the study to the most highly contrasted events while eliminating noise. The time-series so obtained contains both the primaries and main multiples generated in the subwell zone.

The subsequent iteration keeps the initial layer boundaries, but calculates the impedance value associated with each layer in order to obtain a synthetic trace that coincides relatively well with the real upgoing trace. Since the absolute value of the impedance is known at the recording depth —using density and sonic log measurements— a series of reflection coefficients can be based on this reference value.

Certain constraints can be imposed in the light of the local geology. These may be either fixed (reliable information) or introduced with an associated interval of error (uncertain information). The constraints are taken into consideration at each iteration, with an error bracket being calculated for each reflection coefficient. The final result is attained when the error is within the limits of the layer thicknesses.

The two examples presented in Figs. 6.19 and 6.20 are taken from a study carried out in the Niger delta, where an undercompacted shaly formation is at the origin of a fault system that cuts through the succession right up into the Akata shales. High pressures can propagate along the faults to adjacent or overlying reservoirs into blocks that are bounded by pairs of faults. The variation in compaction which is reflected in the decrease of acoustic impedance with depth is therefore gradual in the case of sub-horizontal reflectors. On the contrary, the compaction may increase very rapidly with depth across fault boundaries which correspond to steeply dipping reflectors.

Subwell prediction

In the example illustrated in Fig. 6.19, a well was drilled into the centre of a large anticline far away from any growth faults. On the basis of correlations with neighbouring wells in the region, the overpressured zone beneath this well was predicted to occur at 3720 m.

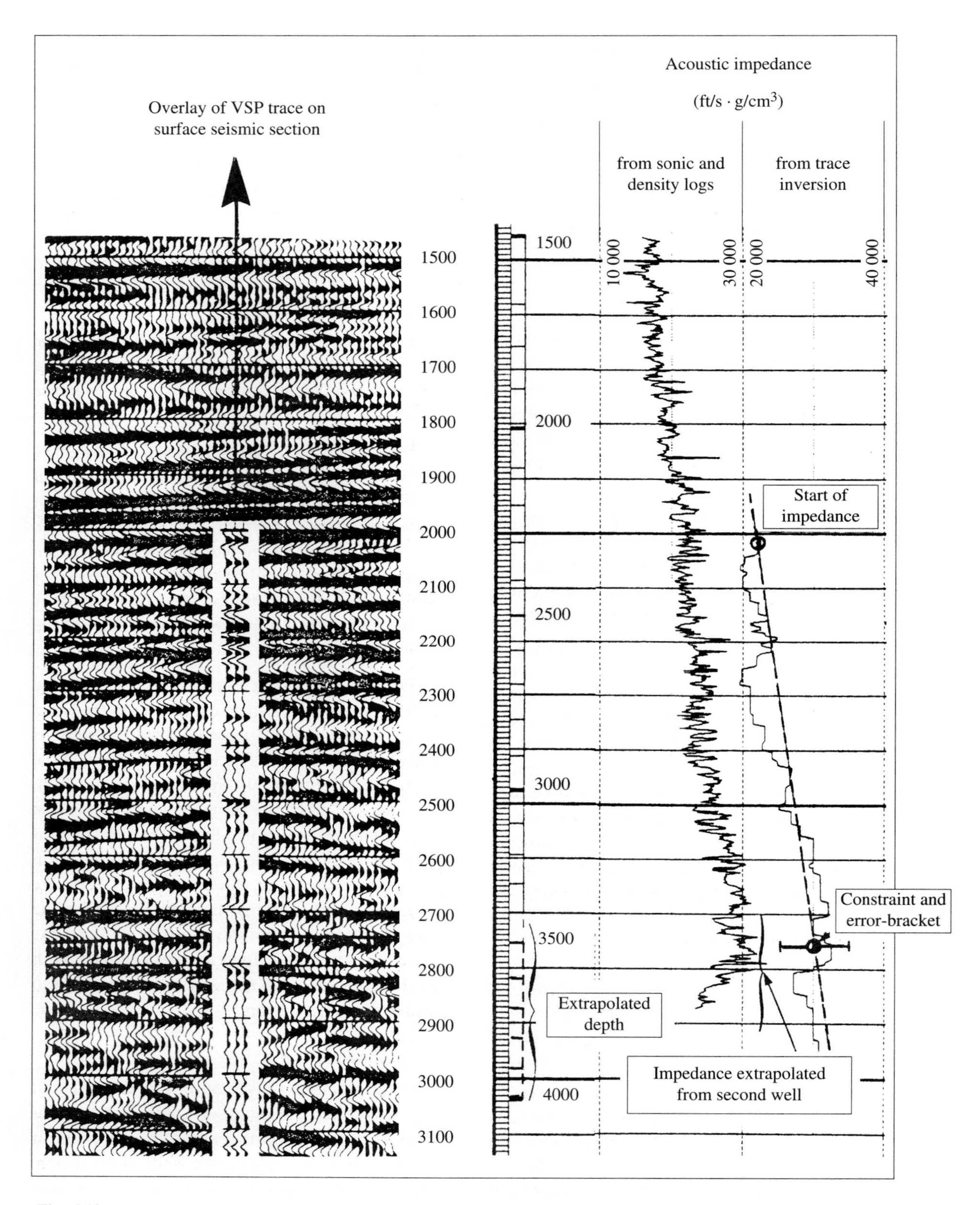

Fig. 6.19 Well logs and vertical seismic profile (VSP) used for prediction of overpressure in sub-horiontal formations.

A Vertical Seismic Profile was recorded, even though the well bottom was at a total depth of 3220 m. The inverted trace used in this study was recorded farther up in the well, at a depth of 2300 m.

The following constraints were applied for inversion of the acoustic impedance log:

- 22 000 ft/s · g/cm^3 at the recording depth of 2300 m.
- 30 000 ± 5000 ft/s · g/cm^3 at the well bottom (at 2.75 s).

The large error-bracket associated with this value is due to the fact that the well was not drilled all the way to the bottom, and the impedance was extrapolated from the overall compaction curve.

Inversion was carried out using a model with 50 layers situated in the 2-3 s TWT interval. On the calculated acoustic impedance log, an abrupt jump can be see at 2.81 s which corresponds to an extrapolated depth of 3580 m.

In the light of this information and other measurements made during the earlier drilling, the well was drilled to a final depth of 3580 m whilst paying attention to the effects of overpressure predicted in this zone.

A second well was then drilled alongside the previous one (in the same environmental conditions), using a procedure that enabled drilling through the overpressured zone.

Figure 6.19 presents the complete log of acoustic impedance obtained from density measurements and sonic logging. The impedance log predicted with the help of seismic data is very close to that obtained from well logging, clearly showing the main geological units despite the choice of a model with a limited number of layers.

The overpressured zone predicted at 2.81 s corresponds to a transition zone indicated on density and sonic logs at 2.8-2.85 s. In such a context, the identification of overpressured zones is relatively easy since there are no clear discontinuities that might be attributed to changes in lithology. In fact, a distinct impedance trend is seen in the centre of this zone of the Niger delta; the only perturbation comes from the presence of overpressure fronts. Identification is also facilitated by the fact that the overpressure fronts and all the reflectors are horizontal, which means that they are positioned at the correct times on the VSP trace.

Detection of overpressured zones with steeply dipping pressure fronts

The studied well was drilled to a depth of 2100 m in the recent part of the Niger delta (Fig. 6.20). The abnormal pressure regime encountered in this area is probably associated whith a fault system that is actually traversed by the well. However, no clear indication of poor compaction has been found.

A VSP was recorded from 2100 to 1000 m in order to evaluate the possibility of predicting arrivals in under-compacted zones of the type already encountered elsewhere.

An inversion procedure was applied to the trace recorded at 1060 m. The initial model used for inversion is based on 50 layers between 0.8 and 1.8 s, using an initial impedance of 28 000 ft/s · g/cm^3 at 0.8 s and 30 000 ± 15 000 ft/s · g/cm^3 at 1.8 s.

Figure 6.20 shows the VSP upgoing waves, the acoustic impedance log obtained from well logs (sonic and density logs) and the acoustic impedance log calculated from inversion of the 1060 m trace. In addition, this figure presents a comparison between the recorded VSP trace and two synthetic traces derived from surface seismic and logging surveys of acoustic impedance.

The inverted trace displays numerous discontinuities. At the top, the weak and strong impedances at 0.91 and 0.96 s, respectively, can be correlated with the sonic and density log profiles. Below this, the two main discontinuous at 1.02 and 1.22 s appear to be offset with respect to the corresponding discontinuities on the VSP (upgoing waves at 1.07 and 1.28 s). These discontinuities appear as oblique seismic arrivals on the VSP upgoing waves, which are superposed onto horizontal upgoing waves associated with flat reflectors.

The impedance profile obtained from well logs shows no clear compaction trend. However, given that the formations traversed by the well are the classic shaley to sandy deposits of the Niger delta, the sharp jumps at 1.07 and 1.28 s (at 1500 and 1850 m) can be attributed to the passage into overpressured zones. This hypothesis is supported

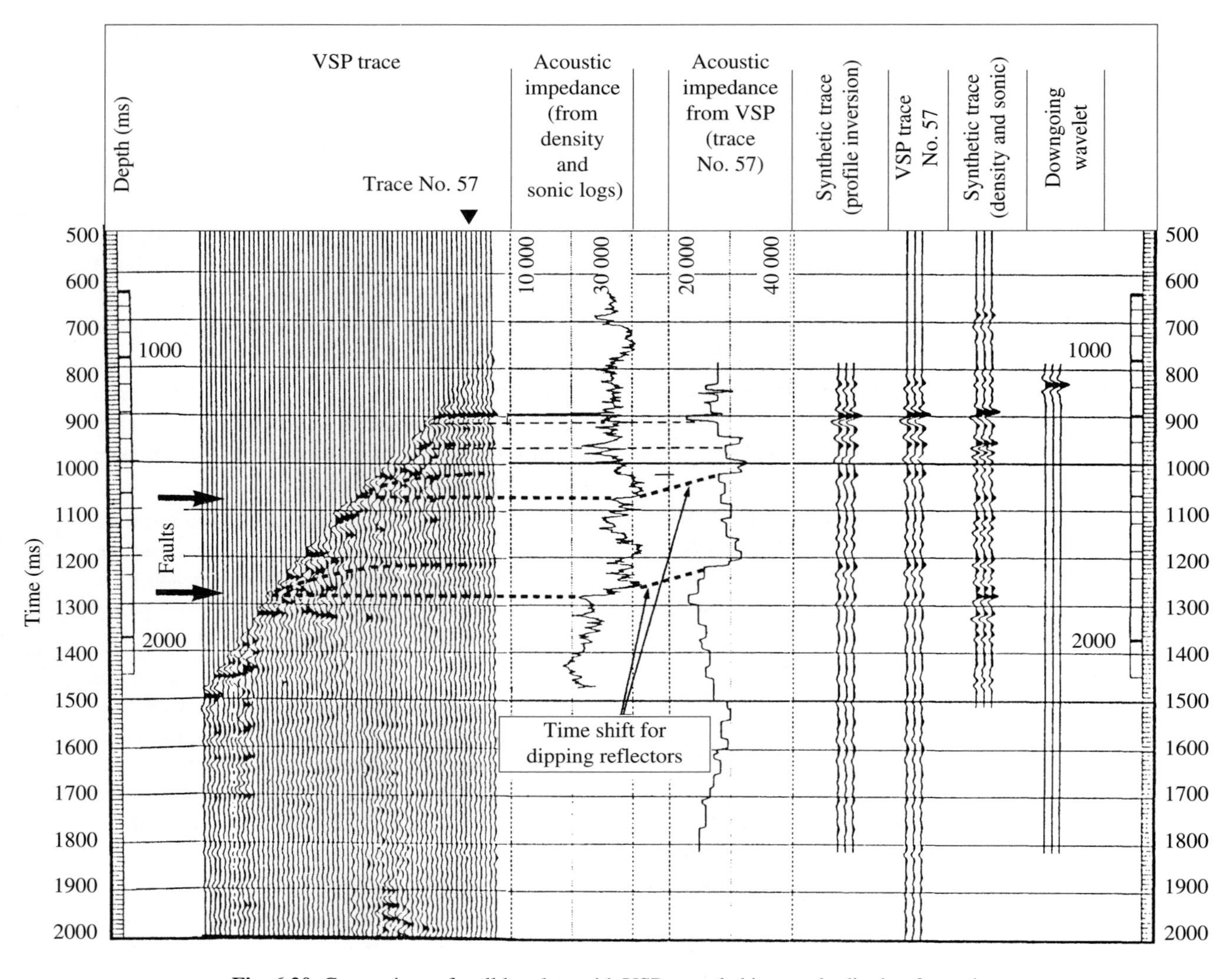

Fig. 6.20 Comparison of well log data with VSP recorded in steeply dipping formations.

by the dipmeter log on Fig. 6.21, which shows the occurrence of two fault-type discontinuities at the same depths as indicated by VSP. On the impedance log obtained from inversion, the horizontal reflectors occur at the expected times, whereas the steeply dipping reflectors are shifted towards lower time values (by an amount depending on the dip). It is evident that this type of one dimensional modelling cannot correctly locate the dipping events. Nevertheless, when dipping events are strongly contrasted, they become superposed onto the tabular horizontal layering in a one dimensional model. In this way, the dips of such events can be evaluated and the real value of the time shift can be approximated. Using the inverted traces in the studied well as a basis for prediction, it appears that the most likely position for the overpressure front corresponds to the sharp decrease in impedance at 1.22 s. On the other hand, the corresponding reflector has an apparent dip and could be located at 1.28 s. The strong dip at this level has been confirmed by the dipmeter log, which shows a fault plane dipping at an angle of 40° to the Southeast. The onset of under-compaction at this level is further supported by the presence of a thick massive shale unit just beneath the fault.

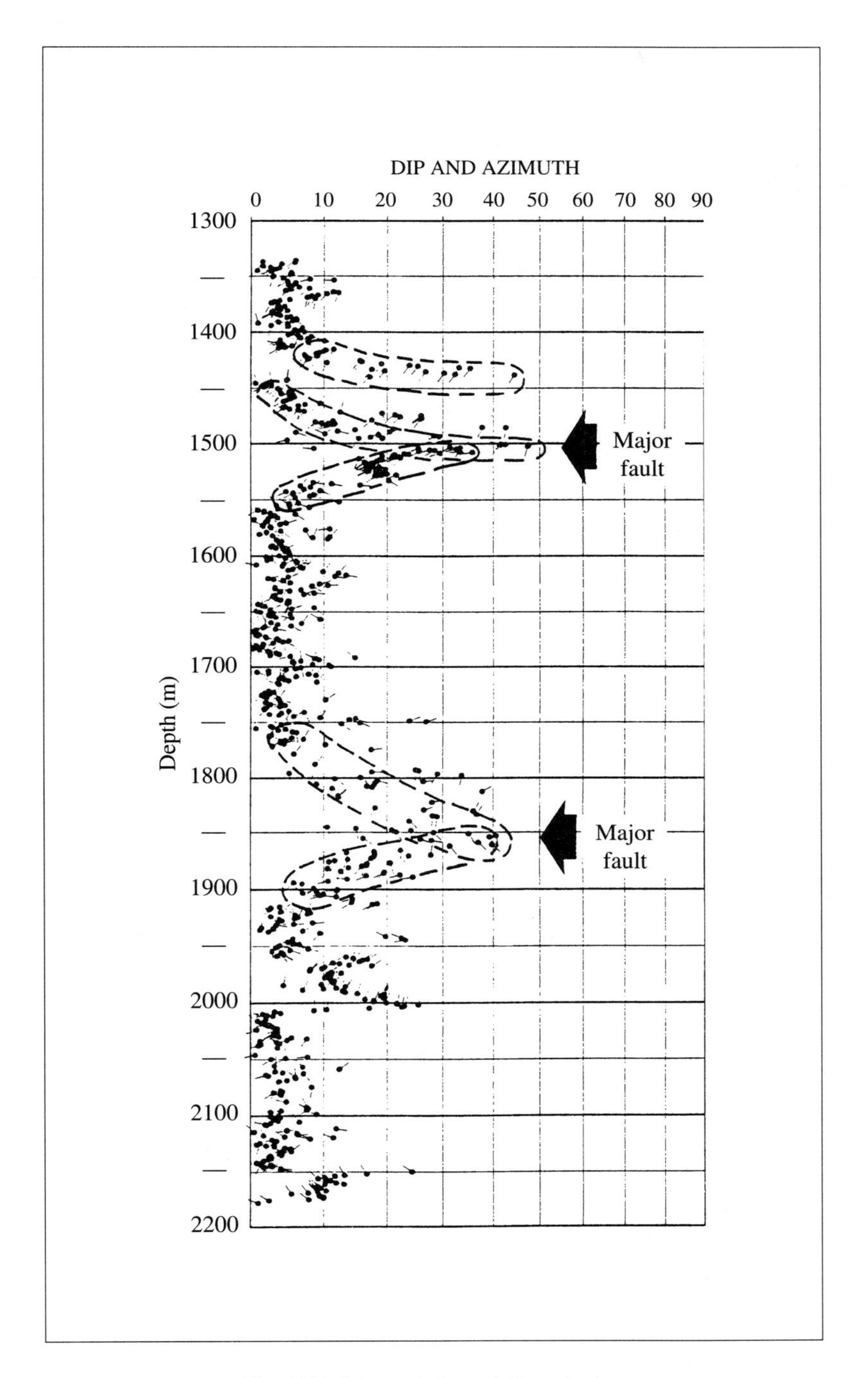

Fig. 6.21 Interpretation of dipmeter log.

Use of wireline logs in the interpretation of seismic lines across salt-bearing formations

(P. Renoux, 1989)

Renoux (1989) has carried out a comparative study on three Permo-Triassic evaporitic sequences of Western Europe in terms of their seismic and well log responses. This author shows the advantages of using acoustic and density logs (see Chapter Four: Synthetic seismograms) in understanding the seismic response of different evaporitic units, and points out the absolute necessity of using the density log to calculate sets of reflectivity coefficients. Furthermore, it is important to evaluate the resolving power as well as the behaviour of acoustic impedance ($\rho/\Delta t$) as a function of lithology

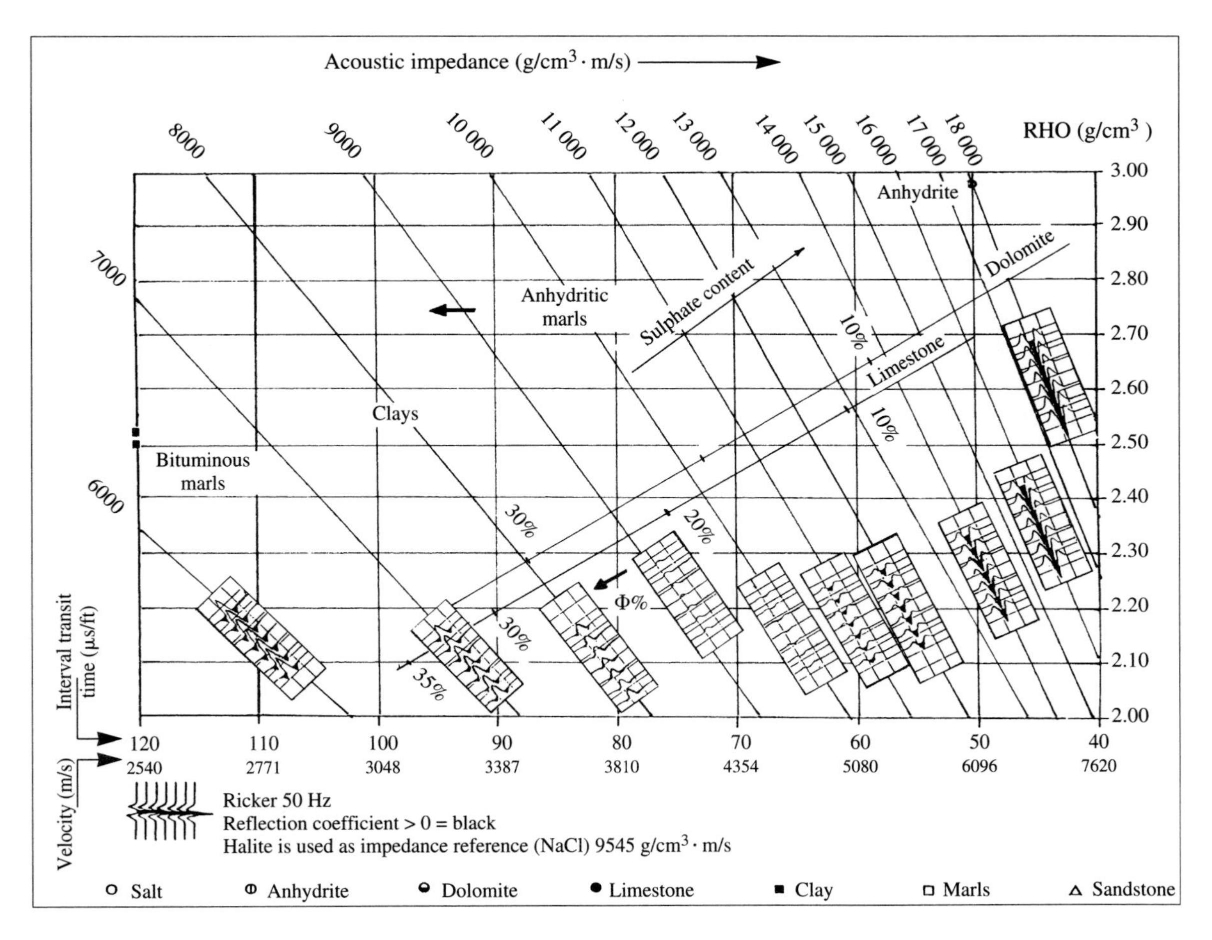

Fig. 6.22 Crossplot showing acoustic impedance as a function of density and velocity.

before using the seismic signature to distinguish different sedimentary sequences.

This study of acoustic impedance behaviour is based on the construction of Density-Velocity crossplot, where the plotted impedance curves enable an improved understanding of the seismic response of various lithologies according to their position on the chart.

Although salt-bearing formations display very different characteristics amongst themselves, they are generally clearly identifiable by logging techniques.

In principle, the conditions of deposition are such that major impedance contrasts are commonly linked to the juxtaposition of halite (NaCl, with a very low density of 2.16) with other formations such as anhydrite which usually follow in the sequence.

However, the sedimentary environment may be perturbed during the accumulation of anhydrite by the arrival of shaley deposits which lead to the formation of anhydritic marls instead of pure anhydrite. This produces impedances which are intermediate between those of anhydrite and salt, thus preventing the detection of salt layers on seismic sections.

Zechstein formations

The Zechstein formations encountered in well A, which were deposited in a shallow-water marine environment, make up a typical salt-bearing succession containing halite at the base followed by anhydrite and dolomite towards the top.

The variations in acoustic impedance are principally related to the large density variations observed in these formations, whereas the velocity variation (Δt) only introduce a low frequency component. The (ρ-Δt- $\rho/\Delta t$) crossplot shows a clear separation of lithologies (Fig. 6.22).

As shown by the logs on Fig. 6.23, the boundaries of the different formations are clearly defined. After stratigraphic deconvolution to achieve a zero-phase seismic signal, the picking of reflections on seismic sections is relatively easy in a medium displaying strong contrasts in acoustic impedance.

A 3-m-thick salt layer (S_1) occurs within the Z_2 anhydrite unit, thus giving rise to a strong impedance contrast which appears equally well on seismic sections.

It is noteworthy that the strong impedance contrast at the top of the Werra Salt introduces a positive "ghost" phase onto the VSP stacked trace as well as the surface seismic section. The top of the salt should actually be picked at the minimum amplitude.

Matching the different reflectors with the well logs (Fig. 6.24) thus enables an improved knowledge of the sedimentary units deposited in this context over wide areas of the Zechstein Basin, i.e.:

(1) the Werra Salt is relatively homogeneous,

(2) the top of the Z_1 unit is characterised by a strong reflector in contact with the anhydrite,

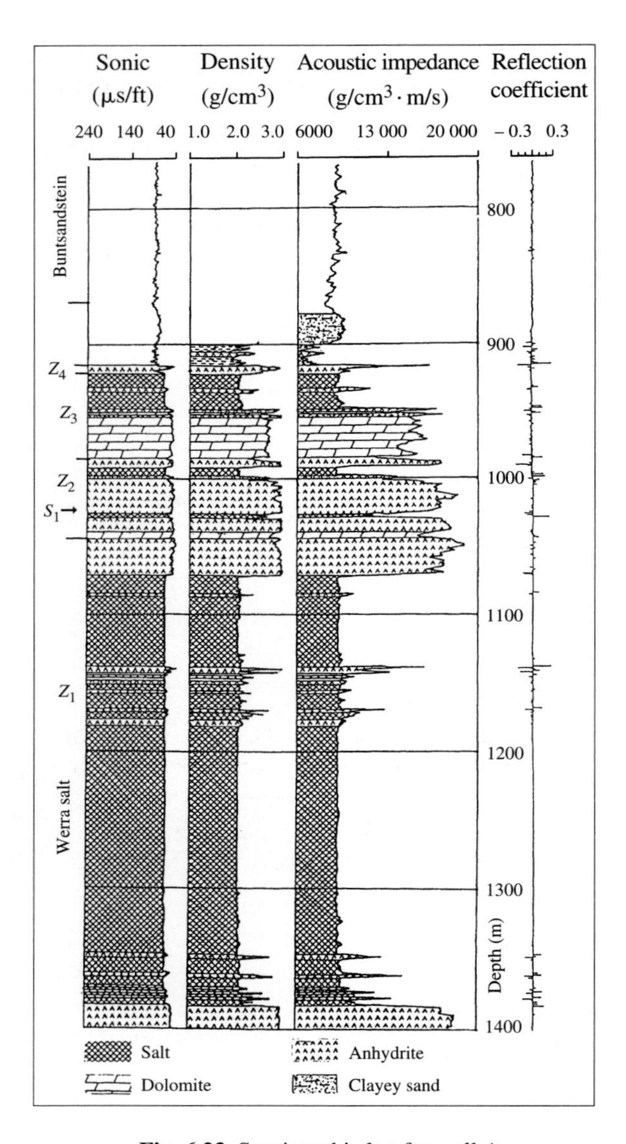

Fig. 6.23 Stratigraphic log for well A.

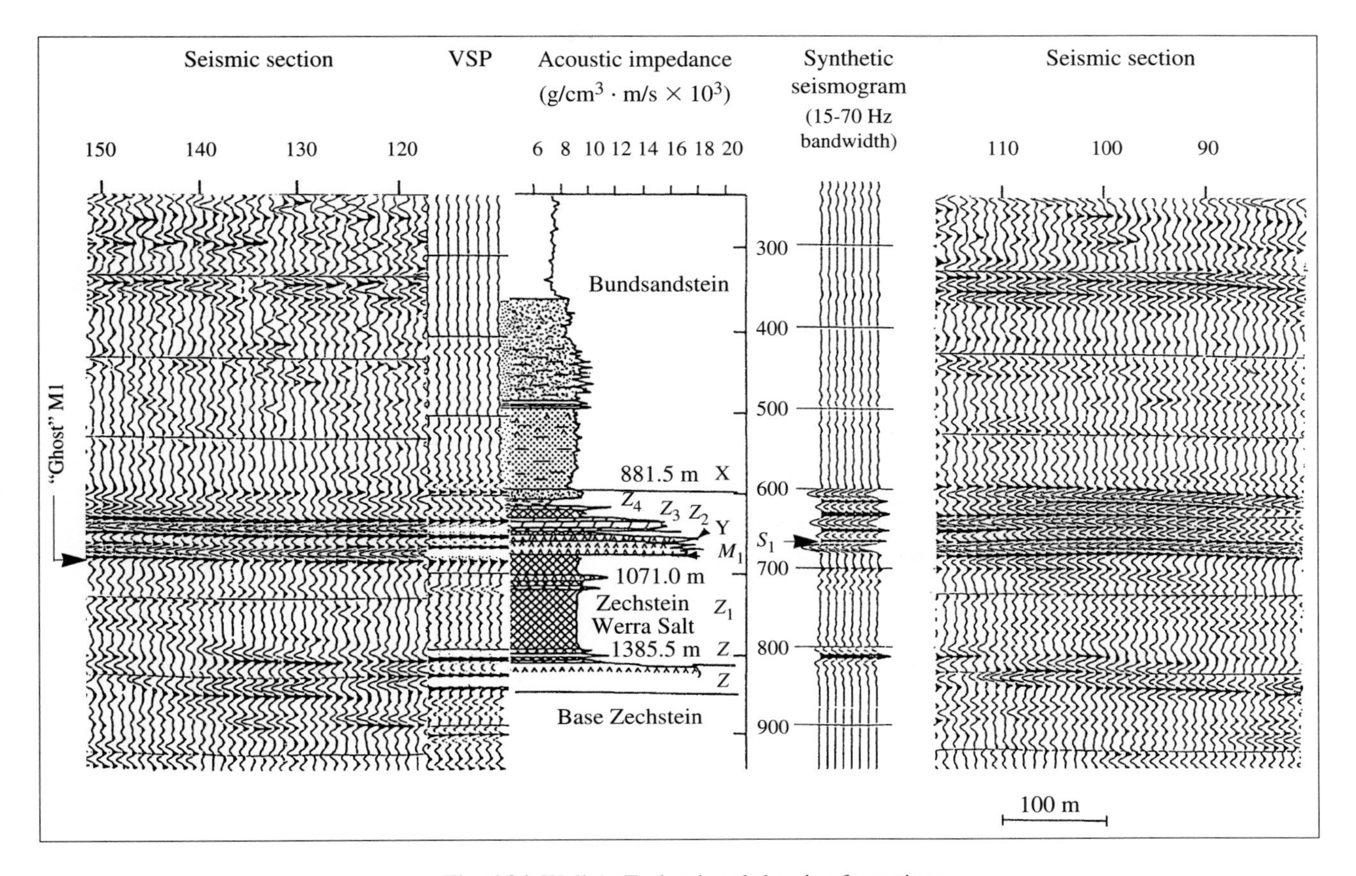

Fig. 6.24 Well A: Zechstein salt-bearing formations.

(3) the anhydrite and dolomite-bearing units Z_2, Z_3 and Z_4 show strong amplitudes,

(4) although the Bundsandstein displays a chaotic facies, it nevertheless shows some reflectors related to anhydritic interbeds.

Muschelkalk Series

The lagoonal facies (sebkha-type sedimentary environment) of the Middle Muschelkalk is intercalated between the transgressive deposits of the Lower Muschelkalk (covered by more or less bituminous marly source rocks) and the carbonates of the Upper Muschelkalk.

The ρ-Δt-$\rho/\Delta t$ crossplot presented in Fig. 6.25 clearly shows three lithological groups corresponding to successive sedimentary environments; the influence of density is just as important as velocity in the calculation of acoustic impedance.

Unlike well A, the top of the upper salt-bearing layer (S_2) is characterised by an increasing impedance gradient due to the presence of a brecciated unit about 10 m thick situated above the anhydrite. The resulting attenuation of contrast prevents any identification of the upper boundary of the salt (Fig. 6.26).

The reflections related to dolomites at the top and base of the limestone are well marked, showing amplitudes that are linked to variations in shale content (*cf.* shaley limestone interval *Sh*) and/or porosity. These variations may bring about the disappearance of the reflections, as can be seen on the seismic section presented in Fig. 6.27.

A fine-scale fitting of the seismic data can be achieved by means of synthetic seismograms, using appropriate wavelets, as illustrated by the comparison of S_1 and S_2 on Fig. 6.28. In the case of the limestones, it is necessary to introduce a wider band wavelet (synthetic seismogram S_1) than that used for the anhydrite interbedded within the salt (synthetic seismogram S_2).

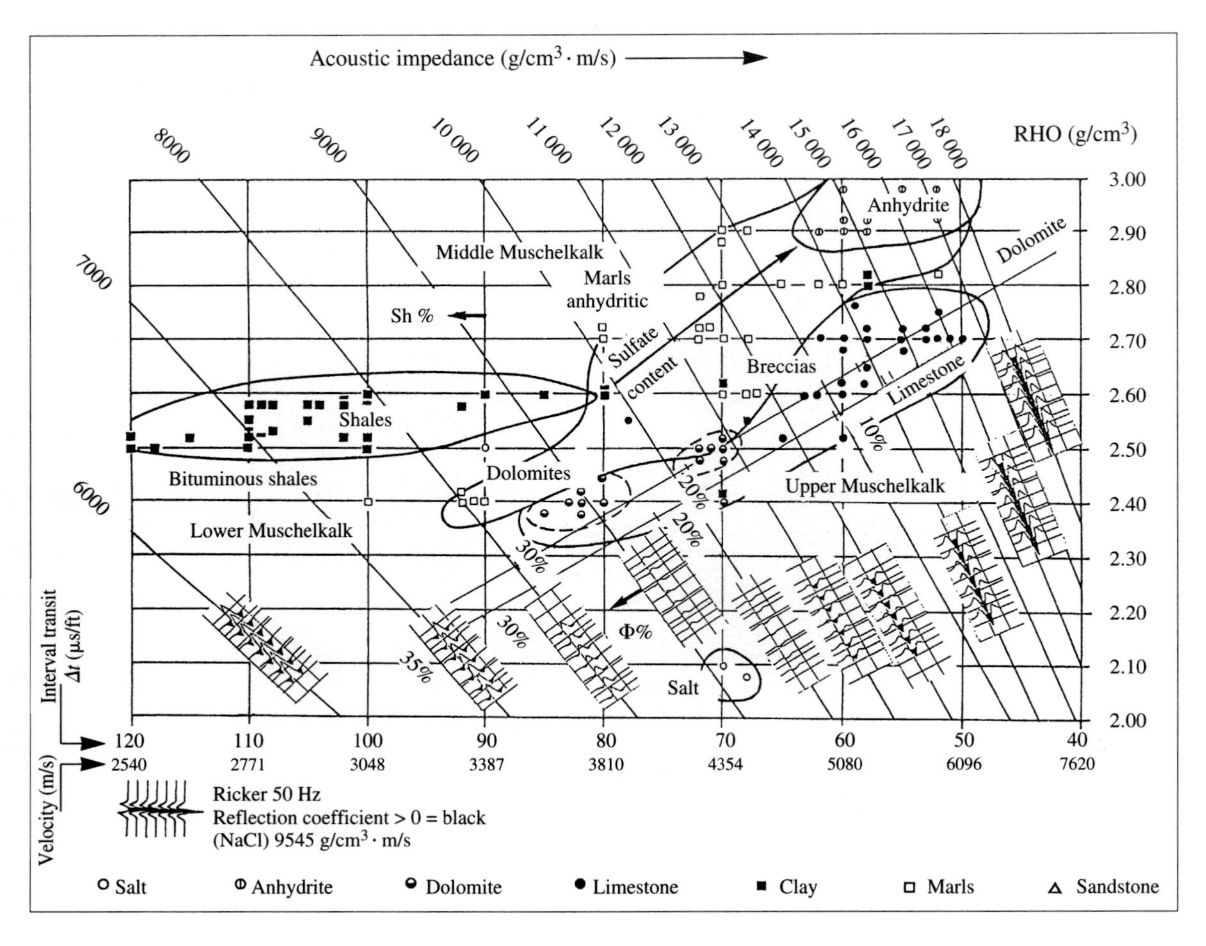

Fig. 6.25 Density-velocity-impedance crossplot for well B.

The vertical resolution of the seismic survey in this case can attain five meters. For thinner beds, amplitude anomalies due to the effects of tuning may be noted on the seismic sections (*cf.* Fig. 6.29; marl layer between anhydrite R_2 and upper anhydrite in well C).

Taking account of the stratigraphic column, the seismic facies are representative of the rock-types present in the succession (provided that the beds are sufficiently thick). Thus, medium amplitudes are observed for the limestone beds and strong amplitudes for the anhydrite. The dolomite beds yield amplitudes that vary as a function of petrophysical properties.

While there is a very good stratigraphic correlation between wells B and C at the base of the salt and above the upper anhydrite unit (beds are flattened with respect to the top of the Upper Muschelkalk dolomite), the lack of correlation in the intermediate strata is shown up by the seismic section presented in Fig. 6.30). Although the R_2 anhydrite unit in well C drapes an erosion surface picked out by breccias, the R_1 reflector in well B corresponds to an anhydrite layer whithin the salt which is totally separate from R_2.

Keuper Series

The halite deposits of Keuper age in well F are intercalated between gypsiferous shales (Lower Keuper) and alternating layers of anhydritic marl with anhydrite.

The absence of reflectors on the seismic section is due to the fact that the lithologies are homo-

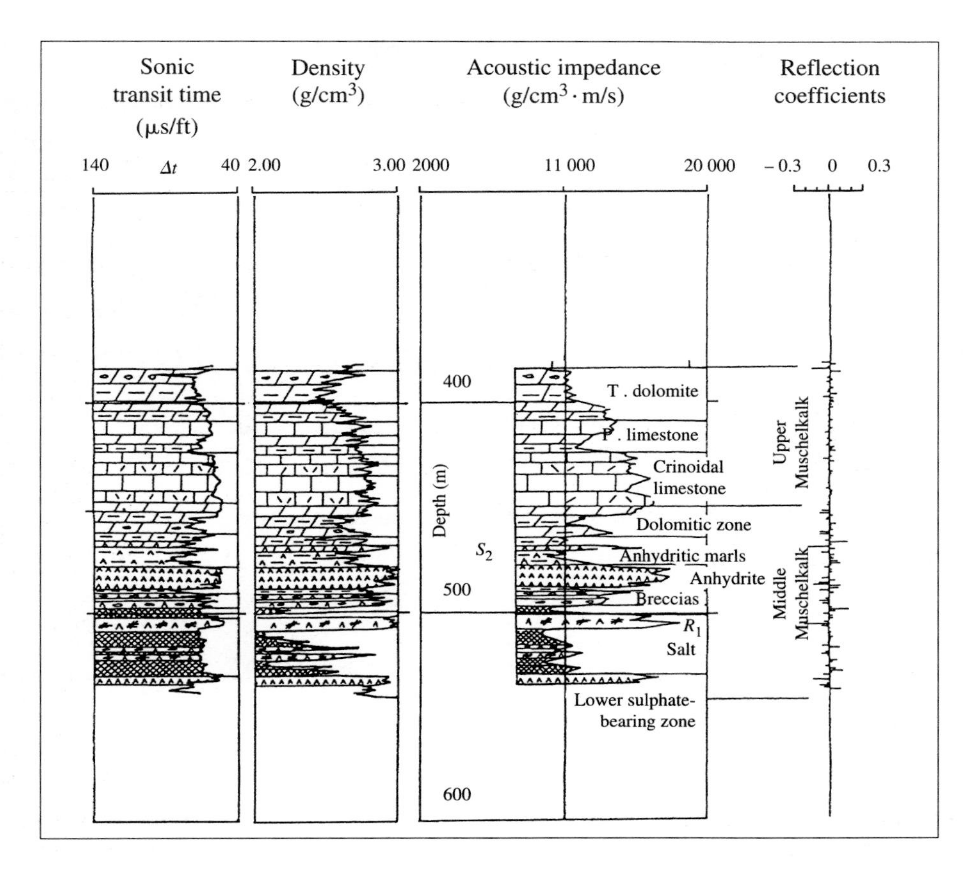

Fig. 6.26 Well log data for well B.

geneous in acoustic impedance (Fig. 6.31). The lower and upper contacts of the salt are weakly apparent on the seismic section of well F (*cf.* Fig. 6.32) and die out progressively away from this site. As a matter of fact, the decrease in impedance contrast corresponds to marls which are richer in sulphates and therefore of higher density.

On the contrary, the dolomite-anhydrite sequence of the Middle Keuper, which is here intercalated between clays of much lower impedance, can be clearly identified on the seismic section. However, the dolomites and anhydrites occuring within this sequence cannot be distinguished from each other due to their similar impedances.

The Rhaetic sands, which locally form hydrocarbon reservoirs, can be picked out relatively well above the clays of the Upper Keuper.

Conclusion

All three groups of evaporites studied here show that the geophysical detection of halite layers is essentially linked to the presence of transitions into sulphate-bearing beds (anhydrite or anhydritic marls). In fact, the supply of shale into the sedimentary basin — by causing an attenuation in impedance contrast — has led to facies which are unfavourable for the development of seismic markers. On the other hand, the presence of anhydrite enables a very clear imaging of erosional effects at the top of the salt, as seen in the case of the Muschelkalk.

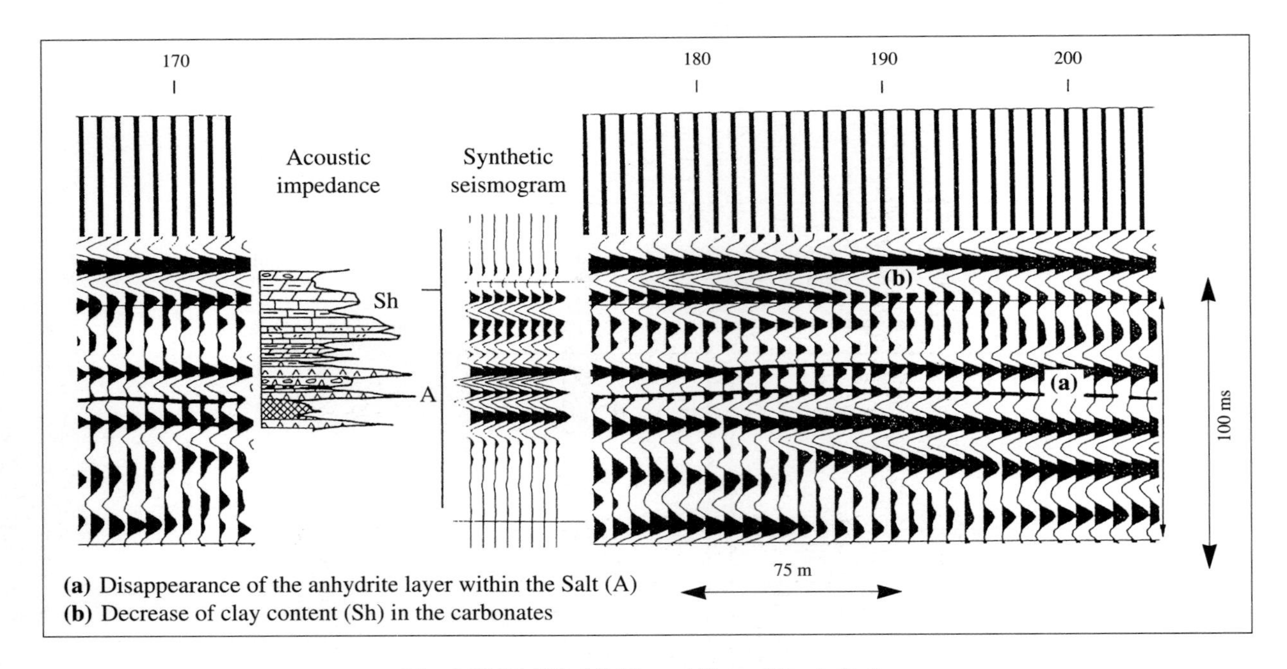

Fig. 6.27 Well B: Middle and Upper Muschelkalk.

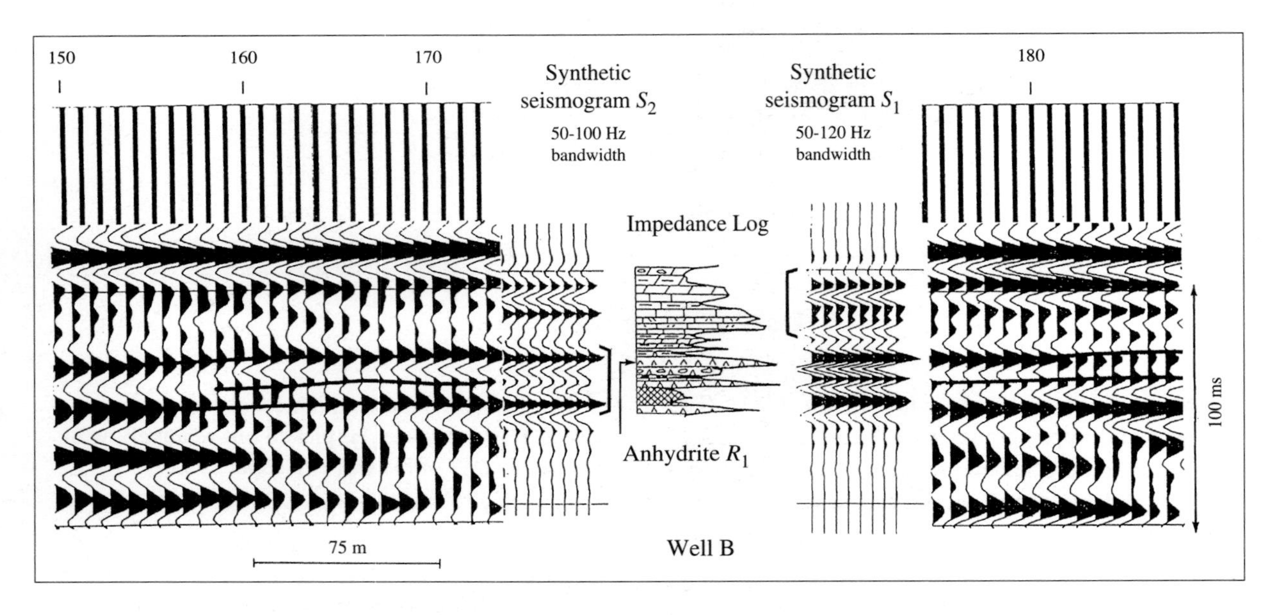

Fig. 6.28 Matching of borehole and surface seismic data.

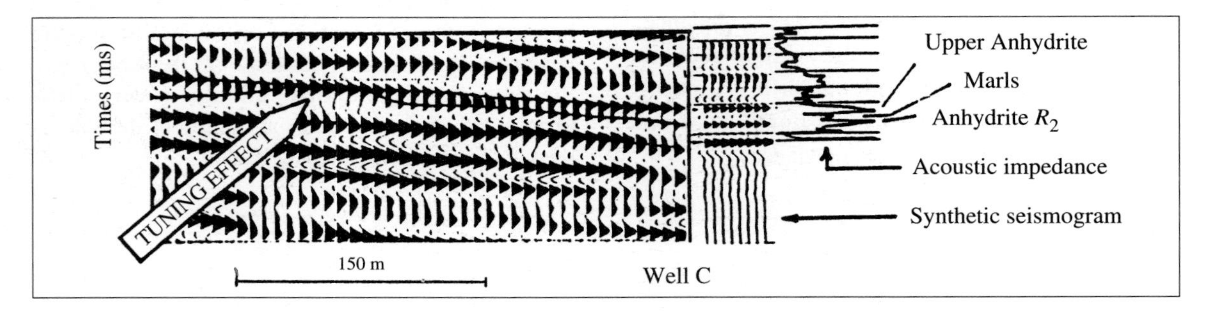

Fig. 6.29 "Tuning" effect.

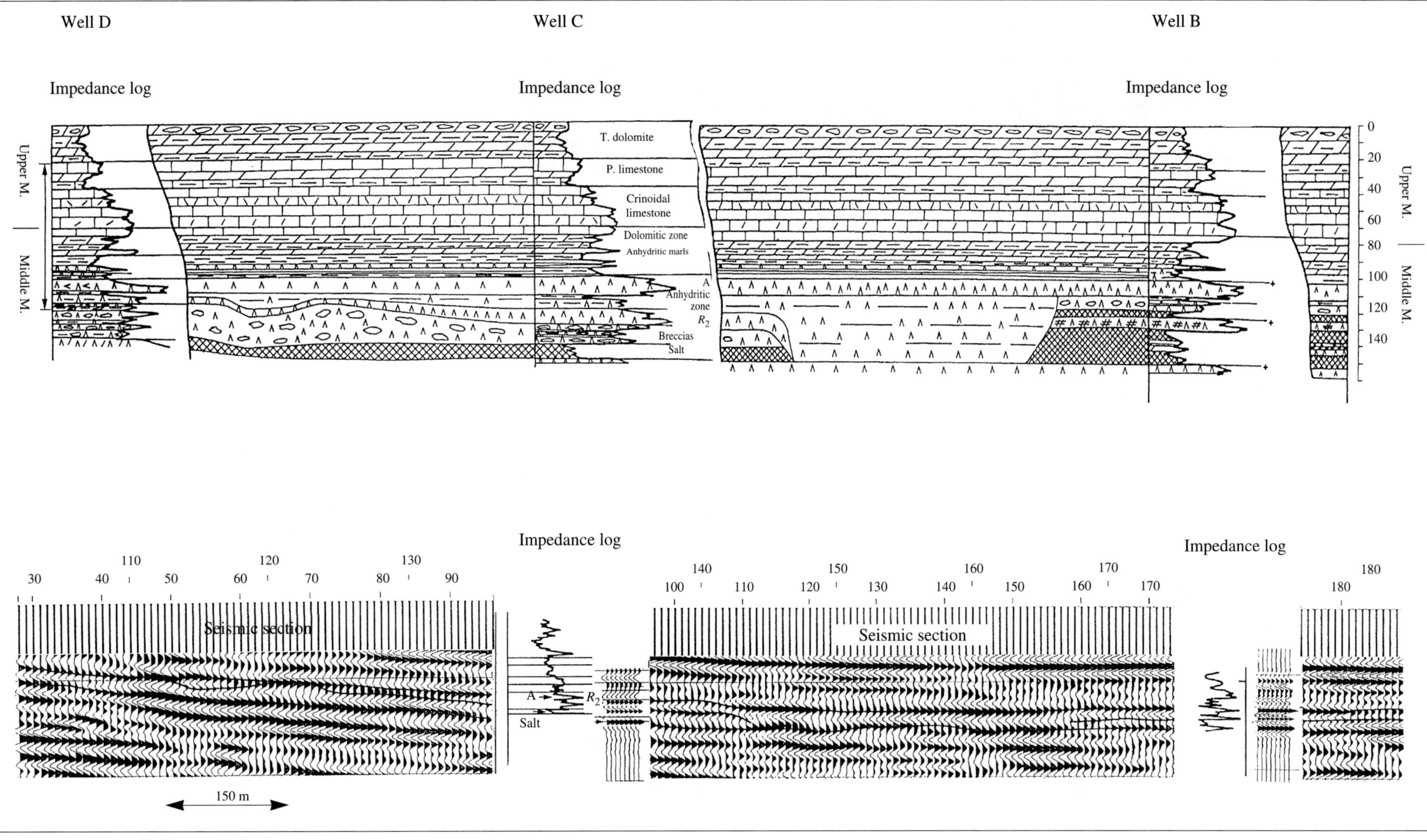

Fig. 6.30 Logs through Middle and Upper Muschelkalk formations, with seismic sections referenced to the top of the dolomites.

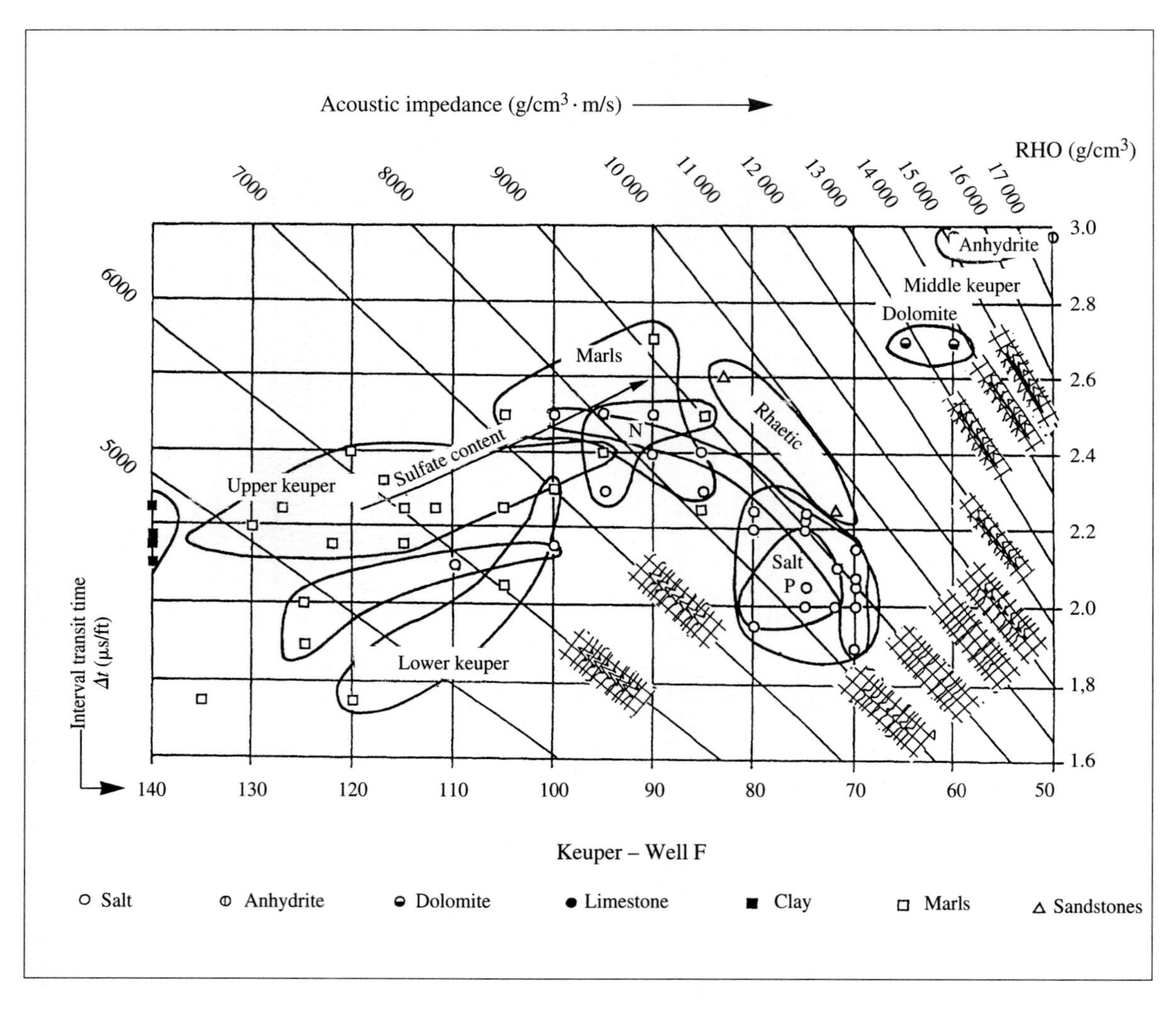

Fig. 6.31 Density-velocity-impedance crossplot for well F.

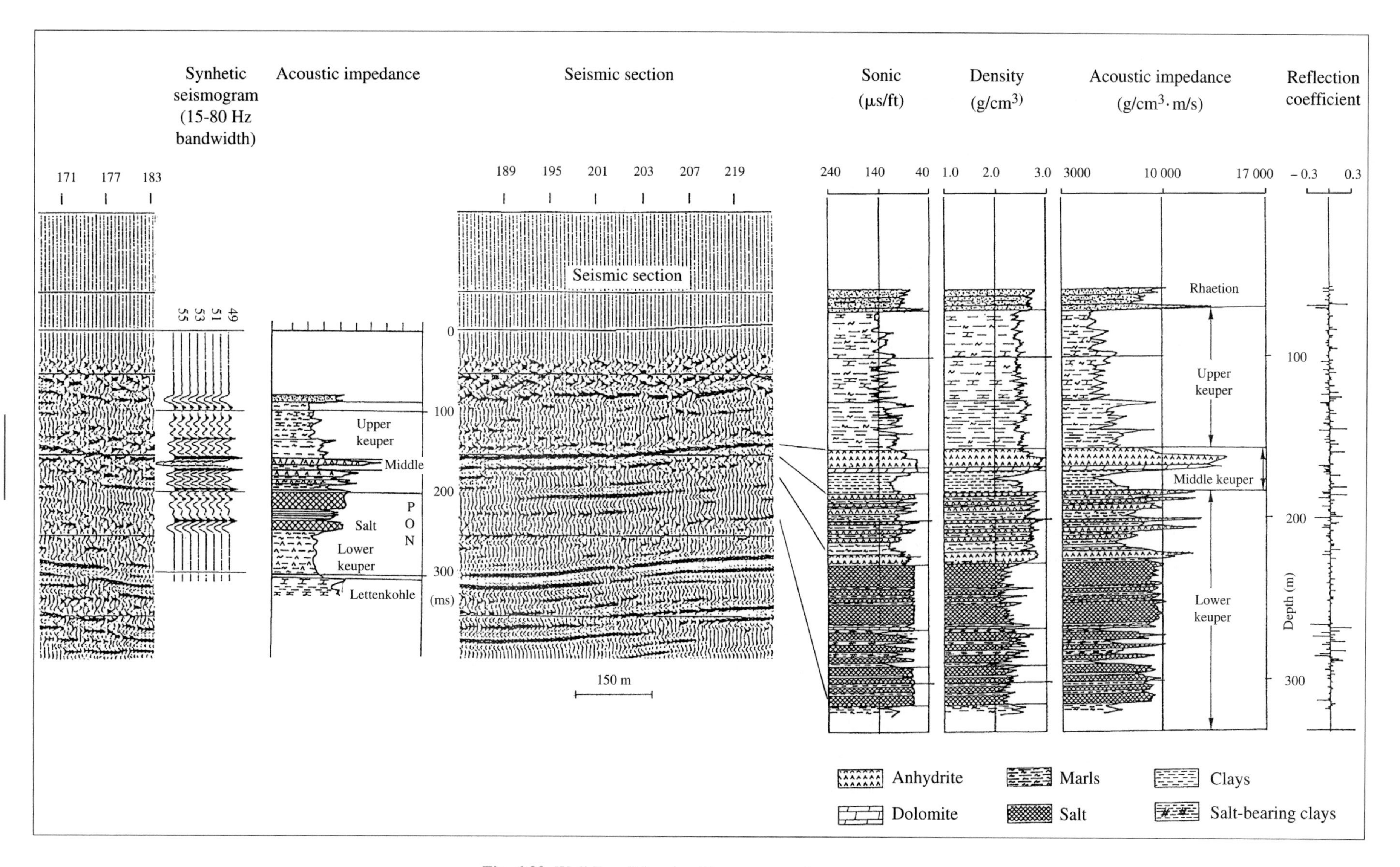

Fig. 6.32 Well F: salt-bearing Keuper succession.

Applications of VSP and dipmeter surveys in the development of a structural model

(J.L. Mari, P. Gavin and F. Verdier, 1990)

This case study presents the respective contributions of dipmeter logs and VSP's to build a structural model. The dipmeter survey helped in the implementation of the VSP, which then provided a geological model with a lateral range of investigation of about 350 m around the axis of the well. The model was checked against the dipmeter data, thus yielding detailed information at the well site.

Context of operation of borehole seismic survey

The AZ 08 well was drilled on the Northeast flank of the Auzance structure (North Aquitaine Platform of Southwest France), which is a N 130°-trending anticline of irregular shape occurring within the Triassic. Numerous faults are seen to affect the structural surface (Fig. 6.33).

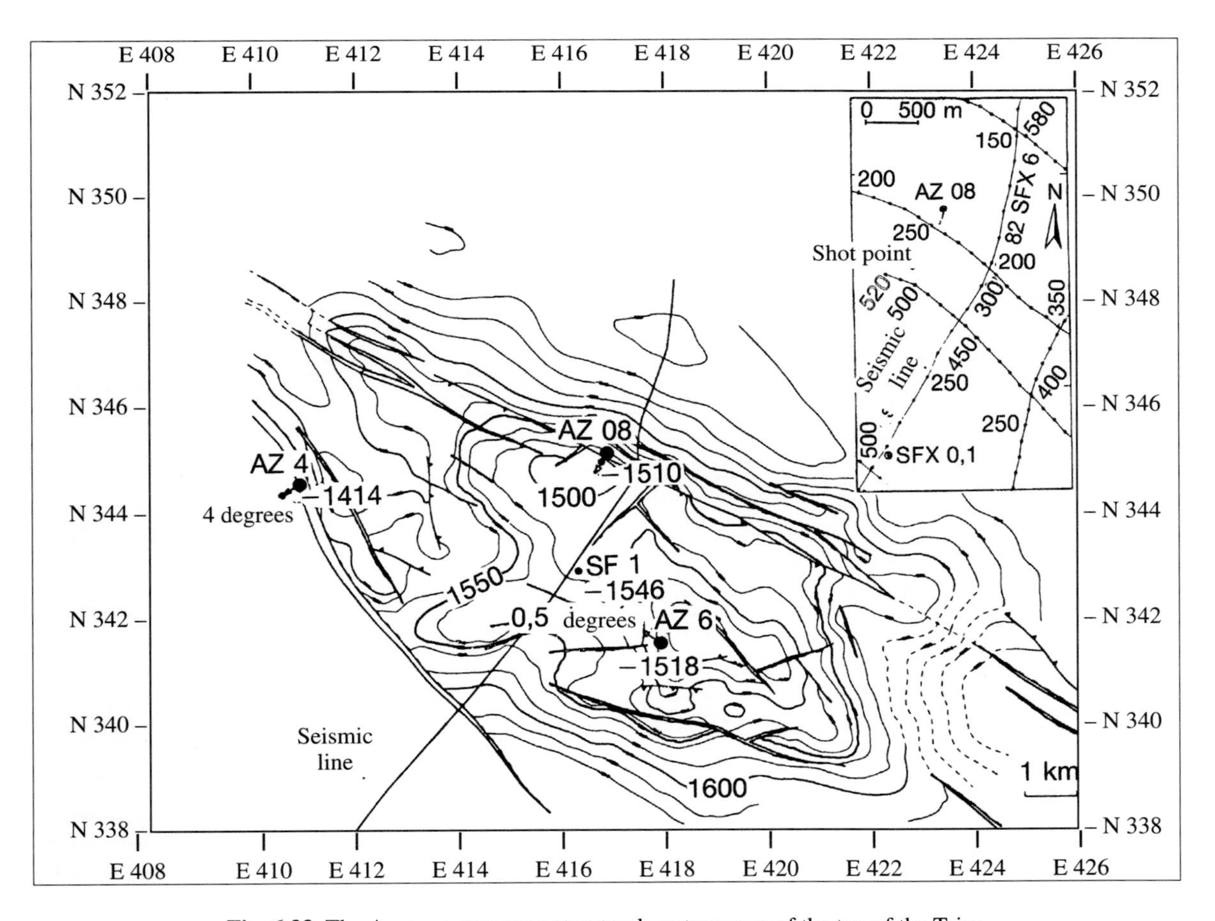

Fig. 6.33 The Auzance structure: structural contour map of the top of the Trias.

The drilling was carried out into the flexured part of the structure, in a zone affected by seismic noise that could not be interpreted in detailed from the surface seismics (Fig. 6.34).

The drilling of well AZ 08 started in the Santonian, then traversed the Cenomanian unconformity at 376.5 m and continued through the Portlandian to the base of the Triassic before stopping in the Palaeozoic basement (total depth: 1784 m).

The objective of the drilling was to improve understanding of the Triassic reservoirs and their structural characteristics.

Deformation of the Triassic is of weak amplitude, being characterised by the presence of numerous faults. However, the deformation observed beneath the Cenomanian is extensive and well developed. This contrast between the post-Cenomanian sequence and its underlying strata can be explained by the regional tectonic history:

(1) An extensional tectonic regime was established up to the end of Jurassic times, leading to the development of normal faults.

(2) The basal Cenomanian unconformity is identified by the truncation of the underlying Jurassic succession, while the Cenomanian is itself transgressive onto the erosion surface so formed.

(3) During the Eocene, a compressive tectonic regime (Pyrenean phase) gave rise to the observed large-scale structure of the Cretaceous layers. This phase led to the reactivation of normal faults as reverse faults and brougth about the structural inversion that is distinctly seen on the seismic profile.

Vertical Seismic Profile

A Vertical Seismic Profile (VSP) was recorded with an offset of 654 m from the wellhead in order to attempt an estimate of the lateral extent and continuity of horizons encountered during drilling. During VSP acquisition, the source was positioned with the help of the dipmeter log in such a way that the plane containing the source and the well-axis was maintained perpendicular to the dip of the structure (dip direction: N 30° E).

The 48 measure points for the VSP were spaced every 15 m recorded between depths of 1050 and 1755 m. A vertical vibrator was used as the source.

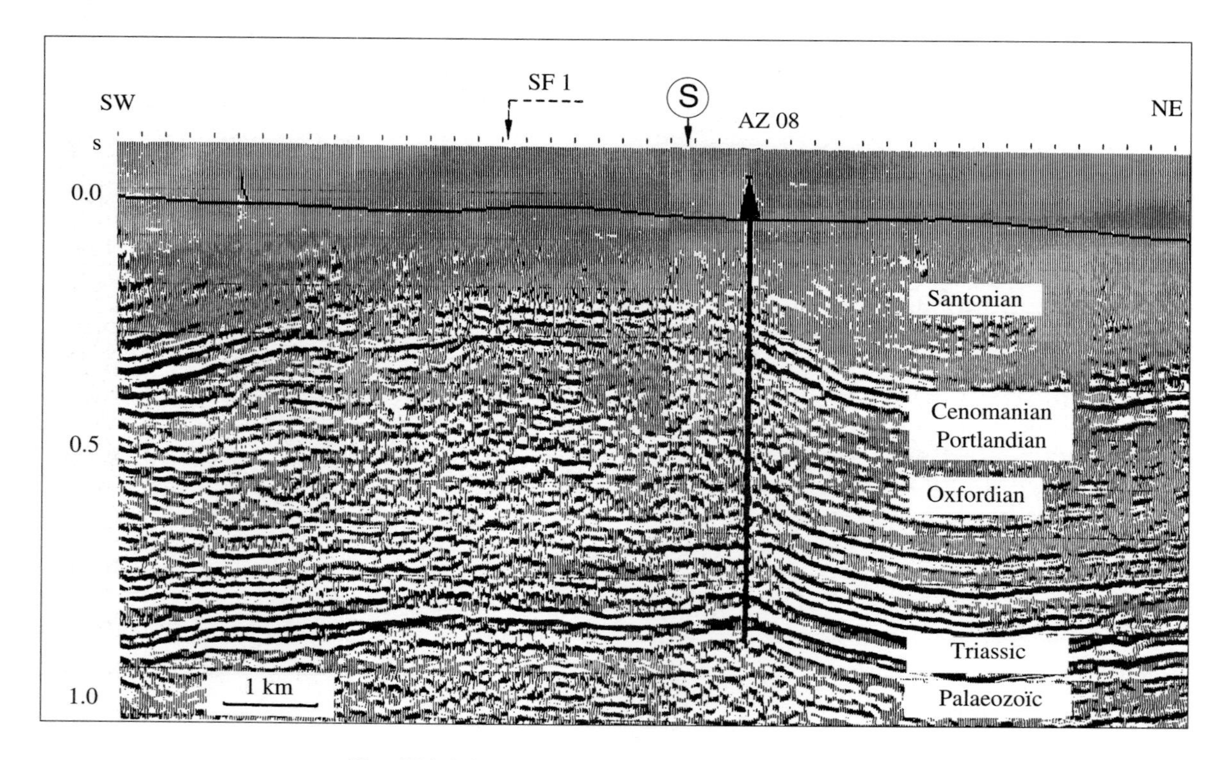

Fig. 6.34 Seismic section in the vicinity of well AZ 08.

The sampling rate was 2 ms, with a recording time of 2 s after correlation. Pre-processing of the data consisted of correlation, editing, stacking of single traces at each depth reading and the reorientation of horizontal components. This final step was carried out to compensate for rotation of the tool and obtain a seismic recording in the plane containing the source and the well-axis.

Despite the large offset of the source in relation to the depth of the well geophone, it can be seen that the compressional waves are mainly recorded along the *Z* component, whereas the shear waves are recorded along *X*. All the seismic sections are presented on the basis of preserved amplitude (Fig. 6.35).

Wave separation was achieved using an algorithm based on spectral matrix filtering with constraints, according to the method proposed by Mari and Gavin (1990). The object of this procedure is to extract both the upgoing and downgoing *P* and *SV* waves on the basis of multi-component data (i.e. *X* and *Z*):

Wave separation was carried out in three steps:

(1) Estimation of the wave vectors S_P and S_{SV} associated with the downgoing *P* and *SV* waves.

(2) Extraction of downgoing *P* and *SV* wavesets using least-squares method to project each component (*Z* and *X*) onto the wave vectors S_P and S_{SV}.

(3) Analysis of the residuals (R_Z and R_X) obtained by subtracting *Z* and *X* from the initial *P* and *SV* wavefields associated with each component.

The residual VSP sections are mainly composed of upgoing waves and noise, with a minor proportion of downgoing *P* waves on the *X* component and *SV* waves on the *Z* component (Fig. 6.36).

The R_Z section shows three *P*-type reflections which are clearly visible in depth (denoted A, B and C on figure). An offset of the markers B and C can be observed on the reflection time-distance curve, suggesting the presence of a fault. The marker A however, appears not to have been faulted. An upgoing shear wave produced by conversion of the downgoing compressional wave at depth of the C marker can be followed continuously on the R_X section.

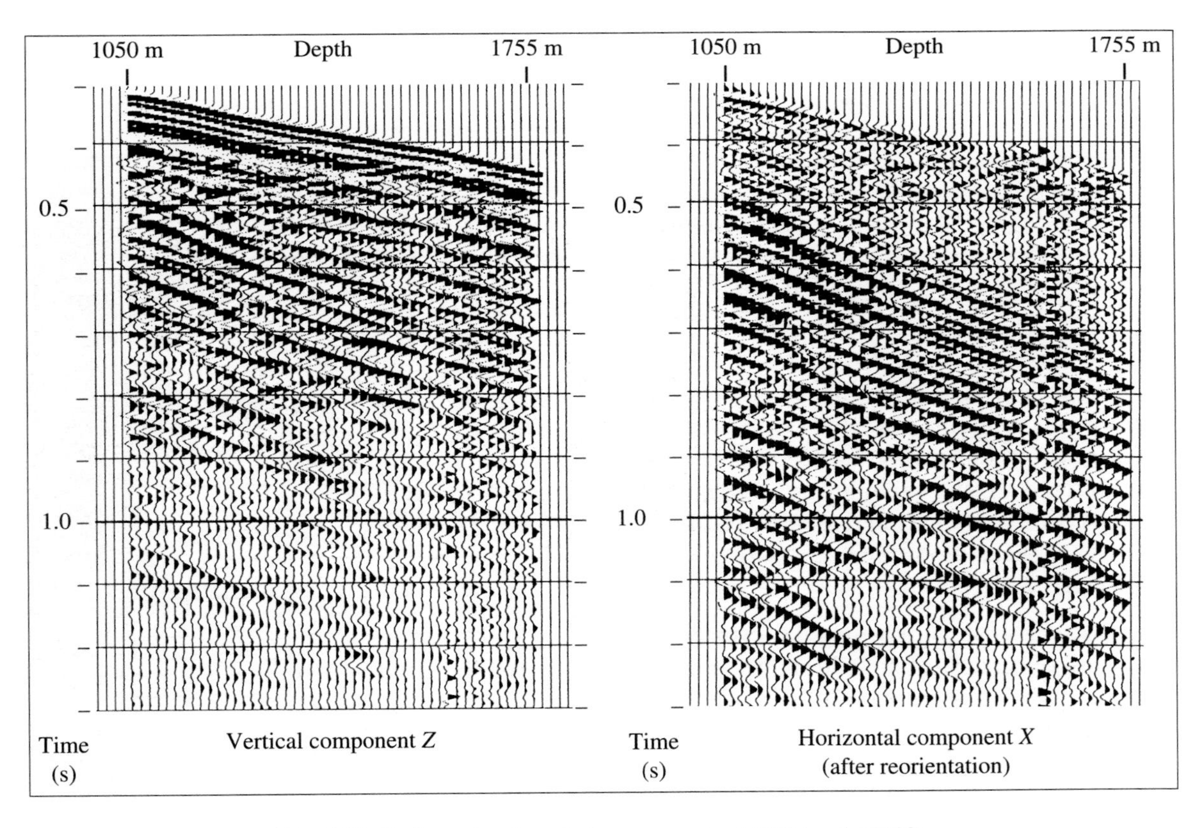

Fig. 6.35 Horizontal and vertical components of VSP (well AZ 08).

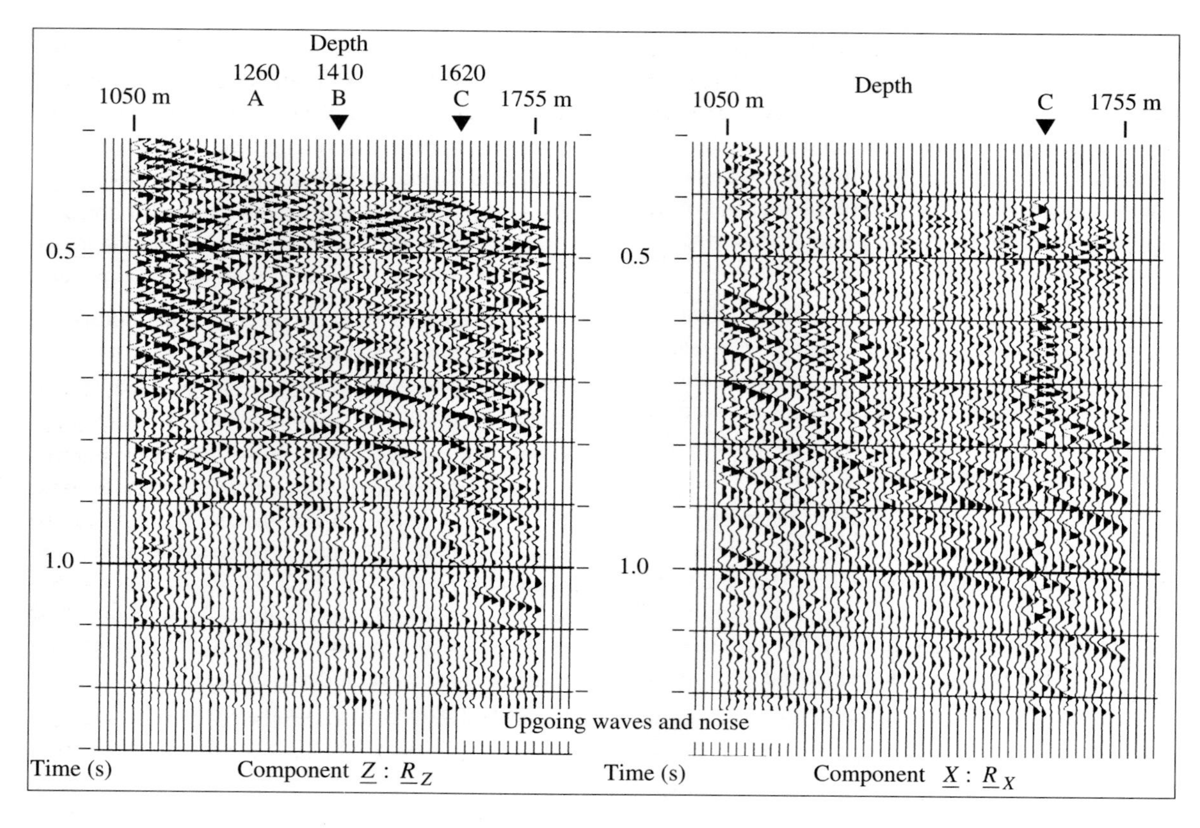

Fig. 6.36 Residual VSP sections.

Because the well was drilled into a complex structure, the interpretation was carried out using charts (time-depth curve vs. dip) and the different reflected-arrival picked-times. An estimation of the velocity field is provided by the time of the first arrival. The mean velocity varies from 3700 to 4000 m/s over the depth range 1200 to 1600 m. Knowing the average velocity and the depth of the marker in the well, it is possible to estimate its dip and also the position of the reflection point (*Xm*, *Zm*) attained by the seismic ray, assuming a homogeneous medium. Interpretation charts are used to characterise the time-distance curve at different dip values.

Figure 6.37 shows the chart used to estimate the dip of marker C located at 1620 m. The average velocity of 4000 m/s can be evaluated by picking the first arrival times. The plotted points correspond to reflected wave arrival times picked from the R_Z section. The estimated dip of marker C is 20-25°.

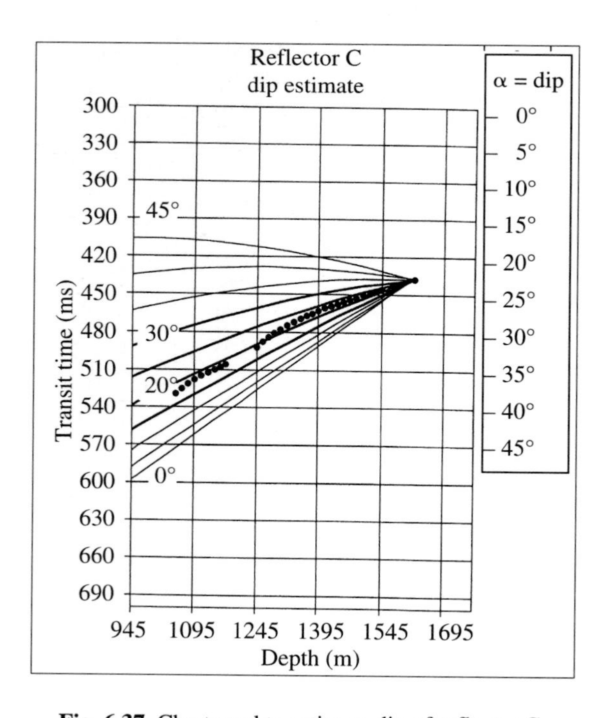

Fig. 6.37 Chart used to estimate dip of reflector C.

The error involved in dip determination can be assessed directly from charts. It is linked not only to the error in picking the wave reflected from the marker but also the position of the geophone.

For geophone positions in the depth range 1050-1755 m, the average error on the dip is 5° for a picking error estimated at about half the wave cycle (12 ms).

Interpretation

The structural model was obtained by migrating the reflected arrivals picked from upgoing wavesets of the VSP point by point, using equations to calculate the coordinates of the reflection points (*Xm*, *Zm*) after estimation of the velocities and dips for each marker (A, B and C).

Charts allowed estimates to be made of the dips of various reflectors; the values so obtained provide structural information over a lateral range of several hundred meters:

Marker	Depth (m)	Velocity (m/s)	Seismic dip (degrees)
A	1260	3750	10
B	1410	3800	10 to 15
C	1620	4000	20 to 25

In the reflection zone investigated by the VSP, marker A is not faulted. Markers B and C are faulted at distances of 200 and 300 m from the well, respectively.

These markers correspond to the following three horizons encountered in the well:

- A top of the Lower Oxfordian at 1260 m,
- B top of the Toarchian at 1420 m,
- C top of the upper reservoir zone in the Triassic at 1620 m.

The dipmeter log shows regular and shallow dip angles (5-10°) for formations in the depth range 0-1200 m. After passing through a highly tectonized zone at 1650-1685 m, the dips become more regular and tend towards more moderate values (10-15°). In the well, the reflectors A, B and C dip, respectively, 10-12°, 14-18° and 20-30° towards the Northeast. These results are in good agreement with the seismic dips (see Fig. 6.38).

Conclusion

VSP's enable the determination of the structural dip of major markers and provide detailed information on the discontinuities affecting these markers at a range of several hundred metres from the well. By correlating well log data with lateral extrapolations provided by the analysis of VSP sections, it is possible to develop a consistent geological model of the well vicinity.

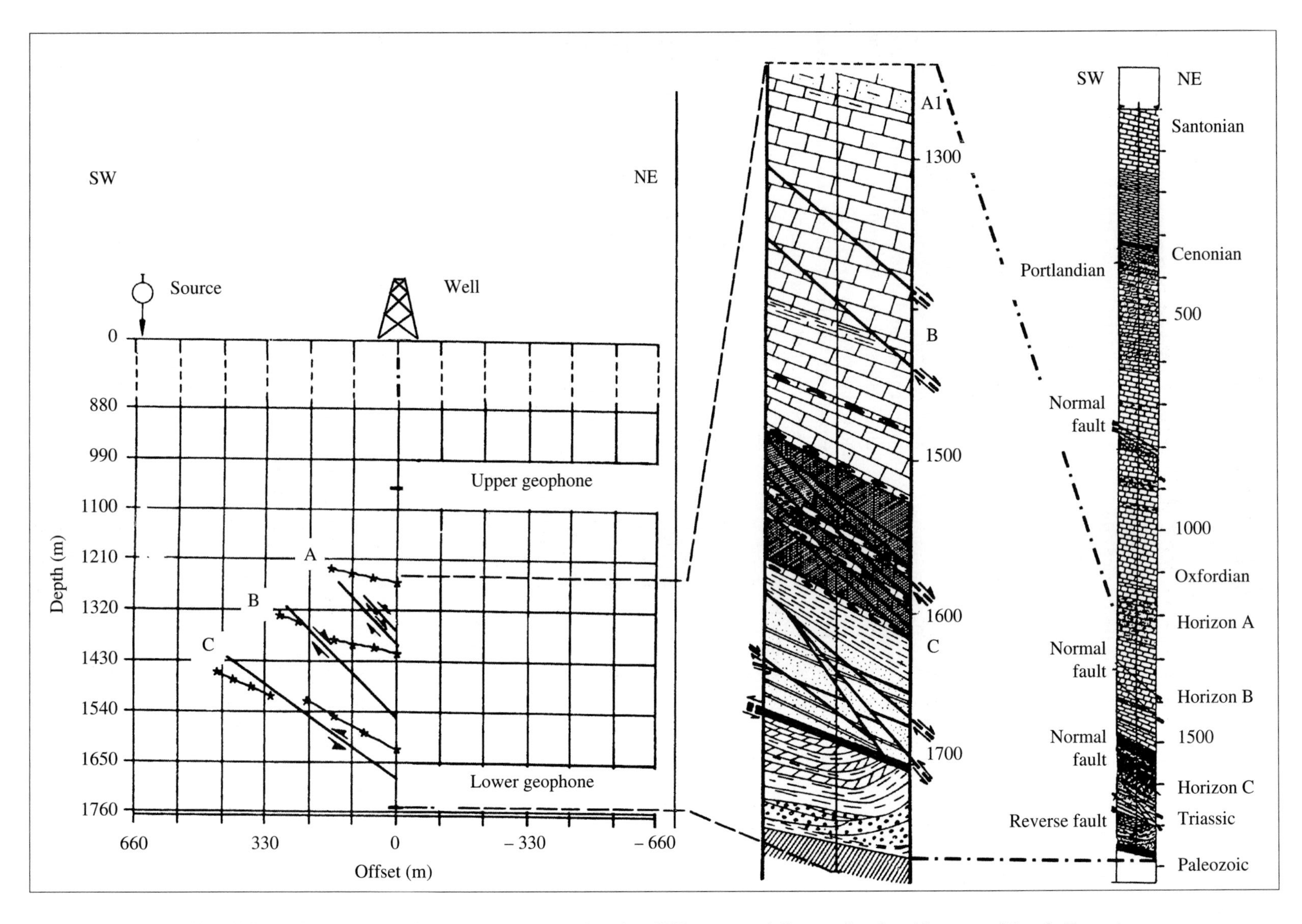

Fig. 6.38 Geological model of the Auzance structure based on VSP survey and dipmeter logging. (*Courtesy of Gaz de France*)

REFERENCES

GENERAL TEXTS

Desbrandes, R. (1982), *Diagraphies dans les sondages.* Éditions Technip, Paris.

Hardage, B.A. (1985), "Vertical Seismic Profiling". Part A: Principles; Vol. 14 A, *Geophysical Press.*

Hossin, A. (1983), *Manuel de Diagraphies Différées – Appareils et interprétation.* Cours de l'ENSPM.

Mari, J.L. and F. Coppens (1989), *La sismique de puits (Seismic Well Surveying),* 1991. Éditions Technip, Paris.

Mari, J.L., F. Coppens, P. Gavin and E. Wicquart (1992), *Traitement des diagraphies acoustiques (Full wave form acoustic data processing,* 1994). Éditions Technip. Paris.

1st part: Mari, J.L. and F. Coppens (1992), "Application de techniques issues de l'intelligence artificielle au pointe des diagraphies acoustiques". *Revue IFP*, Vol. 47, No. 1, pp. 3 to 28.

2nd part: Gavin, P. and J.L. Mari (1992), "Séparation des ondes en diagraphie acoustique". *Revue IFP*, Vol. 47, No. 2, pp. 155 to 178.

3rd part: Mari, J.L. and E. Wicquart (1992), "Caractérisation d'un réservoir par diagraphies acoustiques obtenues avec un outil dipolaire". *Revue IFP*, Vol. 47, No. 4, pp. 443 to 464.

4th part: Gavin, P. and J.L. Mari (1992), "Détermination de pendage par diagraphie acoustique". *Revue IFP*, Vol. 47, No. 5, pp. 103 to 124.

Serra, O. (1979), "Diagraphies Différées – Bases de l'interprétation". Tome 1: "Acquisition des données diagraphiques". *Bull. Cent. Rech. Explor. Prod Elf-Aquitaine.*

Serra, O. (1986), "Fundamental of well – Log Interpretation". 1: The Acquisition of Logging Data (1984). 2: The Interpretation of Logging Data (1986). *Elsevier (Developments in Petroleum Science 15B).*

Sheriff, R.E. (1989), *Geophysical Methods.* Prentice Hall.

White, J.E. (1983), *Underground Sound – Application of seismic waves.* Elsevier Science Publishers B.V., Amsterdam.

JOURNALS

Géologues (Revue Officielle de l'Union Française des Géologues). *Diagraphies Différées.* No. 99, March-April 1993. No 100-101, October 1993.

Revue de l'Institut Français du Pétrole. Éditions Technip, Paris.

The Log Analyst (An International Journal of Society of Professional Well Log Analysts).

Reprint Volumes of Society of Professional Well Log Analysts.

ACOUSTIC WELL LOGGING

Arditty, P.C., G. Arens and P. Staron (1984), "Improvements of formation properties and evaluation through processing and interpretation results of the EVA tool records". *SEG 54th Annual Meeting expanded abstracts.*

Astbury, S. and M.H. Worthington (1986), "The analysis and interpretation of full waveform sonic data – Part I: dominant phases and shear wave velocity". *First Break,* 4 (4), pp. 7-16.

Astbury, S. and M.H. Worthington (1986), "The analysis and interpretation of full waveform sonic data – Part II: multiples, mode conversions and reflections". *First Break,* 4 (6).

Biot, M.A. (1952), "Propagation of elastics waves in a cylindrical bore containing fluid". *Journal of Applied physics 223,* pp. 997-1005.

Bourbié, Th., O. Coussy and B. Zinszner (1986), *Acoustique des milieux poreux.* Éditions Technip, Paris.

Brian, E.H. (1989), "Imaging of near-borehole structure using full-waveform sonic data". *Geophysics,* 54, Vol. 6, pp. 747-757.

Brian, E.H. and F.M. William (1987), "V_P/V_S in unconsolidated oil sands: Shear from Stoneley". *Geophysics,* 52, Vol. 4, pp. 502-513.

Chen, S.T. (1988), "Shear wave logging with dipole source". *Geophysics,* 53, Vol. 5, pp. 659-667.

Cheng, C.H. and M.N. Toksoz (1981), "Elastic wave propagation in a fluid-filled borehole and synthetic acoustic log". *Geophysics,* 46, pp. 1042-1053.

Fortin, J.P., N. Rehbinder and P. Staron (1991), "Reflection imaging around a well with the Eva full-waveform tool". *The log analyst,* Vol. 32, No. 3, pp. 271-278.

Goetz, J.F., L. Dupal and J. Bowler (1979), "An investigation into discrepancies between sonic log and seismic check shot velocities". "*APEA*".

Krief, M., J. Garat, J. Stellingwerff and J. Ventre (1989), "A new petrophysical interpretation using the velocities of *P* and *S* waves (Full wave form Sonic)". *12ème coll. int. de Diagraphies,* (SAID).

Liu, O.Y. (1984), "Stoneley wave derived Δt shear log". *25th SPWLA Logging Symposium,* paper ZZ.

Mari, J.L., F. Coppens, P. Gavin and E. Wicquart (1992), *Traitement des diagraphies acoustiques.* Éditions Technip, Paris.

Morris, C.F., P.M. Little and W. Letton (1984), "A new sonic array tool for full waveform logging". Showed at the *59th annual conference of the Society of Petroleum Engineers of AIME,* Houston, Texas, September, pp. 16-19.

Paillet, F.L. and R. Turpening (1984), "Borehole and surface to borehole seismic applications in fracture characterisation". *SEG 54th Annual Meeting expanded abstracts.*

Raymer, L.L. and J.S. Gardner (1980), "An improved sonic transit time - to - porosity transform". *SPWLA Logging Symposium (Paper P).*

Schmitt, D.P. (1984), "Simulation numérique de diagraphies acoustiques, propagation d'ondes dans des formations cylindriques axisymétriques radialement stratifiées incluant des milieux élastiques et/ou poreux saturés". *Thèse pour le titre de docteur.* Université scientifique et médicale de Grenoble.

Stewart, R.R., Ph.D. Huddleston and Tze Kong Kan (1984), "Seismic versus sonic velocities: A vertical seismic profiling study". *Geophysics,* Vol. 49, No. 8, pp. 1153-1168.

Summers, G.C. and R.A. Broding (1952), "Continuous Velocity Logging". *Geophysics,* 17, pp. 598-614.

Vogel, C.B. (1952), "A seismic logging method". *Geophysics*, 17, pp. 579-586.

Williams, D.M., J. Zemanek, F.A. Angona, C.L. Dennis and R.L. Caldwell (1984), "The long spacing acoustic logging tool". *25th SPWLA Logging Symposium,* paper T.

Zemanek, J., F.A. Angona, D.M. Williams and R.L. Caldwell (1984), "Continuous acoustic shear wave logging". *25th SPWLA Logging Symposium,* paper U.

Zhang Jinzhong and C.H. Cheng (1984), "Numerical studies of body wave amplitudes in waveform logs". *9th international formations evaluation transactions,* paper 14.

PERMEABILITY-FRACTURING

Becquey, M., J.O. Bernet-Rollande and S. Nicoletis (1989), "Microsismicité induite par un arrêt d'injection". SAID, *12th colloque international de Diagraphies.*

Beydoun, W.B., C.H. Cheng and M.N. Toksoz (1985), "Detection of open fractures with vertical seismic profiling". *Journal of Geophysical Research,* Vol. 90, No. B6, pp. 4557-4566.

Biot, M.A. (1956), "Theory of propagation of elastic waves in a fluid-saturated porous solid: I – Low frequency range; II – Higher frequency range". *The Journal of the Acoustical Society of America,* Vol. 28, 2, pp. 168-178 and 179-191.

Chang, S.K., H.L. Liu and D.L. Johnson (1988), "Low frequency tube waves in permeable rocks". *Geophysics,* Vol. 53, 4, pp. 519-527.

Gaudiani, P. and J.L. Mari (1993), "Acoustics on a real scale model: application to fractured media". *Revue de l'Institut Français du Pétrole,* Vol. 48, No. 4, pp. 347-357.

Hardin, E.L., C.H. Cheng, F.L. Paillet and J.D. Mendelson (1987), "Fracture characterization by means of attenuation and generation of tube waves in fractured cristalline rock at Minor Lake, New Hampshire". *Jour. of Geoph. res.,* 92, pp. 7989-8006.

Hou, K., A. Brie and R.A. Plumb (1987), "A new method for fracture identification using array sonic tools". *Journal of Petroleum Technology,* 39, pp. 677-688.

Huang, C.F. and J.A. Hunter (1981), "The correlation of «tube-wave» events with open fractures in fluid-filled boreholes: current research". *Part A, Geological Survey of Canada,* paper 81-1A, pp. 361-376.

Lebreton, F., J.P. Sarda, F. Trocqueme and P. Morlier (1978), "Logging tests in porous media to evaluate the influence of their permeability on acoustic waveforms". *Trans. 19th, SPWLA Annual Logging Symposium,* paper Q.

Liu, O.Y. (1985), "Fracture evaluation using borehole sonic velocity measurements". *Paper SPE,* No. 14399, Las Vegas.

Mathieu, F. and M.N. Toksoz (1984), "Application of full waveform acoustic logging to the estimation of reservoir permeability". *SEG 54th annual meeting expanded abstracts.*

Morris, R.L., D.R. Grine and T.E. Arkfeld (1964), "Using compressional and shear acoustic amplitudes for the location of fractures". *Journal of Petroleum Technology,* 16, pp. 623-632.

Rosenbaum, J.H. (1974), "Synthetic microseismogramms logging in porous formation". *Geophysics,* Vol. 39, pp. 14-32.

Sibbit and Faivre (1985), "The Dual Laterolog response in fractured rocks". *Congrès international SPWLA, June 1985,* Article T.

VERTICAL SEISMIC PROFILING (VSP)

Coppens, F. (1982), "Étude et comparaison de quelques filtres à deux dimensions". *Rapport IFP* No. 29 933.

Coppens, F. (1985), "First arrival picking on common offset trace collection for automatic estimation of static corrections". *Geophysical Prospecting,* 33, pp. 1212-1231.

Coppens, F. and J.L. Mari (1985), *Quelques considérations sur la sismique de gisement.* Cours ENSPM.

Dankbaar, J.W.M. (1985), "Separation of *P* and *S* waves". *Geophysical Prospecting,* 33, No. 7, pp. 970-986.

Dankbaar, J.W.M. (1987), "Vertical seismic profiling – Separation of *P* and *S* waves". *Geophysical Prospecting,* 35, No. 7, pp. 803-814.

Devaney, A.J. and M.L. Oristaglio (1986), "A plane-wave decomposition for elastic wave fields applied to the separation of *P* waves and *S* waves in vector seismic data". *Geophysics,* 51, No. 2, pp. 419-423.

Dillon, P.B. and R.C. Thomson (1984), "Offset source VSP surveys and their image reconstruction". *Geophysical Prospecting,* 32, No. 5, pp. 790-811.

Dillon, P.B. (1985), "VSP migration using the Kirchhoff integral". *Expanded abstracts of the 55th Ann. Int. SEG meeting,* Washington, 6-10 Oct., paper No. BHG 1.2, pp. 19-22.

Dillon, P.B. and V.A. Collyer (1985), "On timing VSP first arrival". *Geophysical Prospecting,* 33, pp. 1174-1194.

Grivelet, P.A. (1985), "Inversion of vertical seismic profiles by iterative modeling". *Geophysics,* 50, pp. 924-930.

Guerendel, P. and J. Laurent (1984), "Study of Borehole seismic tool coupling: application to the Geolock H sonde". *Congrès EAEG,* London, Rapport IFP No. 32 238.

Hardage, B.A. (1985), "Vertical Seismic Profiling". Part A: Principles, Vol. 14 A, *Geophysical Press.*

Kennett, P. and R.L. Ireson (1981), "The VSP as an interpretation tool for stuctural and stratigraphic analysis". *43rd Meeting of EAEG.*

Lamer, A. (1982), *Filtrages sismiques par moyenne et antimoyenne.* Thèse de Doctorat d'État, Université Pierre et Marie Curie, Paris VI. Éditions Technip, Paris.

Mace, D. and P. Lailly (1986), "Solution of VSP one dimensional inverse problem". *Geophysical Prospecting,* 34, pp. 1 002-1 021.

Mammo, T. (1987), "Wave field separation method in VSP". *A review: Bolletino di Geofisica Teorica ed applicata,* Vol. 29 , No. 116 , pp. 275-307.

Mari, J.L. and F. Coppens (1989), *La sismique de puits.* Éditions Technip, Paris.

Mari, J.L., F. Coppens, F. Glangeaud and P. Durand (1990), "Trace pair filtering for separation of upgoing and downgoing waves in vertical seismic profiles". *Revue de l'Institut Français du Pétrole,* Vol. 45, No. 2, pp. 181-203.

Mari, J.L., F. Coppens and E. Blondin (1987), "Le profil SVP: Sismique au voisinage du puits". *Revue de l'Institut Français du Pétrole,* Vol. 42, No. 3, pp. 317-325.

Mari, J.L. and F. Glangeaud (1990), "Spectral matrix applied to VSP processing". *Revue de l'Institut Français du Pétrole,* Vol. 45, No. 3, pp. 417-434.

Mari, J.L. and P. Gavin (1990), "Séparation des ondes *P* et *S* à l'aide de la matrice spectrale avec information a priori". *Revue de l'Institut Français du Pétrole,* Vol. 45, No. 5.

Seeman, B. and L. Horowicz (1983), "Vertical seismic profiling: Separation of upgoing and downgoing acoustic waves in a stratified medium". *Geophysics,* 48, pp. 555-568.

Stewart, R.R., Ph.D. Huddleston and Tze Kong Kan (1984), "Seismic versus sonic velocities: A vertical seismic profiling study". *Geophysics,* Vol. 49, No. 8, pp. 1153-1168.

Tariel, P. and D. Michon (1982), "Comments on VSP processing". *CGG Technical Series,* No. 527.83.02.

Wyatt, K.D. and S.B. Wyatt (1982), "Determination of subsurface structural information using the vertical seismic profile". *Geophysics,* 47, No. 7, pp. 1123-1128.

SYNTHETIC SEISMIC RECORDS

Baranov, V. and G. Kunetz (1960), "Film synthétique avec réflexions multiples – Théorie et calcul pratique". *Geophysical Prospecting,* Vol. 8, pp. 315-325.

Ganley, D.C. (1981), "A method for calculating synthetic seismograms which include the effects of absorption and dispersion". *Geophysics,* Vol. 46, No. 8, pp. 110-1107.

Wuenchel, P.C. (1960), "Seismogram synthesis including multiples and transmission coefficients". *Geophysics,* Vol. 25, pp. 106-129.

DECONVOLUTION-INVERSION

Becquey, M., M. Lavergne and C. Willm (1979), "Acoustic impedance logs computed from seismic traces". *Geophysics,* 449, pp. 1485-1501.

Brac, J., P.Y. Dequirez, F. Hervé, C. Jacques, P. Lailly, V. Richard and D. Tran Van Nhieu (1988), "Inversion with a priori information: an approach to integrated stratigraphic interpretation". Presented at *the 58 th Annual SEG Meeting,* Anaheim.

Dash, B.P. and K.A. Obaidullah (1970), "Determination of signal and noise statistics using correlation theory". *Geophysics,* 35, pp. 24-32.

Duijndam, A.J.W., P. Van Riel and E.J. Kaman (1984), "An iterative scheme for wavelet estimation and seismic section inversion in reservoir seismology". Presented at *the 54th Annual SEG Meeting,* Atlanta.

Frankel, A. and R. W. Clayton (1986), "Finite difference simulations of seismic scattering: implications for the propagation of short-period seismic waves in the crust and models of crustal heterogeneity". *Journal of Geophysical Research,* 91, pp. 6465-6489.

Franklin, J.N. (1970), "Well-posed stochastic extensions of ill-posed linear problems". *J. Math and Appl.,* 31, pp. 682-716.

Gelfand, V. and K. Larner (1984), "Seismic lithologic modeling". *Leading Edge,* 3, pp. 30-35.
Lindseth, R.O. (1979), "Synthetic sonic logs. A process for stratigraphic interpretation". *Geophysics,* 44, No. 1, pp. 3-26.
Lines, L.R. and S. Treitel (1984), "Tutorial: A review of least-squares inversion and its application to geophysical problems". *Geophysical Prospecting,* 32, pp. 159-186.
Mace, D. and P. Lailly (1986), "Solution of the VSP one-dimensional inverse problem". *Geophysical Prospecting,* 34, pp. 1002-1021.
Oldenburg, D.W., S. Levy and K. Stinson (1986), "Inversion of band-limited reflection seismograms: theory and practice". *Proceedings of the IEEE,* 74, 3, pp. 487-497.
Redanz, M., B. Perathoner and J. Fertig (1986), "Wavelets and velocity logs from seismic sections: some experiences and results". *48ème congrès EAEG,* Ostend.
Richard, V. (1986), "High resolution stratigraphic extrapolation: a 1-D pre-stack inverse problem". Presented at *the 48th EAEG Meeting,* Ostend.
Richard, V. and J. Brac (1988), "Wavelet analysis using well log information". Presented at *the 58th SEG Meeting,* Anaheim.
Tarantola, A. and B. Valette (1982), "Inverse problems = quest for information", *J. Geophysics,* Vol. 50, No. 3, pp. 159-170.

ATTENUATION

Aki, K. and P.G. Richards (1980), "Quantitative seismology: Theory and Methods". Vol. II. *W.H. Freeman & Co.*
Aleotti, L. and G.P. Angeleri (1989), "Borehole and surface seismics – An example of improved correlation by studying attenuation effect". *51th EAEG meeting,* Berlin.
Cheng, C.H., M.N. Toksoz and M.E. Willis (1982), "Determination of insitu attenuation from full waveform acoustic logs". *Journal of Geophysical Research,* 87, pp. 5477-5484.
Coppens, F. and J.L. Mari (1984), "L'égalisation spectrale, un moyen d'améliorer la qualité des données sismiques". *Geophysical Prospecting,* 32, pp. 258-281.
Ganley, D.C. and E.R. Kanasewich (1980), "Measurement of absorption and dispersion from check shot survey". *Journal of Geophysical Research,* 85, pp. 5219-5226.
Gelius, L.J. (1987), "Inverse Q-filtering. A spectral balancing technique". *Geophysical Prospecting,* 35, pp. 656-667.
Hauge, P.S. (1981), "Measurements of attenuation from vertical seismic profiles". *Geophysics,* 46, pp. 1 548-1 558.
Mari, J.L. (1989), "Q-Log determination on downgoing wavelet and tube wave analysis in vertical seismic profiles". *Geophysical Prospecting,* 37, pp. 257-277.
Morlier, P. and J.P. Sarda (1971), "Atténuation des ondes élastiques dans les roches poreuses et saturées". *Colloque ARTEP,* Communication No. 6.
Schoemberger, M. and F.K. Levin (1974), "Apparent attenuation due to intrabed multiples". *Geophysics,* 39, pp. 278-291.
Schoemberger, M. and F.K. Levin (1978), "Apparent attenuation due to intrabed multiples (II)". *Geophysics,* 43, pp. 730-737.
Stainsby, S.D. and M.H. Worthington (1985), "Q-estimation from vertical seismic profile data and anomalous variations in the central North Sea". *Geophysics,* 50, pp. 615-626.
Tarif, P. and T. Bourbie (1987), "Experimental comparison between spectral ratio and rise time techniques for attenuation measurement". *Geophysical Prospecting,* 35, pp. 668-680.

OBSERVATIONS CONCERNING AMPLITUDE

Anstey, N.A. (1977), *Seismic Interpretations: The Physical Aspects.* IHRDC, Boston, Mass, USA.
Berkhout, A. J. (1985), "Seismic Resolution: A key to detailed geological information". *World OIL,* pp. 47-51.

Brown, A.R. (1986), *Interpretation of three-dimensional seismic data.* AAPG memoir: 42, Tulsa, Oklahoma, USA.

Dilay, A.J. (1982), "Direct hydrocarbon indicators lied to a canadian gas find". *World Oil,* 195, pp. 149-164.

Domenico, S.N. (1976), "Effect of brine-gas mixture on velocity in an unconsolidated sand reservoir". *Geophysics,* 41, pp. 882-894.

Domenico, S.N. (1977), "Elastic properties of unconsolidated porous sand reservoir". *Geophysics,* 42, pp. 1139-1368.

Ensley, R.A. (1984), "Comparison of *P* and *S* wave seismic data: a new method for detecting gas reservoir". *Geophysics,* 49, pp. 1420-1431.

Gregory, A.R. (1976), "Fluid saturation effects on dynamic elastic properties of sedimentory rocks". *Geophysics,* 41, pp. 895-921.

Hubbert, M.K. (1967), "Application of hydrodynamics to oil exploration". *Proceedings 7th World Petroleum Congress,* V 1B, pp. 59-75.

Lortzer, G.J.M, J.C. Haas and A.J. Berkhout (1988), "Evaluation of existing weighted stacking techniques for amplitude versus offset information". Presented at *the 50th meeting of the EAEG,* The Hague, The Netherlands.

Ostrander, W.J. (1984), "Plane-wave reflection coefficients for gas sands at nonnormal angles of incidence". *Geophysics,* 49, pp. 1637-1648.

Pan, N.D. and G.F. Gardner (1987), "The basic equations of plane elastic wave reflection and scattering applied to AVO analysis". *Seismic Acoustic Laboratory, Univ. of Houston,* Ann. Prog. Rev. 19, pp. 123-139.

Sengbush, R.L. (1983), *Seismic Exploration Methods.* IHRDC, Boston, Mass., USA.

Shuey, R.T. (1985), "A simplification of the Zoeppritz equations". *Geophysics,* 50, pp. 609-614.

Silva, R.L., L.G. Peardon and D.W. March (1989), "Classical and modern attribute analysis – A case study over a known prospect in the North Sea". *51th EAEG meeting,* Berlin.

Taner, M.T., F. Koehler and R.E. Sheriff (1979), "Complex trace analysis". *Geophysics,* 44, pp. 1041-1063.

Vasil'ev, Yu.I. and G.I. Gurvich (1962), "On the ratio between attenuation measurements and propagation velocities of longitudinal and transverse waves". *Bull. Acad. Sci. USSR. Geophys. Ser.,* 12, pp. 1061-1074.

Wides, M.B. (1973), "How thin is a thin Bed". *Geophysics,* 38, pp. 1176-1180.

CASE STUDIES

Blondin, E. and J.L. Mari (1986), "Detection of gas bubble boundary movements". *Geophysical Prospecting,* 34, pp. 73-93.

Brun, S., P. Grivelet and A. Paul (1985), "Prediction of overpressure in Nigeria using Vertical Seismic Profile techniques". *Transactions of 26 th SPWLA Annual Logging Symposium.* Paper J.

De Buyl, M., T. Guidish and F. Bell (1988), "Reservoir description from seismic lithologic parameter Estimation". *Journal of Petroleum Technology,* pp. 475-482.

Mari, J.L., P. Gavin and F. Verdier (1990), "The application of three-component VSP data for the interpretation of the Auzance structure". *Expanded abstracts of 66 th annual SEG meeting.* – Paper BG3.3 –, San Francisco.

Renoux, P. (1989), "Well log and seismic response of Permo-Triassic evaporites Zechstein-Muschelkalk-Keuper". *Special application of EAPG No. 1,* chapter 17, pp. 229-240, Oxford University Press.

Composition MACH 3

60250 Bury

Flashage, impression, reliure Imprimerie CHIRAT
42540 Saint-Just-la-Pendue.

Dépôt légal mai 1997 – N° 3701
Numéro d'éditeur 956